Proceedings of the 9th International Workshop on
Complex Structures, Integrability and Vector Fields

Trends in Differential Geometry, Complex Analysis and Mathematical Physics

Proceedings of the 9th International Workshop on Complex Structures, Integrability and Vector Fields

Trends in Differential Geometry, Complex Analysis and Mathematical Physics

Sofia, Bulgaria 25 – 29 August 2008

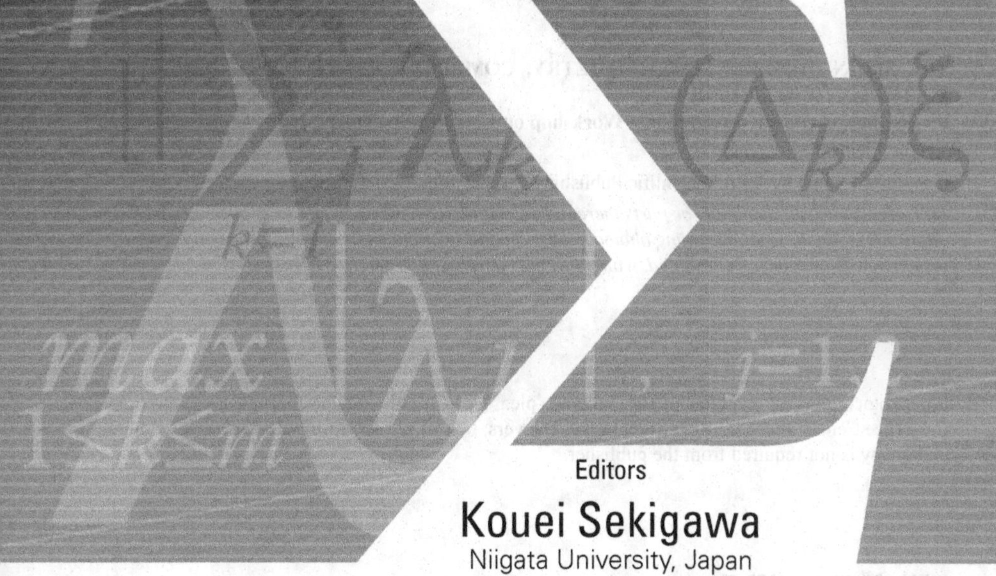

Editors

Kouei Sekigawa
Niigata University, Japan

Vladimir S. Gerdjikov
Stancho Dimiev
Bulgarian Academy of Sciences, Bulgaria

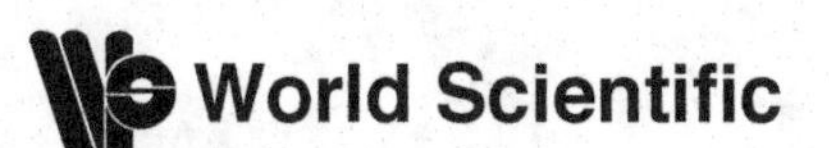

NEW JERSEY • LONDON • SINGAPORE • BEIJING • SHANGHAI • HONG KONG • TAIPEI • CHENNAI

Published by

World Scientific Publishing Co. Pte. Ltd.
5 Toh Tuck Link, Singapore 596224
USA office: 27 Warren Street, Suite 401-402, Hackensack, NJ 07601
UK office: 57 Shelton Street, Covent Garden, London WC2H 9HE

British Library Cataloguing-in-Publication Data
A catalogue record for this book is available from the British Library.

TRENDS IN DIFFERENTIAL GEOMETRY, COMPLEX ANALYSIS AND MATHEMATICAL PHYSICS
Proceedings of the 9th International Workshop on Complex Structures, Integrability and Vector Fields

ISBN-13 978-981-4277-71-6
ISBN-10 981-4277-71-1

Printed in Singapore.

PREFACE

The 9-th International Workshop on Complex Structures, Integrability and Vector Fields was held at Sofia from August 25 to August 29, 2008. Throughout the series of these Workshops we have always been trying to keep intact with the top level achievements in the contemporary Complex Analysis, Differential Geometry and Mathematical Physics without ignoring their classical aspects.

With the years the overlap among different research areas in Mathematics and Physics has been increasing steadily. This reflected on the Workshop by introducing the word *Integrability* in its title, thus including in its programme a new research area deeply related to Mathematical Physics. Many deep and innovative results were reported during the Workshop; most of them are included in the present volume.

A new positive trend in the activity of the Workshop was the increased number of young scientists from South Korea, Poland, Ireland and Bulgaria as well. This may give further new breed and stream to the forthcoming Workshops.

With this volume we are paying deep respect to Professor Pierre Dolbeault from the Pierre et Marie Curie University, Paris. He participated in the 3rd Workshop at St. Konstantin and Elena, nearby Varna, in 1996 and was regarded as a teacher by many of the participants. We also congratulate Professor Julian Ławrynowicz from Łodź University, Poland for his 70-th birthday. Over the long years his regular and active participation has contributed a lot for their success.

Last, but not least, the Editors express their deepest gratitude to Professor T. Oguro for his constant outstanding co-operations and efforts in the preparation of this volume.

The Editors

The 9th International Workshop on Complex Structures, Integrability and Vector Fields

25–29 August 2008 – Bulgaria

ORGANIZING COMMITTEES

ORGANIZERS

University of Niigata, Japan
University of Pisa, Italy
Institute for Nuclear Research and Nuclear Energy of BAS, Bulgaria
Institute of Mathematics and Informatics of BAS, Bulgaria

ORGANIZING COMMITTEE

S. Dimiev (Chairman)	– IMI, Sofia
V. S. Gerdjikov (Vice-Chairman)	– INRNE, Sofia
St. Donev	– INRNE, Sofia
B. Iliev	– INRNE, Sofia
R. Lazov	– IMI, Sofia

SCIENTIFIC COMMITTEE

K. Sekigawa (Chairman)	– Niigata Univ.
H. Hashimoto (Vice-Chairmen)	– Meijyo Univ.
T. Adachi (Vice-Chairmen)	– Nagoya Inst. of Technology
Y. Mathsushita (Vice-Chairmen)	– Shiga Univ.
V. I. Georgiev (Vice-Chairmen)	– Pisa Univ.
S. Dimiev (Vice-Chairmen)	– IMI, Sofia
V. S. Gerdjikov (Vice-Chairmen)	– INRNE, Sofia
D. Alekseevski	– Hulls Univ.
E. Horozov	– Sofia Univ.

PRESENTATIONS

1. K. Sekigawa
   Tangent sphere bundles of constant radii
2. P. Popivanov
   Hypoellipticity, solvability, subellipticity and solutions with stationary singularities of pseudodifferential operators with symplectic characteristics
3. H. Hashimoto
   Deformations of almost complex structures of hypersurfaces of purely imaginary octonions
4. T. Adachi
   A discrete model of Kähler magnetic fields on a complex space form
5. H. Matsuzoe
   Some generalizations of statistical manifolds
6. I. P. Ramadanoff
   Monogenic, hypermonogenic and holomorphic Cliffordian functions
7. J. Lee, J. H. Park, K. Sekigawa
   Four-dimensional TV Bochner flat almost Hermitian manifolds
8. Y. Christodoulides
   Herglotz functions and spectral theory
9. S. Maeda
   A characterization of Clifford minimal hypersurfaces in terms of their geodesics
10. M. J. Hristov
    Some geometric properties and objects related to Bézier curves
11. N. A. Kostov, A. V. Tsiganov
    Lax pair for restricted multiple three wave interaction system, quasi-periodic solutions and bi-hamiltonian structure
12. R. I. Ivanov
    Integrable models for shallow water waves
13. D. Henry
    Persistence of solutions for some integrable shallow water equations

CONTENTS

A DISCRETE MODEL FOR KÄHLER MAGNETIC FIELDS ON A COMPLEX HYPERBOLIC SPACE

*DEDICATED TO PROFESSOR **TOSHIKAZU SUNADA** ON THE OCCASION OF HIS 60TH BIRTHDAY*

TOSHIAKI ADACHI

Department of Mathematics, Nagoya Institute of Technology,
Nagoya, 466-8555, JAPAN
E-mail: adachi@nitech.ac.jp

In this article we propose discrete models of trajectories for Kähler magnetic fields. We consider regular graphs whose edges are colored by 2 colors and compare the asymptotic behaviors of the number of prime cycles on them with those of prime trajectory-cycles on compact quatients of complex hyperbolic spaces.

Keywords: Kähler magnetic fields; Trajectories; Graphs; Prime cycles; Complex hyperbolic spaces.

1. Introduction

A graph (V, E) is a 1-dimensional CW-complex which consists of a set V of vertices and a set E of edges. Graphs are frequently considered as discrete models of Riemannian manifolds of negative curvatures. Paths on a graph correspond to geodesics on a Riemannian manifold, and the adjacency matrix which shows the position of edges corresponds to the Laplace-Bertrami operator.

The aim of this paper is to propose a discrete model for Kähler manifolds with Kähler magnetic fields. On a Kähler manifold M with complex structure J, we can consider Kähler magnetic fields which are constant multiples of the Kähler form $\mathbb{B}_J$. A smooth curve γ parameterized by its arclength is said to be a trajectory for a Kähler magnetic field $\mathbb{B}_k = k\mathbb{B}_J$ if it satisfies $\nabla_{\dot\gamma}\dot\gamma = kJ\dot\gamma$. Clearly, trajectories for a trivial Kähler magnetic

The author is partially supported by Grant-in-Aid for Scientific Research (C) (No. 20540071) Japan Society of Promotion Science.

field $\mathbb{B}_0$ are geodesics. The author has been studying some properties on trajectories in connection with geometry of base manifolds, hence he is interested in introducing their discrete models which correspond to discrete models of geodesics on Riemannian manifolds. In this paper we consider graphs whose edges are colored by 2 colors as such models and study the asymptotic behavior of the number of prime cycles with respect to their lengths.

2. Kähler magnetic fields on complex hyperbolic spaces

In order to explain our standing position, we shall start by recalling some properties of Kähler magnetic fields on a complex hyperbolic space $\mathbb{C}H^n(c)$ of constant holomorphic sectional curvature c. We denote by $\partial\mathbb{C}H^n$ the ideal boundary of a Hadamard manifold $\mathbb{C}H^n$. We say a trajectory γ for $\mathbb{B}_k$ on $\mathbb{C}H^n$ is unbounded in both directions if both of the sets $\gamma([0,\infty))$ and $\gamma((-\infty,0])$ are unbounded. We set $\gamma(\infty) = \lim_{t\to\infty}\gamma(t)$, $\gamma(-\infty) = \lim_{t\to-\infty}\gamma(t) \in \partial\mathbb{C}H^n$ if they exist. For a trajectory γ with bounded image we call it closed if there is a non-zero constant t_c with $\gamma(t+t_c) = \gamma(t)$ for all t. The minimum positive t_c with this property is called the length of γ and is denoted by $\ell(\gamma)$. It is known that every trajectory for a Kähler magnetic field on $\mathbb{C}H^n(c)$ lies on some totally geodesic submanifold $\mathbb{C}H^1(c)$. Moreover, trajectories are classified into three classes according to forces of magnetic fields.

1) If $|k| < \sqrt{|c|}$, then it is unbounded in both directions and satisfies $\gamma(\infty) \neq \gamma(-\infty)$.
2) When $k = \pm\sqrt{|c|}$, it is also unbounded in both directions but satisfies $\gamma(\infty) = \gamma(-\infty)$.
3) If $|k| > \sqrt{|c|}$, then it is closed of length $2\pi/\sqrt{k^2+c}$.

We see this fact more precisely. On the unit tangent bundle UM of a Kähler manifold M, a Kähler magnetic field $\mathbb{B}_k$ induces a flow $\mathbb{B}_k\varphi_t : UM \to UM$ which is defined by $\mathbb{B}_k\varphi_t(v) = \dot\gamma_v(t)$ with a trajectory γ_v whose initial vector is v. This flow is called a Kähler magnetic flow. For a trivial Kähler magnetic field $\mathbb{B}_0$, this flow $\varphi_t = \mathbb{B}_0\varphi_t$ is the geodesic flow. For a complex hyperbolic space $\mathbb{C}H^n$, Kähler magnetic flows are classified into three congruence classes; hyperbolic flows, horocycle flows and rotation flows.

Proposition 2.1 ([1]). *Let $M = \Gamma\backslash\mathbb{C}H^n(c)$ be a Riemannian quotient of a complex hyperbolic space of constant holomorphic sectional curvature c.*

A Kähler magnetic flow $\mathbb{B}_k\varphi_t : UM \to UM$ *is hyperbolic if and only if* $|k| < \sqrt{|c|}$. *It is congruent to the geodesic flow: For such k, there is a diffeomorphism* f_k *on* UM *with* $f_k \circ \mathbb{B}_k\varphi_t = \varphi_{\sqrt{|c|-k^2}\, t/\sqrt{|c|}} \circ f_k$.

In view of this result the feature of trajectories depends on the relation of forces of magnetic fields and sectional curvatures of the base manifold. Though the author gave some comparison theorems in [2,3], it is not enough to make clear this relationship. This also carries us to a study of discrete models for trajectories.

3. Graphs of Kähler type

Let $G = (V, E)$ be a graph. The ends of each edges are vertices. We call G oriented if each edge is oriented. For an oriented edge $e \in E$ we denote the starting vertex and the end vertex of e by $o(e)$, $t(e)$ and call them the origin and terminus of e, respectively. When all edges are not oriented, we call G a non-oriented graph. In this case we regard both two ends of an edge to be its origins and terminuses. When the origin and the terminus of an edge coincide we say this edge to be a loop. If there are two edges e_1, e_2 with $o(e_1) = o(e_2)$ and $t(e_1) = t(e_2)$ we call them multiple edges. If a graph does not have loops and multiple edges, we say this graph simple. The set E of edges of a simple graph hence is a subset of $V \times V \backslash \{(v, v) \mid v \in V\}$. When a simple graph is non-oriented, we consider $(v, w) \in E$ if and only if $(w, v) \in E$ and identify them. Given a simple graph $G = (V, E)$ with finite cardinality $\sharp(V)$ of V, we define its adjacency matrix $A_G = (a_{vw})$ by

$$a_{vw} = 1, \text{ if } (v, w) \in E, \quad a_{vw} = 0, \text{ if } (v, w) \notin E.$$

When we treat graphs it is usual to suppose there are no "hairs". That is, on a non-oriented graph for each vertex v there are at least two edges emanating from v, and on an oriented graph for each vertex v there are edges e_1, e_2 with $o(e_1) = v = t(e_2)$.

For a graph $G = (V, E)$, an element $c = (e_1, e_2, \ldots, e_n) \in E \times E \times \cdots \times E$ is said to be an n-step path if it satisfies $t(e_i) = o(e_{i+1})$. We put $\ell(c) = n$ and call it the length of c. An n-step path c is called closed if its origin $o(c) = o(e_1)$ and its terminus $t(c) = t(e_n)$ coincide. When G is a non-oriented graph, we say a path $c = (e_1, \ldots, e_n)$ has backtrackings if there is i_0 $(1 \leq i_0 \leq n-1)$ with $e_{i_0} = e_{i_0+1}$. When c is closed and $e_1 = e_n$ we also say that it has a backtracking. For a non-oriented graph it is usual to consider only paths without backtrackings. When c is an n-step closed path, for a divisor k of n, we say an n-step path c to be n/k-folded if $e_{i+k} = e_i$,

$i = 1, \ldots, n$, where indices are considered modulo n. If an n-step path is not k-folded for every divisor k of n, it is called prime. We call two closed n-step paths $c = (e_1, \ldots, e_n)$ and $c' = (e'_1, \ldots, e'_n)$ are congruent to each other if there is k with $e'_i = e_{i+k}$, $i = 1, \ldots, n$. A congruence class of closed paths is called a cycle. Clearly, prime cycles correspond to images of closed geodesics.

If we consider the universal covering of a graph as a CW-complex, it is a graph which is so-called a tree, which does not have closed paths. We can consider its ideal boundary as the set of limit points of unbounded paths from a base vertex. Hence graphs are considered as discrete models of Riemannian manifolds of negative curvature. For the sake of later use, we here recall graphs which are considered as models of quotients of non-compact symmetric spaces of rank one. A non-oriented graph $G = (V, E)$ is called regular if the order at $v \in V$, which is the cardinality $\sharp\big(o^{-1}(v)\big)$ of the set of edges emanating from v, does not depend on the choice of v. If we consider the order at v as the negativity of curvature at this vertex, we can consider regular graphs as models of quotients of real hyperbolic spaces. A non-oriented graph $G = (V, E)$ is called bipartite if the set of vertices are divided into two subsets $V = V_1 + V_2$ and each edge joins only a vertex in V_1 and a vertex in V_2. A bipartite graph $(V_1 + V_2, E)$ is called (m, n)-regular if the order at $v \in V_1$ is m and the order at $w \in V_2$ is n. In the case that $m = n$ it is an n-regular bipartite graph. From the viewpoint of sectional curvature, this can be seen as a discrete model of a quotient of a complex hyperbolic space and a quotient of a quaternionic hyperbolic space.

We now consider graphs where we can study models of trajectories for Kähler magnetic fields. It is needless to say that complex structures are real even dimensional objects and on contrary graphs are 1-dimensional. Hence, if we intend to use graphs as discrete models of Kähler manifolds we need to add some structures on graphs. We shall say a simple oriented/non-oriented graph (V, E) to be of *Kähler type* if edges are colored by 2-colors. More precisely, E is divided into 2 subsets as $E = E_1 + E_2$ satisfying that $o^{-1}(v) \cap E_i \neq \emptyset$ for $i = 1, 2$. We call edges in E_1 *principal*, and those in E_2 *auxiliary*. In order to get rid of the existence of hairs, we suppose at each vertex $v \in V$ that $\sharp\big(o^{-1}(v) \cap E_i\big) \geq 2$ for non-oriented graphs of Kähler type and that $t^{-1}(v) \cap E_i \neq \emptyset$ for oriented graphs of Kähler type if we need. We here note that the line graph of a graph of Kähler type is not necessarily a bipartite graph. A graph (V, E) of Kähler type is called *regular* if both of its principal graph (V, E_1) and its auxiliary graph (V, E_2) are regular. More clearly we say a graph of Kähler type is (m, n)-regular if its principal graph is m-regular and its auxiliary graph is n-regular. We here show figures of some regular non-oriented graphs of Kähler type.

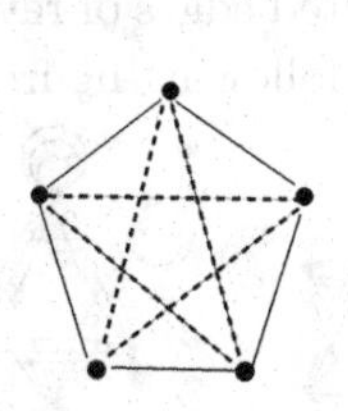

Fig. 1. $K_5(2,2)$

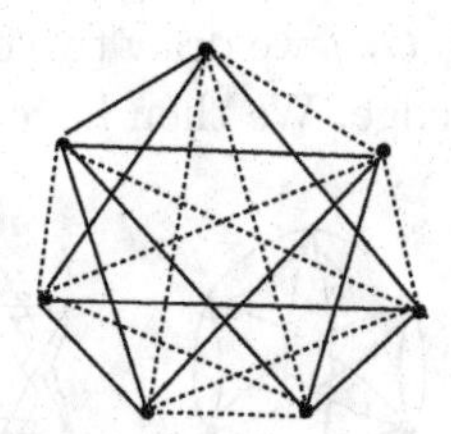

Fig. 2. $K_7(3,3)$

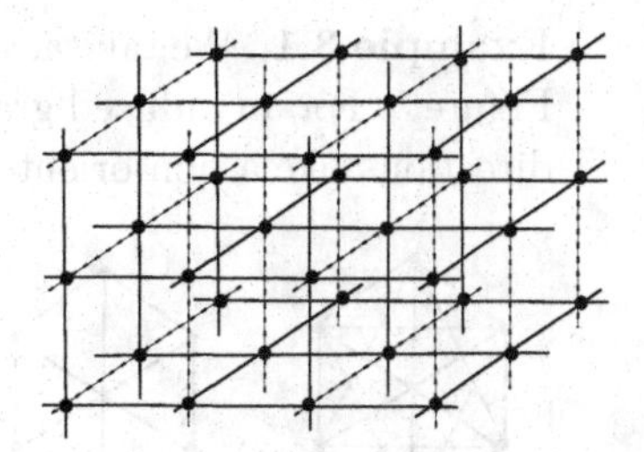

Fig. 3. complex Euclid

On a graph $(V, E_1 + E_2)$ of Kähler type we can consider models of trajectories for Kähler magnetic fields in the following manner. Let a, b be relatively prime positive integers. An n-step path $c = (e_1, \dots, e_n)$ is said to be an (a, b)-path if

i) $a + b$ is a divisor of n,
ii) $e_i \in E_1$, $i \equiv 1, 2, \dots, a \quad (\mathrm{mod}\ a + b)$,
iii) $e_i \in E_2$, $i \equiv a+1, a+2, \dots, a+b\ (\mathrm{mod}\ a + b)$.

We regard (a, b)-paths correspond to trajectories for a Kähler magnetic field $\mathbb{B}_{b/a}$. This means that an a-step path on the principal graph $G_{1,0} = (V, E_1)$ is bended under the force of a magnetic field and turns to an (a, b)-path on G. But as orders of vertices of the auxiliary graph $G_{0,1} = (V, E_2)$ are not 1, an a-step path on $G_{1,0}$ can be seen as bended many ways. In order to get rid of bifurcations of charged particles, we consider the auxiliary graph $G_{0,1}$ stochastically. That is, for $G_{0,1}$ we take the stochastic adjacency matrix $P = (p_{u,v})$ which is given as

$$p_{vw} = \begin{cases} 1/\sharp\big(o^{-1}(v)\big), & (v, w) \in E_2, \\ 0, & (v, w) \notin E_2. \end{cases}$$

Figure 4 shows a part of a graph of Kähler type. By use of this adjacency matrix, we consider as follows: Two $(3, 1)$-paths $(\mathrm{OABCD}), (\mathrm{OABCH})$ are 4-steps and of weight $1/2$, and the path (DEFGO) is of weight $1/3$.

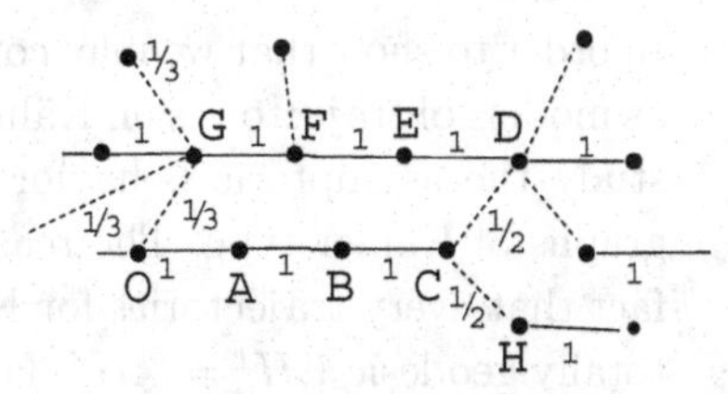

Fig. 4. weight of colored paths

We should note that colored paths on a graph $G = (V, E)$ of Kähler type induces oriented graphs. If we set $E_{a,b}$ to be the set of all $(a + b)$-step

(a, b)-paths on G, we obtain an oriented graph $G_{a,b} = (V, E_{a,b})$ which may have loops and multiple edges.

Example 3.1. We take a colored complete graph $K_6(3, 2)$ which is given in Figure 5. For an induced graph $G_{a,b}$ we draw two oriented edges of reversed directions like a non-oriented edge. We then have the following figures.

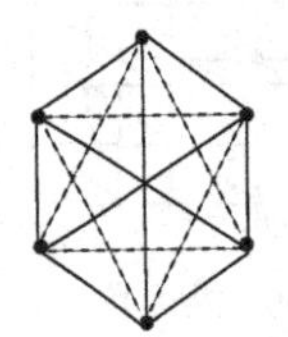

Fig. 5. $K_6(3, 2)$

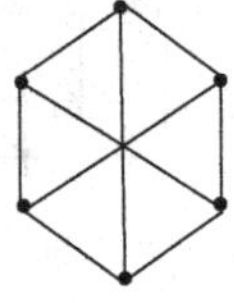

Fig. 6. G_0

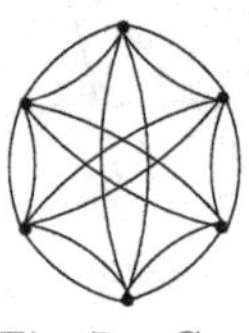

Fig. 7. $G_{1,1}$

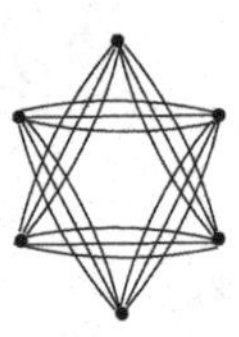

Fig. 8. $G_{2,0}$

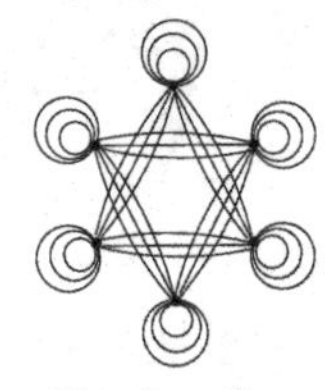

Fig. 9. $G_{2,1}$

We call a graph $G = (V, E)$ irreducible if for arbitrary $v, w \in V$ there is a path c with $o(c) = v$ and $t(c) = w$. We shall call a graph of Kähler type (a, b)-irreducible if the induced graph $(V, E_{a,b})$ is irreducible. In view of Example 3.1 we see a graph $K_6(3, 2)$ is $(1, 1)$-irreducible but is $(2, 1)$-reducible.

We say a graph $G = (V, E_1 + E_2)$ of Kähler type bipartite if V is divided into two subsets as $V = V_1 + V_2$ and both of principal graph $(V_1 + V_2, E_1)$ and auxiliary graph $(V_1 + V_2, E_2)$ are bipartite. For a bipartite graph of Kähler type we see the induced graphs $G_{a,b}$ with odd $a+b$ are also bipartite graphs and induced graphs $G_{a,b}$ with even $a+b$ are not connected. We call a bipartite graph $(V, E_1 + E_2)$ of Kähler type $((m, m'), (n, n'))$-regular if bipartite graphs (V, E_1) and (V, E_2) are (m, m')-regular and (n, n')-regular, respectively. In the case that $m = m'$ and $n = n'$ it is an (m, n)-regular bipartite graph of Kähler type.

4. Counting prime cycles

In order to show that we may consider (a, b)-paths on graphs of Kähler type as models of trajectories on Kähler manifolds of negative curvature, we shall study the asymptotic behavior of the number of prime cycles on regular graphs of Kähler type. The reason why we take regular graphs lies on the fact that every trajectories for Kähler magnetic fields on $\mathbb{C}H^n$ lies on some totally geodesic $\mathbb{C}H^1 = \mathbb{R}H^2$. For a graph G and a positive T, we denote by $\mathfrak{P}_G(T)$ the set of all prime cycles on G of length not greater than T. Given two functions $f, g : (0, \infty) \to \mathbb{R}$ or $f, g : \mathbb{N} \to \mathbb{R}$ we call them asymptotic to each other and denote by $f \sim g$ if they satisfy $\lim_{T\to\infty} f(T)/g(T) = 1$. It is well-konwn that the asymptotic behavior of the cardinality of the set $\mathfrak{P}_G(T)$

closely related with the maximum eigenvalue of the adjacency matrix of G. When a graph of Kähler type is regular, then its derived graphs are also regular. Applying Perron-Frobenius theorem, we obtain the following.

Theorem 4.1. *Let $G = (V, E)$ be a $(p+1, q+1)$-regular graph of Kähler type with $p \geq 2$ and $q \geq 1$. If the induced graph $G_{a,b}$ is irreducible and is not bipartite, then the asymptotic behavior of the number of prime cycles is*

$$\sharp(\mathfrak{P}_{G_{a,b}}(T)) \sim e^{h_{a,b}(G)T} \,/\, h_{a,b}(G)T \quad (T \to \infty).$$

Here

$$h_{a,b}(G) = \frac{\{\log(p+1) + (a-1)\log p + (b-1)\log\bigl(q/(q+1)\bigr)\}}{a+b}.$$

Theorem 4.2. *Let $G = (V, E)$ be a $(p+1, q+1)$-bipartite regular graph of Kähler type with $p \geq 2$ and with $q \geq 1$. If it is (a, b)-irreducible (in particular $a+b$ is odd), then the asymptotic behavior of the number of prime cycles is*

$$\sharp(\mathfrak{P}_{G_{a,b}}(2T)) \sim e^{2h_{a,b}(G)T} \,/\, h_{a,b}(G)T \quad (T \to \infty).$$

Here

$$h_{a,b}(G) = \frac{\{\log(p+1) + (a-1)\log p + (b-1)\log\bigl(q/(q+1)\bigr)\}}{a+b}.$$

We compare these results with the result on trajectories for Kähler magnetic fields. We say two trajectories γ_1, γ_2 are congruent to each other if there is t_0 with $\gamma_2(t+t_0) = \gamma_1(t)$ for all t, and say a congruence class of closed trajectories to be a cycle. A pair $\mathfrak{p} = ([\gamma], \ell(\gamma))$ of a cycle $[\gamma]$ and its length is called a prime cycle. We denote by $\mathfrak{P}_k(T; M)$ the set of all prime cycles for $\mathbb{B}_k$ on a Kähler manifold M whose lengths are not greater than T. We should note that the set $\mathfrak{P}_k(M)$ of all prime cycles for $\mathbb{B}_k$ on M is identified with the set of prime cycles of Kähler magnetic flow $\mathbb{B}_k\varphi_t$. When the magnetic flow $\mathbb{B}_k\varphi_t$ is hyperbolic, the generating function $\zeta_{\mathbb{B}_k\varphi_t}$ which is called the zeta function for $\mathbb{B}_k\varphi_t$ is defined by

$$\zeta_{\mathbb{B}_k\varphi_t}(s) = \prod_{\mathfrak{p}\in\mathfrak{P}_k(M)} \frac{1}{1 - e^{-s\ell(\mathfrak{p})}}.$$

By Proposition 2.1 we have the following.

Proposition 4.1. *Let $M = \Gamma\backslash\mathbb{C}H^n(c)$ be a compact quotient of a complex hyperbolic space. The zeta functions of the geodesic flow φ_t and a Kähler magnetic flow $\mathbb{B}_k\varphi_t$ with $|k| < \sqrt{|c|}$ for M are related as*

$$\zeta_{\mathbb{B}_k\varphi_t}(s) = \zeta_{\varphi_t}\bigl(\sqrt{|c|/(|c|-k^2)}\; s\bigr).$$

Therefore if we denote the topological entropy of $\mathbb{B}_k\varphi_t$ by $h_k(M)$ which coincides with $n\sqrt{|c|-k^2}$, it satisfies the following:

(1) *$\zeta_{\mathbb{B}_k\varphi_t}(s)$ converges absolutely and is holomorphic on $\Re(s) > h_k(M)$,*
(2) *$\zeta_{\mathbb{B}_k\varphi_t}(s)$ has a nowhere vanishing meromorphic extension on a domain containing $\Re(s) \geqq h_k(M)$,*
(3) *$\zeta_{\mathbb{B}_k\varphi_t}(s)$ has a simple pole at $s = h_k(M)$, and is holomorphic on $\Re(s) \geqq h_k(M)$.*

As a direct consequence we can conclude the following.

Corollary 4.1. *On a compact quotient $M = \Gamma\backslash\mathbb{C}H^n(c)$ of a complex hyperbolic space, if $|k| < \sqrt{|c|}$ then the asymptotic behavior of the number of prime cycles for $\mathbb{B}_k$ is as follows:*

$$\sharp(\mathfrak{P}_k(T;M)) \sim e^{h_k(M)T}/h_k(M)T \quad (T \to \infty).$$

We are interested in the constants $h_{a,b}(G)$ and $h_k(M)$. It is clear that the topological entropy $h_k(M)$ of $\mathbb{B}_k\varphi_t$ and that $h(M)$ of geodesic flow on UM are related as $h_k(M) = \sqrt{1-(k^2/|c|)}\,h(M)$. If we consider a trajectory segment $\gamma : [0,\ell] \to \mathbb{C}H^n(c)$ for $\mathbb{B}_k$ on the universal covering of M, we see the distance $d(\gamma(0),\gamma(\ell))$ between two end points of γ satisfies

$$\frac{\sinh\bigl(\sqrt{|c|-k^2}\,\ell/2\bigr)}{\sqrt{|c|-k^2}} = \frac{\sinh\bigl(\sqrt{|c|}\,d(\gamma(0),\gamma(\ell))/2\bigr)}{\sqrt{|c|}}.$$

Thus one can see that for about Kähler magnetic fields the relation on topological entropies and the relation between arc-lengths and distances of trajectory segments are closely related to each other. Next we consider $\pi_1(V,E_1)$ covering graph $\widetilde{G}$ of G. On this graph for an $(a+b)$-step (a,b)-path c we may consider the distance between $o(c)$ and $t(c)$ is equal to a because we can consider an a-step path on the principal graph is bended by a magnetic field to this path. On the other hand, for regular graphs it is known that their zeta functions are essentially the same as the Ihara zeta function associated with a cocompact torsion-free discrete subgroup of $PSL_2(\mathbb{Q}_P)$ over p-adic numbers. Therefore, the asymptotic behavior of the number of prime cycles on the principal graph $G_{1,0}$ is given as follows: If $G_{1,0}$ is irreducible and is not bipartite, then

$$\sharp(\mathfrak{P}_{G_{1,0}}(T)) \sim e^{h(G_{1,0})T}/h(G_{1,0})T \quad (T \to \infty),$$

and if $G_{1,0}$ is irreducible and is bipartite

$$\sharp(\mathfrak{P}_{G_{1,0}}(2T)) \sim e^{2h(G_{1,0})T}/h(G_{1,0})T \quad (T \to \infty),$$

where $h(G_{1,0}) = \log(p+1)$. Since the entropy $h_{a,b}(G)$ approximately equals to $\frac{1}{1+(b/a)}h(G_{1,0})$, the relation of entropies and the relation between steps and distances of colored paths have a similar relationship as for trajectories for Kähler magnetic fields. From this point of view, we may say that colored paths on graphs of Kähler type can be regard as discrete models of trajectories for Kähler magnetic fields.

References

1. T. Adachi, *Kähler magnetic flows on a manifold of constant holomorphic sectional curvature,* Tokyo J. Math. 18 (1995), 473–483.
2. T. Adachi, *A comparison theorem on crescents for Kähler magnetic fields,* Tokyo J. Math. 28 (2005), 289–298.
3. T. Adachi, *A comparison theorem on sectors for Kähler magnetic fields,* Proc. Japan Acad. Sci. 81 Ser. A (2005), 110–114.
4. T. Adachi and T. Sunada, *Twisted Perron-Frobenius theorem and L-functions,* Journal of Functional Analysis 71 (1987), 1–46.
5. Y. Ihara, *On discrete subgroups of the two by two projective linear group over p-adic field,* J. Math. Soc. Japan 18 (1966), 219–235.
6. J.P. Serre, *Tree,* Springer New York, 1980.
7. T. Sunada, *Fundamental groups and Laplacians — Number theoretc methods in geometry —*, in Japanese, Kinokuniya 1988.
8. A. Terras, *A stroll through the garden of graph zeta functions,* preprint, http://math.ucsd.edu/˜aterras/newbook.pdf.

INTEGRABILITY CONDITION ON THE BOUNDARY PARAMETERS OF THE ASYMMETRIC EXCLUSION PROCESS AND MATRIX PRODUCT ANSATZ

BOYKA ANEVA

Institute for Nuclear Research and Nuclear Energy,
Bulgarian Academy of Sciences,
blvd. Tzarigradsko chaussee 72, 1784 Bulgaria
E-mail: blan@inrne.bas.bg

We consider the question whether with the constraint from the Bethe ansatz solution on the ASEP boundary parameters the matrix product ansazt fails to produce the stationary state. We discuss applicability of the matrix product ansatz from the point of view of the boundary algebra which reveals deep symmetry properties of the ASEP. We argue that the Bethe ansatz integrability constraint is the defining condition for a finite dimensional representation of the boundary algebra when the system reaches a steady state satisfying detailed balance.

Keywords: Stochastic system; Nonequilibrium physics; Steady state; Boundary algebra; Orthogonal polynomials

1. Introduction

The asymmetric simple exclusion process (ASEP) has become a paradigm in nonequilibrium physics [1,2] due to its simplicity, rich behaviour and wide range of applicability. It is an exactly solvable model of an open many-particle stochastic system interacting with hard core exclusion. Introduced originally as a simplified model of one dimensional transport for phenomena like hopping conductivity [3] and kinetics of biopolymerization [4], it has found applications from traffic flow [5] to interface growth [6], shock formation [7], hydrodynamic systems obeying the noisy Burger equation, problems of sequence alignment in biology [8]. At large time the ASEP exhibits relaxation to a steady state, and even after the relaxation it has a nonvanishing current. An intriguing feature is the occurrence of boundary induced phase transitions [9] and the fact that the stationary bulk properties strongly depend on the boundary rates.

The ASEP dynamics is governed by a master equation for the probability distribution $P(s_i, t)$ of a stochastic variable $s_i = 0, 1$, at a site $i = 1, 2, \ldots, L$ of a linear chain. In the set of occupation numbers $(s_1, s_2, \ldots, s_L)$ specifying a configuration of the system $s_i = 0$ if a site i is empty, or $s_i = 1$ if the site i is occupied. On successive sites particles hop with probability $g_{01}dt$ to the left, and $g_{10}dt$ to the right. The event of hopping occurs if out of two adjacent sites one is a vacancy and the other is occupied by a particle. The symmetric simple exclusion process is the lattice gas model of particles hopping between nearest-neighbour sites with a constant rate $g_{01} = g_{10} = g$. The asymmetric simple exclusion process with hopping in a preferred direction is the driven diffusive lattice gas of particles moving under the action of an external field. The process is totally asymmetric if all jumps occur in one direction only, and partially asymmetric if there is a different non-zero probability of both left and right hopping. The number of particles in the bulk is conserved and this is the case of periodic boundary conditions. In the case of open systems, the lattice gas is coupled to external reservoirs of particles of fixed density and additional processes can take place at the boundaries. Namely, at the left boundary $i = 1$ a particle can be added with probability αdt and removed with probability γdt, and at the right boundary $i = L$ it can be removed with probability βdt and added with probability δdt. The master equation

$$\frac{dP(s,t)}{dt} = \sum_{s'} \Gamma(s, s')P(s', t)$$

can be mapped to a Schrödinger equation in imaginary time

$$\frac{dP(t)}{dt} = -HP(t)$$

for a quantum Hamiltonian with nearest-neighbour interaction in the bulk and single-site boundary terms. Through the mapping the equivalence to the integrable spin 1/2 XXZ quantum spin chain with anisotropy $\Delta = -(q + q^{-1})/2$, $q = g_{01}/g_{10}$ and most general non diagonal boundary terms was obtained [10] which allowed for the derivation of exact results for the ASEP using Matrix Product Approach (MPA) and Bethe Ansatz (BA).

Bethe Ansatz solution was recently achieved [11] for the XXZ chain Hamiltonian and subsequently for the ASEP [12] provided the model parameters satisfy a condition which in terms of the ASEP notations reads

$$(q^{\frac{1}{2}(L+2k)} - 1)(\alpha\beta - q^{\frac{1}{2}(L-2k-2)}\gamma\delta) = 0 \tag{1}$$

with $|k| \leq L/2$. Given k, the first factor zero in (1) assumes q to be a root of unity which is unacceptable for the ASEP. The second factor zero imposes a relation between the bulk and the boundary parameters and it is believed [10,13] that with this constraint the MPA fails to produce the stationary state. Quite remarkably, however, for generic q, one can satisfy the constraint [12] by choosing k to be $k = -L/2$, which, as commented in [13] implies additional symmetries for the ASEP.

Emphasizing the dependence of the steady state bulk behaviour on the boundary rates we consider the boundary Askey-Wilson algebra (as well as the tridiagonal one) which reveals hidden symmetries of the ASEP. We discuss the consequences of the algebraic properties on the applicability of the MPA in relation to the BA integrability condition.

2. Matrix Product State Ansatz (MPA)

The idea is that the steady state properties of the ASEP can be obtained exactly in terms of matrices obeying a quadratic algebra [7,14]. For a given configuration $(s_1, s_2, \ldots, s_L)$ the stationary probability is defined by the expectation value $P(s) = \langle w|D_{s_1} D_{s_2} \ldots D_{s_L}|v\rangle / Z_L$, where $D_{s_i} = D_1$ if a site $i = 1, 2, \ldots, L$ is occupied and $D_{s_i} = D_0$ if a site i is empty and $Z_L = \langle w|(D_0 + D_1)^L|v\rangle$ is the normalization factor to the stationary probability distribution. The operators $D_i, i = 0, 1$ satisfy the quadratic (bulk) algebra

$$D_1 D_0 - q D_0 D_1 = x_1 D_0 - D_1 x_0, \qquad x_0 + x_1 = 0 \tag{2}$$

with boundary conditions of the form

$$\begin{aligned} (\beta D_1 - \delta D_0)|v\rangle &= x_0|v\rangle, \\ \langle w|(\alpha D_0 - \gamma D_1) &= -\langle w|x_1. \end{aligned} \tag{3}$$

and $\langle w|v\rangle \neq 0$. We stress the one parameter dependence of the MPA due to $x_0 = -x_1 = \zeta$ with $0 < \zeta < \infty$. In most known applications it is restricted to the choice $\zeta = 1$. In our opinion the relation $x_0 + x_1 = 0$ implies an abelian symmetry with a conserved quantity $D_0 + D_1$, following from $D_0 \to D_0 + x_0$, $D_1 \to D_1 + x_1$.

The MPA is an efficient method to evaluate exactly all the relevant physical quantities such as the mean density at a site i, the correlation functions, the current J through a bond between sites i and $i+1$. Exact results for the ASEP with open boundaries were obtained within the MPA through the relation of the stationary state to q-Hermite [16] and Al-Chihara polynomials [17] in the case $\gamma = \delta = 0$ and to the Askey-Wilson polynomials [18] in the general case. Finite dimensional representations [10,15] have been considered too and they simplify calculations. Due to a constraint on the model parameters they define an invariant subspace of the infinite matrices and give exact results only on some special curves of the phase diagram. The MPA was readily generalized to many-species models and to dynamical MPA [19].

In the general case of incoming and outgoing particles at both boundaries with all boundary parameters nonzero there are four operators βD_1, $-\delta D_0$, $-\gamma D_1$, αD_0 and one needs an addition rule to form two linear independent boundary operators acting on the dual boundary vectors. A solution to the boundary problem can be obtained by using the bulk $U_q(su(2))$ symmetry algebra written as a deformed $(u,-u)$-algebra ($u<0$ and $0<q<1$)

$$[N, A_{\pm}] = \pm A_{\pm} \qquad [A_-, A_+] = uq^N - uq^{-N} \tag{4}$$

with a central element $Q = A_+A_- - u(q^N + q^{1-N})/(1-q)$. The representations, labeled by the values of the Casimir $Q(\kappa) = -u(q^\kappa + q^{1-\kappa})(1-q)$ for a fixed parameter κ, are finite dimensional and in a basis $|n,\kappa\rangle$ given by

$N|n,\kappa\rangle = (\kappa+n)|n,\kappa\rangle$, $A_-|n,\kappa\rangle = r_n|n-1,\kappa\rangle$, $A_+|n,\kappa\rangle = r_{n+1}|n+1,\kappa\rangle$,

where $r_n^2 = ((1-q^n)(-uq^\kappa + uq^{1-n-\kappa})/(1-q)$. ($|0,\kappa\rangle$ is the vacuum with $r_0 = 0$.) The dimension $l+1$ of the spin κ-representation is defined by the condition

$$-uq^\kappa + uq^{-l-\kappa} = 0 \tag{5}$$

for some $n = l$. Given the $U_q(su(2))$ generators $A_\pm$ and N we can present the boundary operators in the form

$$\begin{aligned}
\beta D_1 - \delta D_0 &= -\frac{x_1\beta}{\sqrt{1-q}} q^{N/2} A_+ - \frac{x_0\delta}{\sqrt{1-q}} A_- q^{N/2} \\
&\quad - \frac{x_1\beta q^{1/2} + x_0\delta}{1-q} q^N - \frac{x_1\beta + x_0\delta}{1-q}, \\
\alpha D_0 - \gamma D_1 &= \frac{x_0\alpha}{\sqrt{1-q}} q^{-N/2} A_+ + \frac{x_1\gamma}{\sqrt{1-q}} A_- q^{-N/2} \\
&\quad + \frac{x_0\alpha q^{-1/2} + x_1\gamma}{1-q} q^{-N} + \frac{x_0\alpha + x_1\gamma}{1-q}.
\end{aligned} \tag{6}$$

Separating the shift parts from the boundary operators and denoting the corresponding rest parts by A and A^* we can prove that the operators A and A^* defined by

$$\begin{aligned}
A &= \beta D_1 - \delta D_0 + \frac{x_1\beta + x_0\delta}{1-q}, \\
A^* &= \alpha D_0 - \gamma D_1 - \frac{x_0\alpha + x_1\gamma}{1-q},
\end{aligned} \tag{7}$$

and their q-commutator $[A, A^*]_q = q^{1/2}AA^* - q^{-1/2}A^*A$ form a closed linear algebra

$$[[A, A^*]_q, A]_q = -\rho A^* - \omega A - \eta,$$
$$[A^*, [A, A^*]_q]_q = -\rho^* A - \omega A^* - \eta^*, \tag{8}$$

where the operator-valued structure constants depend on $q, \alpha, \beta, \gamma, \delta$ and the $su_q(2)$ Casimir Q. In particular

$$-\rho = x_0 x_1 \beta\delta q^{-1}(q^{1/2} + q^{-1/2})^2,$$
$$-\rho^* = x_0 x_1 \alpha\gamma q^{-1}(q^{1/2} + q^{-1/2})^2. \tag{9}$$

Relations (8) are the well known Askey-Wilson (AW) relations [20,21] for the shifted boundary operators A, A^*, from which the defining relations of a tridiagonal algebra follow as well. This is an associative algebra with a unit generated by a (tridiagonal) pair of operators A, A^* [22,23], $A \to tA + c$, $A^* \to t^*A^* + c^*$ where t, t^*, c, c^* are some scalars.

The AW algebra (8) possesses important properties that allow to obtain its ladder representations, spectra, overlap functions. Namely, there exists a basis (of orthogonal polynomials) f_r according to which the operator A is diagonal and the operator A^* is tridiagonal. There exists a dual basis f_p in which the operator A^* is diagonal and the operator A is tridiagonal. The overlap function of the two basis $\langle s|r\rangle = \langle f_s^*|f_r\rangle$ is expressed in terms of the Askey-Wilson polynomials. Let $p_n = p_n(x; a, b, c, d|q)$ denote the nth AW polynomial [25] depending on four parameters a, b, c, d, with $p_0 = 1$, $x = y + y^{-1}$, $0 < q < 1$ and a three term recurrence relation

$$xp_n = b_n p_{n+1} + a_n p_n + c_n p_{n-1}, \qquad p_{-1} = 0. \tag{10}$$

We need only the explicit form of the matrix elements b_n

$$b_n = \frac{(1 - abq^n)(1 - acq^n)(1 - adq^n)(1 - abcdq^{n-1})}{a(1 - abcdq^{2n-1})(1 - abcdq^{2n})}. \tag{11}$$

After rescaling $A \to \frac{1}{\beta}A$, $A^* \to \frac{1}{\alpha}A^*$ we have found the explicit form of the infinite dimensional representation (and the dual one) for the boundary operators. Let $\mathcal{A}$ denote the matrix satisfying (10) in the basis $(p_0, p_1, p_2, \dots)$. *Result*: There exist a representation π in the space of Laurent polynomials with basis $(p_0, p_1, p_2, \dots)^t$ with respect to which $\pi(D_1 - \frac{\delta}{\beta}D_0)$ is diagonal with eigenvalues

$$\lambda_n = \frac{1}{1-q}\left(bq^{-n} + dq^{n-1}\right) + \frac{1}{1-q}(1 + bd) \tag{12}$$

and $\pi(D_0 - \frac{\gamma}{\alpha}D_1)$ is tridiagonal

$$\pi(D_0 - \frac{\gamma}{\alpha}D_1) = \frac{1}{1-q}b\mathcal{A}^t + \frac{1}{1-q}(1+ac). \tag{13}$$

The dual representation π^* has a basis $p_0, p_1, p_2, \ldots$ with respect to which $\pi^*(D_0 - \frac{\gamma}{\alpha}D_1)$ is diagonal with eigenvalues

$$\lambda_n^* = \frac{1}{1-q}\left(aq^{-n} + cq^n\right) + \frac{1}{1-q}(1+ac) \tag{14}$$

and $\pi^*(D_1 - \frac{\delta}{\beta}D_0)$ is tridiagonal

$$\pi^*(D_1 - \frac{\delta}{\beta}D_0) = \frac{1}{1-q}a\mathcal{A} + \frac{1}{1-q}(1+bd). \tag{15}$$

The choice $\langle w| = h_0^{-1/2}(p_0, 0, 0, \ldots)$, $|v\rangle = h_0^{-1/2}(p_0, 0, 0, \ldots)^t$ (h_0 is a normalization) as eigenvectors of the diagonal matrices $\pi(D_1 - \frac{\delta}{\beta}D_0)$ and $\pi^*(D_0 - \frac{\gamma}{\alpha}D_1)$ yields a solution to the boundary equations which uniquely relates a, b, c, d to $\alpha, \beta, \gamma, \delta$. Namely

$$a = \kappa_+^*(\alpha, \gamma), \qquad b = \kappa_+(\beta, \delta), \qquad c = \kappa_-^*(\alpha, \gamma), \qquad d = \kappa_-(\beta, \delta),$$

where

$$\kappa_\pm^{(*)}(\nu, \tau) = \frac{-(\nu - \tau - (1-q)) \pm \sqrt{(\nu - \tau - (1-q))^2 + 4\nu\tau}}{2\nu}.$$

These are the functions of the parameters which define the phase diagram of the ASEP. They have been used in previously known MPA applications where have always been taken for granted. It is remarkable that here they follow from the AW algebra representations.

The condition for the representation to be finite dimensional is $b_n = 0$ for some $n = n_f$, where $n_f = 1, 2, \ldots$ is the dimension of the representation. From the explicit form of b_n it follows that the representation is finite dimensional if any of the factors in the numerator in (11) is zero, in particular, with our normalization $abcd = \frac{\gamma\delta}{\alpha\beta}$, if the condition holds

$$\alpha\beta = q^{n_f - 1}\gamma\delta. \tag{16}$$

Since we are using $U_q(su(2))$ besides the constraint (16) there is one more constraint (given by eq. (5)) which is the first factor in (1). Comparison with the second factor in (1) (with the value k from the first factor) suggests that it defines a finite dimensional representation of the MPA boundary operators of dimension $n_f = L$. Finite dimensional representations of the boundary algebra imply finite dimensional matrices D_0 and D_1. The examples considered in [15] correspond to the condition $ab = \kappa_+^*\kappa_+ = q^{1-L}$, which defines an L-dimensional representation of the boundary algebra according to the first factor in the nominator of b_n. A special case of the tridiagonal algebra with $c = d = 0$ is the open system with only incoming (outgoing) particles at the left (right) boundary when the (shifted) diagonal elements are of the form $q^{\pm n}$ and the tridiagonal operators satisfy the recurrence relation for either q-Hermite or Al-Chihara polynomials. A nontrivial special case of an open system with four boundary parameters related to the representations above with $b = 0$ and $c = 0$, was considered in [26] for the MPA solution of the weakly ASEP. The constraint $\alpha\beta = q^{L-1}\gamma\delta$ was found to describe the detailed balance when the steady state is a Bernoulli measure at density, the same for both reservoirs.

3. Detailed Balance and MPA

Detailed balance (DB) means that the probability $P(\{s\})$ for a transition from a configuration $\{s\}$ to another configuration $\{s'\}$ is equal to the probability of the reverse transition. Thus for the event of hopping between site i and $i+1$ one has

$$P(s_1, \dots, 0, 1, \dots, s_L) = q^{-1} P(s_1, \dots, 1, 0, \dots, s_L), \tag{17}$$

for an event at the left boundary -

$$P(1, s_2, \dots, s_L) = \frac{\alpha}{\gamma} P(0, s_2, \dots, s_L) \tag{18}$$

and correspondingly at the right boundary

$$P(s_1, \dots, s_{L-1}, 1) = \frac{\delta}{\beta} P(s_1, \dots, s_{L-1}, 0). \tag{19}$$

Starting from a given configuration and using eqs. (17) and (18) one can always calculate the weights of all configurations by removing particles at the left boundary and the consistency with eq. (19) [26] requires $\alpha\beta = q^{L-1}\gamma\delta$ which is the DB condition. As can be readily verified the above DB relations with P expressed as matrix elements of the type $\langle w|D_{s_1} D_{s_2} ... D_{s_L}|v\rangle$ are consistent with the limit $x_0 = -x_1 = 0$ of the MPA defining relations (2) and (3). The presence of the parameter dependent linear terms in the bulk algebra is due to the boundary processes and these quadratic-linear algebraic relations provide recursive expressions for matrix elements of the steady weights, the current, the correlation functions. The quadratic algebra induces a reordering property and an element of length L can be brought in a linear combination of elements of length $L-1$ with positive coefficients. Hence one can compute all matrix elements of length L if $\langle w|D_0^k D_1^{L-k}|v\rangle$,

$k = 0, 1, \ldots, L$, and all $(L-1)$- elements are known. From the boundary conditions (3) and after reordering one has

$$\beta\langle w|D_0^k D_1^{L-k}|v\rangle - q^{L-k-1}\delta\langle w|D_0^{k+1}D_1^{L-k-1}|v\rangle = x_0 P_{L-1},$$
$$-q^k\gamma\langle w|D_0^k D_1^{L-k}|v\rangle + \alpha\langle w|D_0^{k+1}D_1^{L-k-1}|v\rangle = -x_1 P'_{L-1},$$

where P_{L-1} and P'_{L-1} are positive linear combinations of matrix elements of length $L-1$. The conclusion is straightforward: If $\alpha\beta - q^{L-1}\gamma\delta \neq 0$, then the recursions are consistent. If $\alpha\beta - q^{L-1}\gamma\delta = 0$, then in order that a matrix element of length L exists, the zero determinant $x_0\alpha P_{L-1} - x_1 q^{L-k-1}\delta P'_{L-1} = 0$ implies $x_0 = -x_1 = 0$. (All matrix elements of length $L-k$, with $k = 0, 1, \ldots, L-1$ are nonzero if $\alpha\beta - q^{L-k-1}\gamma\delta = 0$ for some $\alpha, \beta, \gamma, \delta$.) This has the consequence that the current

$$J = \frac{\langle w|(D_0+D_1)^{i-1}(D_1D_0 - qD_0D_1)(D_0+D_1)^{L-i-1}|v\rangle}{Z_L}$$

vanishes and the probabilities satisfy the DB condition. Thus with the constraint on the boundary parameters even though the boundary processes are present they become irrelevant and the nonequilibrium behaviour of the system is no longer maintained. The MPA in the limit $x_0 = x_1 = 0$ produces a stationary state whose probability weights satisfy DB.

The boundary operators of the detailed balance case satisfy an AW algebra with $\eta = \eta^* = 0$ and $\rho = \rho^*$, $\omega \simeq Q$. Without writing the explicit form of the representation, related now to q-Hahn polynomials, we note that in the limit $x_0 = -x_1 = 0$ from the boundary equations it follows $b = d$ and $a = c$, with $b = \sqrt{-\frac{\beta}{\delta}}$ and $a = \sqrt{-\frac{\gamma}{\alpha}}$. Inserting it in the condition $abcd = a^2b^2 = q^{1-L}$ we obtain the DB condition. Formally $a^2b^2 = q^{1-L}$ for some parameters $a' = -a^2, b' = -b^2$, with $L = 1$, resembles the limit $q \to 1$ and we recover the detailed balance case at equal density $\rho_{a'} = \rho_{b'}$, with $\rho_{a'} = \frac{\alpha}{\alpha+\gamma}$, $\rho_{b'} = \frac{\beta}{\beta+\delta}$ [27]. The system will be in a product measure state with uniform density and zero current when $a'b' = 1$, valid for the one dimensional representation. We can identify the one dimensional representation of the boundary operators with $\langle w|D_1|v\rangle$ and $\langle w|D_0|v\rangle$, which are non zero in the limit $x_0 = -x_1 = 0$. A non vanishing matrix element of length L can be formed as the product of single-site matrix elements $\langle w|D_1|v\rangle^{L-k}$ and $\langle w|D_0|v\rangle^k$. Given the DB condition $\alpha\beta = q^{L-1}\gamma\delta$, we can always find corresponding $\alpha'\beta' = \gamma'\delta'$ which define the one dimensional representation of the boundary algebra and hence determine the steady state as a Bernoulli product measure at equal density for both reservoirs.

4. Conclusion

The MPA defining algebraic relations (2) and boundary conditions (3) provide solvable recursions for the stationary state, the current, the correlation functions of an open L-site nonequilibrium system for all values of the parameters in the range $0 \leq \alpha, \beta, \gamma, \delta, \leq 1$, $\alpha\gamma \neq 0$, $\beta\delta \neq 0$, $0 < q < 1$ and $x_0 = -x_1 = \zeta > 0$. The recursions remain valid with final dimensional matrices of dimension L determined by (one of) the conditions $\kappa_+(\alpha, \gamma)\kappa_+(\beta, \delta) = \kappa_+(\alpha, \gamma)\kappa_-(\beta, \delta) = q^{1-L}$, which also define L dimensional representations of the boundary AW algebra. Exceptional subrange of parameters is given by the constraints $\alpha\beta = q^{L-1}\gamma\delta$ and $\zeta = 0$ when the MPA produces a stationary state satisfying detailed balance.

In the BA condition (1) the first factor zero with $k = -L/2$ defines a representation of dimension $L+1$ of the bulk $U_q(su(2))$ symmetry. The second factor zero defines a finite L-dimensional representation of the boundary AW algebra. The BA condition relates a spin $|k| = L/2$ representation of the $U_q(su(2))$ symmetry to a finite L representation of the boundary algebra. The Bethe integrability condition on the boundary parameters corresponds to finite dimensional representations of the MPA matrices which constrain the parameters of the open system to a range when the system will eventually reach a steady state satisfying detailed balance. For arbitrary values of the ASEP parameters the BA condition is automatically satisfied by the spin $|k| = L/2$ representation of the bulk $U_q(su(2))$. The symmetry behind this property is the boundary algebra, which assures the exact solvability and allows for employing Bethe ansatz or applying matrix product ansatz with no restriction on the physics of the system.

Acknowledgments

The author is grateful to the organizers for the invitation to participate in the workshop.

References

1. V. Privman (ed), *Nonequilibrium Statistical Mechanics in one dimension*, Cambridge University Press, 1997.
2. G.M. Schuetz, *Phase Trans. and Crit. Phen.* **19**, Academic Press, London, 2000.
3. P.M. Richards, *Phys. Rev* **B16**, 1393 (1977).
4. J.T. Macdonald, J.H. Gibbs and A.C. Pipkin, Biopolymers **6**, 1 (1968).
5. M. Schreckenberg, A. Schadschneider, K. Nagel and N. Ito, Phys. Rev. **E51**, 2939 (1995).

6. J. Krug and H. Spohn, in *Solids far from equilibrium*, eds. C. Godreche, Cambridge University Press, 1991.
7. B. Derrida, Physics Reports, **301**, 65 (1998).
8. R. Bundschuh, Phys. Rev **E65**, 031911 (2002).
9. J. Krug, Phys. Rev. Lett, **67**, 1882, (1991).
10. F.H. L. Essler and V. Rittenberg, J. Phys. **A29**, 3375 (1996).
11. R.I. Nepomechie, J. Stat. Phys. **111**, 1363 (2003); J. Phys. **A 37**, 433 (2004).
12. J. de Gier and F.H.L. Essler, Phys. Rev. Lett. **95**, 240601 (2005); J. Stat. Mech. P12011 (2006).
13. J. de Gier and F.H.L. Essler, J. Stat. Mech., P12011 (2006).
14. B. Derrida, M.R. Evans, V. Hakim and V. Pasquier, J. Phys. **A26**, 1493 (1993).
15. K. Malick and S. Sandow, J. Phys. **A30**, 4513 (1997).
16. R.A. Blythe, M.R. Evans, F. Colaiori and F.H.L. Essler, J. Phys. **A33**, 2313 (2000).
17. T. Sasamoto, J. Phys. **A32**, 7109 (1999).
18. M. Uchliama, T. Sasamoto, M. Wadati, J. Phys. **A37**, 4985 (2004).
19. R.B. Stinchcombe and G.M. Schuetz, Phys. Rev. Lett. **75**, 140 (1995).
20. A.S. Zhedanov, Theor Math Fiz **89** 190 (1991).
21. Y.A. Granovskii and A.S. Zhedanov, J. Phys. A **26**(1993) L357.
22. P. Terwilliger, Lec. Notes in Math. **1883**, 225 (2006).
23. P. Terwilliger, math. QA/0307016.
24. L. Dolan and M. Grady, Phys. Rev. D **3**, 1587 (1982).
25. R.A. Askey, J.A. Wilson, Mem. Amer. Math. Soc. 54 (319) (1985).
26. C. Enaud and B. Derrida, J. Stat. Phys. **114**, 537 (2004).
27. B. Derrida, J.L. Lebowitz, J.R. Speer, J. Stat. Phys. **107**, 599 (2002).

REMARKS ON THE DOUBLE-COMPLEX LAPLACIAN

L. APOSTOLOVA

Institute of Mathematics and Informatics,
of the Bulgarian Academy of Sciences,
Sofia 1113, Bulgaria
E-mail: liliana@math.bas.bg

These remarks are devoted to a basic properties of the double-complex Laplace operator $\Delta_+ U(z,w) = \frac{\partial^2 U(z,w)}{\partial z^2} + i\frac{\partial^2 U(z,w)}{\partial w^2}$, where z, w are complex variables and $U(z,w)$ is a double-complex value function. A decomposition into two complex operators is given. An initial problem for the double-complex Laplace operator on the whole space $\mathbb{C}(1,j)$ is solved with a D'Alember's type formula. An initial problem on the cartesian product of two unit squares in $\mathbb{C}$ for the zero eigenvalue problem for the double-complex Laplace operator is treated with the Fourier method of separating of the variables in double-complex context. The corresponding system of two equations of second order and the corresponding operator of fourth order with real coefficients are obtained.

Keywords: Double-complex number; Fundamental solution; Double-complex Laplace operator; Ultrahyperbolic system; Double-complex Laplace operator.

1. Double-complex Laplace operator

Let $\mathbb{C}(1,j)$ be the commutative algebras of the arranged couple of complex numbers (z,w) written also as $z + jw$, $z, w \in \mathbb{C}$ with an unit j having the property $j^2 = i$, i being the imaginary unit in $\mathbb{C}$. This is a commutative algebras without division, as the numbers $(z,w) : z^2 - iw^2 = 0$ are divisors of the zero in $\mathbb{C}(1,j)$. Actually, $(z+jw)(z-jw) = z^2 - j^2w^2 = z^2 - iw^2 = 0$ and $z + jw$ is a divisor of the zero in the case $z^2 - iw^2 = 0$.

We shall consider the functions of double complex variables with values double-complex variables, i.e. functions of the type $f : U \to \mathbb{C}(1,j)$, where U is an open subset in the algebra $\mathbb{C}(1,j)$.

In papers [1,5] the following differential operator of second order with complex coefficients

$$\Delta_+ = \frac{\partial^2}{\partial z^2} + i\frac{\partial^2}{\partial w^2} \tag{1}$$

is obtained. This operator is called a double-complex Laplace operator. Differential operators arising for other generalizations of the numbers are given in [2].

Let us show the factorization of the operator Δ_+ as a composition of two operators of first order with double-complex coefficients. We shall make the following transformations:

$$\begin{aligned}\Delta_+ &= \frac{\partial^2}{\partial z^2} + i\frac{\partial^2}{\partial w^2} = \frac{\partial^2}{\partial z^2} + j^2\frac{\partial^2}{\partial w^2} = \frac{\partial^2}{\partial z^2} - j^6\frac{\partial^2}{\partial w^2} \\ &= \left(\frac{\partial}{\partial z} + j^3\frac{\partial}{\partial w}\right)\left(\frac{\partial}{\partial z} - j^3\frac{\partial}{\partial w}\right) \\ &= \left(\frac{\partial}{\partial z} + ji\frac{\partial}{\partial w}\right)\left(\frac{\partial}{\partial z} - ji\frac{\partial}{\partial w}\right) = \partial\partial^*.\end{aligned}$$

This factorizations shows the appropriate Cauchy-Riemann equations for double-complex functions and the conjugate Cauchy-Riemann equation associated with operator Δ_+.

Now let us consider the double complex Laplace operator $\Delta_+ = \dfrac{\partial^2}{\partial z^2} + i\dfrac{\partial^2}{\partial w^2}$, which is a second order partial differential operator with complex coefficients. We shall obtain the complex symbol of the double-complex Laplace operator

$$\Delta_+ = \frac{1}{4}\left(\frac{\partial}{\partial x} - i\frac{\partial}{\partial y}\right)^2 + \frac{i}{4}\left(\frac{\partial}{\partial u} - i\frac{\partial}{\partial v}\right)^2.$$

To do this we replace respectively $\partial/\partial x$ by ξ_1, $\partial/\partial y$ by ξ_2, $\partial/\partial u$ by ξ_3 and $\partial/\partial v$ by ξ_4. In such a way we obtain the quadratic form

$$\begin{aligned}B(\xi_1,\xi_2,\xi_3,\xi_4) &= \frac{1}{4}(\xi_1 - i\xi_2)^2 + \frac{i}{4}(\xi_3 - i\xi_4)^2 \\ &= \frac{1}{4}(\xi_1^2 - \xi_2^2 - 2i\xi_1\xi_2 + i\xi_3^2 - i\xi_4^2 + 2\xi_3\xi_4),\end{aligned}$$

whose matrix is the following one

$$Q = \frac{1}{4}\begin{pmatrix} 1 & -i & 0 & 0 \\ -i & -1 & 0 & 0 \\ 0 & 0 & i & 1 \\ 0 & 0 & 1 & -i \end{pmatrix}.$$

The quadratic form $B(\xi_1,\xi_2,\xi_3,\xi_4)$ is the complex symbol of the operator Δ_+. The matrix of this quadratic form has rank 2 and so the operator is a degenerate one. We shall note that the matrix Q of the operator Δ_+ is not a Hermitian matrix.

2. Decomposition of the double-complex Laplace operator into two $\overline{\partial}$-operators

We propose a decomposition of the double-complex Laplace operator into two $\overline{\partial}$ operators. Since $-i = \left(\frac{1-i}{\sqrt{2}}\right)^2$ we can present Δ_+ as follows

$$\begin{aligned}\Delta_+ U(z,w) &= \frac{\partial^2 U(z,w)}{\partial z^2} - \left(\frac{1-i}{\sqrt{2}}\right)^2 \frac{\partial^2 U(z,w)}{\partial w^2} \\ &= \left(\frac{\partial}{\partial z} - \frac{1-i}{\sqrt{2}}\frac{\partial}{\partial w}\right)\left(\frac{\partial U(z,w)}{\partial z} + \frac{1-i}{\sqrt{2}}\frac{\partial U(z,w)}{\partial w}\right).\end{aligned}$$

Introducing new variables $z_1 = z - (1-i)/\sqrt{2}w$ and $w_1 = z + (1-i)/\sqrt{2}w$, we get

$$\begin{aligned}\frac{\partial U(z,w)}{\partial \overline{z}_1} &:= \frac{\partial U(z,w)}{\partial z} - \frac{1-i}{\sqrt{2}}\frac{\partial U(z,w)}{\partial w}, \\ \frac{\partial U(z,w)}{\partial \overline{w}_1} &:= \frac{\partial U(z,w)}{\partial z} + \frac{1-i}{\sqrt{2}}\frac{\partial U(z,w)}{\partial w}\end{aligned}$$

and we obtain

$$\Delta_+ U(z_1,w_1) = \frac{\partial}{\partial \overline{z}_1} \circ \frac{\partial U(z_1,w_1)}{\partial \overline{w}_1}.$$

Then the composition of the operators Δ_+ and $\overline{\Delta}_+$ will be equal to the composition of the two real Laplace operators, namely

$$\begin{aligned}PU &= \Delta_+\overline{\Delta}_+ U(x_1,y_1,u_1,v_1) \\ &= \left(\frac{\partial^2}{\partial x_1^2} + \frac{\partial^2}{\partial y_1^2}\right)\left(\frac{\partial^2 U(x_1,y_1,u_1,v_1)}{\partial u_1^2} + \frac{\partial^2 U(x_1,y_1,u_1,v_1)}{\partial v_1^2}\right) \\ &= \frac{\partial^4 U(x_1,y_1,u_1,v_1)}{\partial x_1^2 \partial u_1^2} + \frac{\partial^4 U(x_1,y_1,u_1,v_1)}{\partial x_1^2 \partial v_1^2} + \frac{\partial^4 U(x_1,y_1,u_1,v_1)}{\partial y_1^2 \partial u_1^2} \\ &\quad + \frac{\partial^4 U(x_1,y_1,u_1,v_1)}{\partial y_1^2 \partial v_1^2}.\end{aligned}$$

It is appropriate to assume that in the formula above the function $U(x_1,y_1,u_1,v_1)$ is real valued.

3. D'Alamber's type formula for the initial value problem in the whole space $\mathbb{C}(1,j)$

Let us consider the initial value problem for the equation

$$\Delta_+ F(z,w) = 0 \tag{2}$$

on the whole space $\mathbb{C}(1,j)$, namely we shall consider a function $F(z,w)$: $\mathbb{C}(1,j) \to \mathbb{C}(1,j)$ with the following initial conditions

$$F(z,0) = \varphi_0(z) + j\varphi_1(z), \qquad F'_w(z,0) = \psi_0(z) + j\psi_1(z), \tag{3}$$

on the complex space $\mathbb{C}$. Here $\varphi_0(z), \varphi_1(z)$ and $\psi_0(z), \psi_1(z)$ are holomorphic functions on $\mathbb{C}$. As the Laplace operator Δ_+ has complex coefficients, the problem reduces to two complex-valued problems, namely one for the complex-valued function $F_0(z,w)$ and one for the complex-valued function $F_1(z,w)$.

We can obtain immediately a solution of this problem by D'Alamber's type formula as follows:

$$\begin{aligned} F(z,w) &= \frac{1}{2}\left(\varphi_0\left(z + \frac{1+i}{\sqrt{2}}w\right) + \varphi_0\left(z - \frac{1+i}{\sqrt{2}}w\right)\right) \\ &+ j\frac{1}{2}\left(\varphi_1\left(z + \frac{1+i}{\sqrt{2}}w\right) + \varphi_1\left(z - \frac{1+i}{\sqrt{2}}w\right)\right) \\ &+ \frac{1}{\sqrt{2}(1+i)}\int_{z-(1+i)w/\sqrt{2}}^{z+(1+i)w/\sqrt{2}} \psi_0(t)\,dt \\ &+ j\frac{1}{\sqrt{2}(1+i)}\int_{z-(1+i)w/\sqrt{2}}^{z+(1+i)w/\sqrt{2}} \psi_1(t)\,dt. \end{aligned}$$

In this formula the integration does not depend on the curve along which it is performed, since $\psi_0(z)$ and $\psi_1(z)$ are holomorphic functions. We assume it is performed along the segment between the points $z - (1+i)w/\sqrt{2}$ and $z + (1+i)w/\sqrt{2}$.

Actually, it would be checked that the function $F(z,w)$ defined above, satisfies the equation $\Delta_+ F(z,w) = 0$ as well as the initial conditions

$$F(z,0) = \varphi_0(z) + j\varphi_1(z), \qquad F_w(z,0) = \psi_0(z) + j\psi_1(z).$$

Therefore this formula gives a solution of the given initial problem in the whole space $\mathbb{C}(1,j)$.

4. Real representation of the equation $\Delta_+ u = 0$

Now we shall reduce the partial differential equation of second order with complex coefficients and one unknown function

$$\Delta_+(f+ig) = \frac{\partial^2(f+ig)}{\partial z^2} + i\frac{\partial^2(f+ig)}{\partial w^2} = 0,$$

to a system of two partial differential equations of second number with real coefficients and two unknown function f and g. Indeed, if we separate the real and the imaginary parts in the written above equations we obtain the system

$$\begin{cases} \dfrac{\partial^2 f}{\partial x^2} - \dfrac{\partial^2 f}{\partial y^2} - 2\dfrac{\partial^2 f}{\partial u \partial v} - \dfrac{\partial^2 g}{\partial u^2} + \dfrac{\partial^2 g}{\partial v^2} + 2\dfrac{\partial^2 g}{\partial x \partial y} = 0 \\ \dfrac{\partial^2 f}{\partial u^2} + \dfrac{\partial^2 f}{\partial v^2} - 2\dfrac{\partial^2 f}{\partial x \partial y} - \dfrac{\partial^2 g}{\partial x^2} - \dfrac{\partial^2 g}{\partial y^2} - 2\dfrac{\partial^2 g}{\partial u \partial v} = 0. \end{cases} \tag{A}$$

Denoting

$$A(f) = \frac{\partial^2 f}{\partial x^2} - \frac{\partial^2 f}{\partial y^2} - 2\frac{\partial^2 f}{\partial u \partial v}, \qquad B(f) = \frac{\partial^2 f}{\partial u^2} - \frac{\partial^2 f}{\partial v^2} - 2\frac{\partial^2 f}{\partial x \partial y},$$

the system can be written as

$$\begin{aligned} Af - Bg &= 0, \\ Bf + Ag &= 0. \end{aligned} \tag{B}$$

The symbol of this system is the following matrix

$$\sigma \equiv \begin{pmatrix} \sigma(A), & -\sigma(B) \\ \sigma(B), & \sigma(A) \end{pmatrix} = \begin{pmatrix} \xi_1^2 - \xi_2^2 - 2\xi_3\xi_4, & -\xi_3^2 + \xi_4^2 + 2\xi_1\xi_2 \\ \xi_3^2 - \xi_4^2 - 2\xi_1\xi_2, & \xi_1^2 - \xi_2^2 - 2\xi_3\xi_4 \end{pmatrix}.$$

The rank of this matrix over the line $\xi_1 = \xi_2 = \frac{\sqrt{2}}{2}\xi_3$, $\xi_4 = 0$ is equal to 0 and over the line $\xi_1 = \xi_2 = -\frac{\sqrt{2}}{2}\xi_3$, $\xi_4 = 0$ is equal to 1. So it is of variable type.

Then we see that

$$(A^2 + B^2)f = A(A(f)) + B(B(f)) = A(B(g)) - B(A(g)) = 0,$$

as the operators A and B have constant coefficients.

So the system (B) implies the following partial differential equation of fourth order with one unknown function:

$$(A^2 + B^2)f = 0,$$

and analogously

$$(A^2 + B^2)g = 0.$$

Written down in explicit form we get

$$\left(\frac{\partial^2}{\partial x^2} - \frac{\partial^2}{\partial y^2} - 2\frac{\partial^2}{\partial u \partial v}\right)^2 f + \left(\frac{\partial^2}{\partial u^2} - \frac{\partial^2}{\partial v^2} - 2\frac{\partial^2}{\partial x \partial y}\right)^2 f = 0$$

and the same equation for the function g.

The symbol of the operator $A^2 + B^2$ then is

$$(\xi_1^2 - \xi_2^2 - 2\xi_3\xi_4)^2 + (\xi_3^2 - \xi_4^2 - 2\xi_1\xi_2)^2.$$

The system

$$\xi_1^2 - \xi_2^2 + 2\xi_3\xi_4 = 0, \qquad \xi_3^2 - \xi_4^2 - 2\xi_1\xi_2 = 0,$$

has as nonzero real characteristic vectors the following four straight lines in the corresponding hyperplanes:

$$\textbf{I.}\quad \xi_1 = \xi_2 = \frac{\sqrt{2}}{2}\xi_3, \quad \xi_4 = 0,$$

$$\textbf{II.}\quad \xi_1 = \xi_2 = \frac{\sqrt{2}}{2}\xi_4, \quad \xi_3 = 0,$$

$$\textbf{III.}\quad \xi_3 = \xi_4 = \frac{\sqrt{2}}{2}\xi_1, \quad \xi_2 = 0, \quad \text{and}$$

$$\textbf{VI.}\quad \xi_3 = \xi_4 = \frac{\sqrt{2}}{2}\xi_2, \quad \xi_1 = 0, \quad \text{in} \quad \mathbb{R}^4.$$

Let us note also that the operators A and B are ultra-hyperbolic which is seen by the change of variables $u' = (u+v)/\sqrt{2}$, $v' = (u-v)/\sqrt{2}$ for the operator A and $x' = (x+y)/\sqrt{2}$, $y' = (x-y)/\sqrt{2}$ for the operator B.

The double complex Laplace operator has very interesting properties from the point of view of linear partial differential operators, see also [4]. We also note that the theory of double-complex power series is considered in [3].

Acknowledgements

The author thanks prof. S. Dimiev, acad. P. Popivanov and prof. N. Kutev for many valuable remarks.

References

1. S. Dimiev, Double-complex analytic functions, in: *Proceedings of Bedlevo Conferences, 2006*, 2007.
2. S. Dimiev, Generalized multi-scalar variables and PDE, (to be published).
3. S. Dimiev, P. Furlinska, Double-complex power series, *Buletine de la Societe des sciences et des letters de Lodz, vol. LVII, Recherches sur les deformations*, vol. LII (2007), pp. 79–85.
4. P. Popivanov, *A short note on the double-complex Laplace operator*, In this volume, pp. 195–198.
5. P. Stoev, In: Eds. S. Dimiev, K. Sekigawa. *Topics in Contemporary Diferential Geometry, Complex Analysis and Mathematical Physics*. World Scientific, (2007) pp. 299–307.

GENERALIZATIONS OF CONJUGATE CONNECTIONS

OVIDIU CALIN

Department of Mathematics, Eastern Michigan University,
Ypsilanti, Michigan 48197, USA
E-mail: ocalin@emich.edu

HIROSHI MATSUZOE

Department of Computer Science and Engineering,
Graduate School of Engineering, Nagoya Institute of Technology,
Gokiso-cho, Showa-ku, Nagoya, 466-8555, Japan
E-mail: matsuzoe@nitech.ac.jp

JUN ZHANG

Department of Psychology, University of Michigan,
Ann Arbor, Michigan 48109, USA
E-mail: junz@umich.edu

Generalizations of conjugate connections are studied in this paper. It is known that generalized conjugate connections, and semi-conjugate connections are generalizations of conjugate connections. To clarify relations of these connections, the notion of dual semi-conjugate connections is introduced. Then their relations are elucidated. Local triviality of generalized conjugate connection is also studied.

Keywords: Conjugate connection; Generalized conjugate connection; Semi-conjugate connection; Information geometry; Affine differential geometry.

Introduction

Geometry of conjugate connections is a natural generalization of geometry of Levi-Civita connections from Riemannian manifolds theory. Since conjugate connections arise from affine differential geometry and from geometric theory of statistical inferences, many studies have been carried out in the recent 20 years [1–3].

In this paper, we study generalizations of conjugate connections. Some of such generalizations have been introduced independently. The generelized conjugate connection was introduced in Weyl geometry [4]. The semi-conjugate connection was introduced in affine differential geometry [7]. In order to consider the relations between these connections, we introduce the dual semi-conjugate connection. Then we shall discuss the properties and the relations between these connections.

In the later part of this paper, we concentrate to study generalized conjugate connections. It is known that the generalized conjugate connection is invariant under gauge transformations. Hence, under suitable conditions, the generalized conjugate connection reduces to the standard conjugate connection. This property is called local triviality. Therefore, we give sufficient conditions for the generalized conjugate connection to have a local triviality.

1. Generalizations of conjugate connections

We assume that all the objects are smooth throughout this paper. We may also assume that a manifold is simply connected since we discuss local geometric properties on a manifold.

Let (M, g) be a semi-Riemannian manifold, and ∇ an affine connection on M. We can define another affine connection ∇^* by

$$Xg(Y,Z) = g(\nabla_X Y, Z) + g(Y, \nabla^*_X Z). \tag{1}$$

We call ∇^* the *(standard) conjugate connection* or the *(standard) dual connection* of ∇ with respect to g. It is easy to check that $(\nabla^*)^* = \nabla$.

In this paper, we consider generalizations of these conjugate connections.

Let (M, g) be a semi-Riemannian manifold, ∇ an affine connection on M, and $\widetilde{C}$ a $(0,3)$-tensor field on M. We define another affine connection $\widetilde{\nabla^*}$ by

$$Xg(Y,Z) = g(\nabla_X Y, Z) + g(Y, \widetilde{\nabla^*}_X Z) + \widetilde{C}(X,Y,Z). \tag{2}$$

If the tensor $\widetilde{C}$ vanishes identically, then $\widetilde{\nabla}^*$ is the standard conjugate connection of ∇. Since the metric tensor g is symmetric, using twice Equation (2), we obtain the following proposition.

Proposition 1.1. $g((\widetilde{\widetilde{\nabla}^*})^*_X Y - \nabla_X Y, Z) = \widetilde{C}(X,Y,Z) - \widetilde{C}(X,Z,Y)$.

Next, we shall define generalizations of the aforementioned conjugate connections.

Definition 1.1. Let (M, g) be a semi-Riemannian manifold, ∇ an affine connection on M, and τ a 1-form on M.

(1) The *generalized conjugate connection* [5,6] $\overline{\nabla}^*$ of ∇ with respect to g by τ is defined by

$$Xg(Y,Z) = g(\nabla_X Y, Z) + g(Y, \overline{\nabla}^*_X Z) - \tau(X)g(Y,Z). \qquad (3)$$

(2) The *semi-conjugate connection* [5,7] $\hat{\nabla}^*$ of ∇ with respect to g by τ is defined by

$$Xg(Y,Z) = g(\nabla_X Y, Z) + g(Y, \hat{\nabla}^*_X Z) + \tau(Z)g(X,Y). \qquad (4)$$

(3) The *dual semi-conjugate connection* $\check{\nabla}^*$ of ∇ with respect to g by τ is defined by

$$Xg(Y,Z) = g(\nabla_X Y, Z) + g(Y, \check{\nabla}^*_X Z) - \tau(X)g(Y,Z) - \tau(Y)g(X,Z). \qquad (5)$$

The generalized conjugate connection is introduced in Weyl geometry to characterize Weyl connections [4]. The semi-conjugate connection arises naturally in affine hypersurface theory [7]. The dual semi-conjugate connection is introduced in this paper. As we will see later in this section, the dual semi-conjugate connection has dual property of the semi-conjugate connection.

From Proposition 1.1, $\overline{(\overline{\nabla}^*)}^* = \nabla$ holds for a generalized conjugate connection. On the other hand, this equality does not hold for a semi-conjugate connection $\hat{\nabla}$, or a dual semi-conjugate connection $\check{\nabla}^*$.

To clarify relations among the connections $\overline{\nabla}^*$, $\hat{\nabla}^*$ and $\check{\nabla}^*$, let us recall the projective equivalence relation and the dual-projective equivalence relation of affine connections.

Suppose that ∇ and ∇' are affine connections on a semi-Riemannian manifold (M, g). We say that ∇ and ∇' are *projectively equivalent* if there exists a 1-form τ such that

$$\nabla'_X Y = \nabla_X Y + \tau(Y)X + \tau(X)Y. \qquad (6)$$

We say that ∇ and ∇' are *dual-projectively equivalent* if there exists a 1-form τ such that

$$\nabla'_X Y = \nabla_X Y - g(X,Y)\tau^{\#}, \qquad (7)$$

where $\tau^{\#}$ is the metrical dual vector field, i.e., $g(X, \tau^{\#}) = \tau(X)$.

The readers should not confuse that, even if ∇ and ∇' are projectively (or dual-projectively) equivalent, their dual connections ∇^* and $(\nabla')^*$ may not be dual-projectively (or projectively) equivalent, respectively.

Proposition 1.2. *Let (M, g) be a semi-Riemannian manifold, ∇ an affine connection on M, and ∇^* the standard conjugate connection of ∇ with respect to g. Suppose that an affine connection ∇' is projectively equivalent to ∇ by τ. Then the following relations hold.*

(1) The generalized conjugate connection $\overline{\nabla'}^*$ of ∇' by τ is dual-projectively equivalent to ∇' by τ with respect to g.
(2) The dual semi-conjugate connection $\check{\nabla}'^*$ of ∇' by τ coincides with ∇^*, that is, $\check{\nabla}'^* = \nabla$.
(3) The semi-conjugate connection $\hat{\nabla}'^*$ of ∇' by τ is given by

$$\hat{\nabla}'^*_X Y = \nabla^*_X Y + g(X, Y)\tau^\# + \tau(Y)X + \tau(X)Y.$$

Proof. Form Equations (1), (3) and (6), we obtain

$$\begin{aligned} Xg(Y, Z) &= g(\nabla'_X Y, Z) + g(Y, \overline{\nabla'}^*_X Z) - \tau(X)g(Y, Z) \\ &= g(\nabla_X Y, Z) + g(Y, \overline{\nabla'}^*_X Z) - \tau(Y)g(X, Z). \end{aligned}$$

The last equality implies that $\overline{\nabla'}^*$ is dual-projectively equivalent to ∇^* by τ with respect to g.

On the other hand, from Equations (1), (5) and (6), we obtain

$$\begin{aligned} Xg(Y, Z) &= g(\nabla'_X Y, Z) + g(Y, \check{\nabla}'^*_X Z) - \tau(Y)g(X, Z) - \tau(X)g(Y, Z) \\ &= g(\nabla_X Y, Z) + g(Y, \check{\nabla}'^*_X Z). \end{aligned}$$

This implies that the dual semi-conjugate connection $\check{\nabla}'^*$ by τ coincides with ∇^*.

From Equations (1), (4) and (6), we obtain the third statement. □

Following similar arguments as in Proposition 1.2, we obtain the following proposition.

Proposition 1.3. *Let (M, g) be a semi-Riemannian manifold, ∇ an affine connection on M, and ∇^* the standard conjugate connection of ∇ with respect to g. Suppose that an affine connection ∇' is dual-projectively equivalent to ∇ by τ. Then the following relations hold.*

(1) The generalized conjugate connection $\overline{\nabla'}^*$ of ∇' by τ is projectively equivalent to ∇' by τ.
(2) The semi-conjugate connection $\hat{\nabla}'^*$ of ∇' by τ coincides with ∇^*, that is, $\hat{\nabla}'^* = \nabla$.
(3) The dual semi-conjugate connection $\check{\nabla}'^*$ of ∇' by τ is given by

$$\check{\nabla}'^*_X Y = \nabla^*_X Y + g(X, Y)\tau^\# + \tau(Y)X + \tau(X)Y.$$

2. Generalized conjugate connections and gauge transformations

Here after, we concentrate on generalized conjugate connections.

Let (M, g) be a semi-Riemannian manifold, ∇ an affine connection on M, and ϕ a function on M. We consider a conformal change of the metric $\bar{g} := e^{\phi} g$. Denote by $\overline{\nabla}^*$ the standard conjugate connection of ∇ with respect to the conformal metric $\bar{g}$. Then we obtain

$$\begin{aligned} &X\bar{g}(Y, Z) = \bar{g}(\nabla_X Y, Z) + \bar{g}(Y, \overline{\nabla}^*_X Z) \\ &\iff Xg(Y, Z) = g(\nabla_X Y, Z) + g(Y, \overline{\nabla}^*_X Z) - d\phi(X) g(Y, Z). \end{aligned} \tag{8}$$

This implies that $\overline{\nabla}^*$ is the generalized conjugate connection of ∇ with respect to g by $d\phi$.

Let τ be a 1-form on M. Set

$$(\bar{g}, \bar{\tau}) := (e^{\phi} g, \tau - d\phi). \tag{9}$$

The pair $(\bar{g}, \bar{\tau})$ is called a *gauge transformation* of (g, τ).

Proposition 2.1. *The generalized conjugate connection is invariant under gauge transformations. That is,* $\overline{\nabla}^*$ *is the generalized conjugate connection of* ∇ *with respect to* g *by* τ *if and only if* $\overline{\nabla}^*$ *is the generalized conjugate connection of* ∇ *with respect to* $\bar{g}$ *by* $\bar{\tau}$.

Proof. From Equations (8) and (9), we easily obtain

$$\begin{aligned} &Xg(Y, Z) = g(\nabla_X Y, Z) + g(Y, \overline{\nabla}^*_X Z) - \tau(X) g(Y, Z) \\ &\iff X\bar{g}(Y, Z) = \bar{g}(\nabla_X Y, Z) + \bar{g}(Y, \overline{\nabla}^*_X Z) - \bar{\tau}(X) \bar{g}(Y, Z). \end{aligned}$$ □

We remark that the gauge invariance is an important notion in Weyl geometry. We can discuss Weyl geometry in terms of the generalized conjugate connections [4,8].

3. Local triviality of generalized conjugate connections and equiaffine structures

Suppose that $\overline{\nabla}^*$ is the generalized conjugate connection of an affine connection ∇ by a 1-form τ. Equation (8) implies that $\overline{\nabla}^*$ is reduced to the standard conjugate connection with respect to some conformal metric $\bar{g}$ if $\tau = d\phi$.

Definition 3.1. The generalized conjugate connection $\overline{\nabla}^*$ is called *locally trivial* if there exists a function ϕ such that $\overline{\nabla}^*$ is the standard conjugate connection with respect to some conformal metric $\bar{g} = e^{\phi}g$. That is, the equivalence (8) holds.

In order to elucidate the local triviality of generalized conjugate connections, we consider equiaffine structures on a manifold.

Definition 3.2. Let ∇ be a torsion-free affine connection on M. Let ω be a volume form on M, that is, ω is an n-form on M which does not vanish everywhere. We say that the pair (∇, ω) is an *equiaffine structure* on M if ω is parallel with respect to ∇, that is, $\nabla\omega = 0$. We say that the connection ∇ is *equiaffine*, and the volume form ω is *parallel* with respect to ∇.

Let R be the curvature tensor of ∇, and Ric the Ricci tensor of ∇, i.e., $\mathrm{Ric}(Y,Z) = \mathrm{tr}\{X \mapsto R(X,Y)Z\}$. In local coordinate expressions, the quantities are given by

$$\nabla_{\frac{\partial}{\partial x^i}} \frac{\partial}{\partial x^j} = \sum_{k=1}^{n} \Gamma_{ij}^{k} \frac{\partial}{\partial x^k},$$

$$R\left(\frac{\partial}{\partial x^i}, \frac{\partial}{\partial x^j}\right) \frac{\partial}{\partial x^l} = \sum_{k=1}^{n} R_{lij}^{k} \frac{\partial}{\partial x^k}$$

with

$$R_{lij}^{k} = \frac{\partial \Gamma_{lj}^{k}}{\partial x^i} - \frac{\partial \Gamma_{li}^{k}}{\partial x^j} + \sum_{m=1}^{n} \left(\Gamma_{mi}^{k} \Gamma_{lj}^{m} - \Gamma_{li}^{m} \Gamma_{mj}^{k}\right). \tag{10}$$

Proposition 3.1. *Let ∇ be a torsion-free affine connection on M. Then the following conditions are equivalent.*

(1) ∇ is equiaffine.
(2) Ric is symmetric.
(3) $\displaystyle \frac{\partial}{\partial x^i}\left(\sum_{k=1}^{n} \Gamma_{jk}^{k}\right) = \frac{\partial}{\partial x^j}\left(\sum_{k=1}^{n} \Gamma_{ik}^{k}\right).$

Proof. The statement (1) $\iff$ (2) is Proposition 3.1 in Section 1 by Nomizu and Sasaki [1]. Contracting (10) to get the Ricci tensor, we can easily obtain (2) $\iff$ (3). □

For further information about equiaffine structures, see Zhang's paper [9], or Zhang and Matsuzoe's paper [10].

Proposition 3.2. *Let ∇ be a torsion-free affine connection on M, and $dv = \sqrt{g}dx^1 \wedge \cdots \wedge dx^n$ the standard volume element with respect to g.*

Then the connection is equiaffine if and only if there exists a function f such that

$$\frac{\partial}{\partial x^k}\left(\log\left(f\sqrt{g}\right)\right) = \sum_{j=1}^{n} \Gamma_{kj}^{j}. \tag{11}$$

Proof. Recall that $\sqrt{g}$ is a non-zero function on M. If ∇ is equiaffine, then the condition (3) in Proposition 3.1 holds. This is the integrability condition of f since $\sqrt{g}$ is a non-zero function.

On the other hand, if the formula (11) holds, then the condition (3) in Proposition 3.1 holds. This implies the desired result, and we may say that the function f is just the proportionality function between the parallel form ω and the volume form dv. □

In the following it makes sense to assume τ exact. If M is connected and simply connected, by Poincaré's lemma, there is a function ϕ on M such that $\tau = d\phi$. Since we are considering local properties on a manifold, we have the following result.

Theorem 3.1. *Let (M, g) be a connected and simply connected n-dimensional semi-Riemannian manifold. Let ∇ be a torsion-free affine connection on M, and τ a 1-form on M. Suppose that the generalized conjugate connection $\overline{\nabla}^*$ of ∇ is torsion-free. Consider the following three conditions:*

(1) τ is an exact 1-form.
(2) ∇ is equiaffine.
(3) $\overline{\nabla}^*$ is equiaffine.

Then any two of the above conditions imply the third one.

Proof. In a local coordinate expression, the definition of $\overline{\nabla}^*$ in (3) is written

$$\frac{\partial g_{ij}}{\partial x^k} = \Gamma_{ki,j} + \overline{\Gamma}^*_{kj,i} - \tau_k g_{ij}.$$

Contracting the above equation with g^{ij}, we have

$$\sum_{i,j=1}^{n} g^{ij}\frac{\partial g_{ij}}{\partial x^k} = \sum_{i=1}^{n} \Gamma_{ki}^{i} + \sum_{i=1}^{n} \overline{\Gamma}^{*i}_{\ ki} - n\tau_k.$$

Let $\nabla^{(0)}$ be the Levi-Civita connection with respect to g. Using that

$$\sum_{i=1}^{n} \Gamma^{(0)\,i}_{ki} = \frac{1}{2}\sum_{i,j=1}^{n} g^{ij}\frac{\partial g_{ij}}{\partial x^k},$$

we obtain

$$n\tau_k = \sum_{j=1}^{n} \left\{ \Gamma^j_{kj} + \overline{\Gamma}^{*\ j}_{kj} - 2\Gamma^{(0)\ j}_{kj} \right\}.$$

Differentiating with respect to x^i and then x^k, subtracting, yields

$$n\left(\frac{\partial \tau_k}{\partial x^i} - \frac{\partial \tau_i}{\partial x^k}\right) = \sum_{j=1}^{n} \left\{ \left(\frac{\partial}{\partial x^i}\Gamma^j_{kj} - \frac{\partial}{\partial x^k}\Gamma^j_{ij}\right) + \left(\frac{\partial}{\partial x^i}\overline{\Gamma}^{*j}_{\ kj} - \frac{\partial}{\partial x^k}\overline{\Gamma}^{*j}_{\ ij}\right) \right\}$$
$$- 2\sum_{j=1}^{n} \left(\frac{\partial}{\partial x^i}\Gamma^{(0)\ j}_{kj} - \frac{\partial}{\partial x^k}\Gamma^{(0)\ j}_{ij}\right).$$

The last parenthesis vanishes since the Levi-Civita connection $\nabla^{(0)}$ is equiaffine (it admits a parallel volume element $dv = \sqrt{g}dx^1 \wedge \cdots \wedge dx^n$, or, equivalently, it has a symmetric Ricci tensor). Then

$$n\left(\frac{\partial \tau_k}{\partial x^i} - \frac{\partial \tau_i}{\partial x^k}\right) = \sum_{j=1}^{n} \left\{ \left(\frac{\partial}{\partial x^i}\Gamma^j_{\ kj} - \frac{\partial}{\partial x^k}\Gamma^j_{\ ij}\right) + \left(\frac{\partial}{\partial x^i}\overline{\Gamma}^{*j}_{\ kj} - \frac{\partial}{\partial x^k}\overline{\Gamma}^{*j}_{\ ij}\right) \right\}.$$

If any two of the above parentheses vanish, then the third one must vanish. Applying Proposition 3.1 leads to the desired result. □

From Theorem 3.1 and Proposition 3.1, we obtain the following.

Corollary 3.1. *Suppose that ∇ and $\overline{\nabla}^*$ are torsion-free. Consider the following three conditions:*

(1) τ is an exact 1-form.
(2) ∇ has a symmetric Ricci tensor.
(3) $\overline{\nabla}^*$ has a symmetric Ricci tensor.

Then any two of the above conditions imply the third one.

Let $\omega = f dv$, and $\overline{\omega}^* = \overline{f}^* dv$ be the volume elements associated with the generalized conjugate connections ∇ and $\overline{\nabla}^*$, with $f, \overline{f}^*$ functions, *i.e.*,

$$\nabla\omega = 0, \qquad \overline{\nabla}^*\overline{\omega}^* = 0,$$

where $dv = \sqrt{g}\,dx$ denotes the Riemannian volume element. We shall consider the following notations

$$[\omega] := f, \qquad [\overline{\omega}^*] := \overline{f}^*.$$

Theorem 3.2. *Let $\tau = d\phi$. Then*

$$[\omega][\overline{\omega}^*] = Ce^{n\varphi}, \tag{12}$$

with $C > 0$, constant.

Proof. Applying Proposition 3.2 to formula (11) yields

$$n\tau_k = \frac{\partial}{\partial x^k}\left(\log(f\sqrt{g})\right) + \frac{\partial}{\partial x^k}\left(\log(\bar{f}^*\sqrt{g})\right) - 2\frac{\partial}{\partial x^k}(\log\sqrt{g}).$$

Using $\tau_k = \frac{\partial \phi}{\partial x^k}$, the above relation can be written as an exact equation

$$\begin{aligned}\frac{\partial}{\partial x^k}\left(n\phi - \log(f\sqrt{g}) - \log(\bar{f}^*\sqrt{g}) + 2\log\sqrt{g}\right) = 0 \Longleftrightarrow \\ \frac{\partial}{\partial x^k}\left(n\phi - \log(f\bar{f}^*)\right) = 0 \Longleftrightarrow \\ n\phi - \log(f\bar{f}^*) = c \quad \text{constant} \Longleftrightarrow \\ f\bar{f}^* = Ce^{n\phi},\end{aligned}$$

with $C = e^{-c}$. □

Acknowledgement

The authors are partially supported by the NSF Grant No. 0631541. The second named author is also partially supported by Grant-in-Aid for Encouragement of Young Scientists (B) No. 19740033, Japan Society for the Promotion of Science.

References

1. K. Nomizu and T. Sasaki, *Affine differential geometry: geometry of affine immersions*, Cambridge Univ. Press, 1994.
2. S. Amari and H. Nagaoka, *Method of information geometry*, Amer. Math. Soc., Oxford University Press, Providence, RI, 2000.
3. K. Nomizu and U. Simon, Notes on conjugate connections, in *Geometry and Topology of Submanifolds IV*, eds. F. Dillen and L. Verstraelen, World Scientific, 1992.
4. A. Norden, *Affinely Connected Spaces*, GRMFL, Moscow, 1976.
5. K. Nomizu, Affine connections and their use, in *Geometry and Topology of Submanifolds VII*, ed. F. Dillen, World Scientific, 1995.
6. H. Matsuzoe, Statistical manifolds and its generalization, in *Proceedings of the 8th International Workshop on Complex Structures and Vector Fields*, World Scientific, 2007.
7. S. Ivanov, *J. Geom.* **53**, 89 (1995).
8. H. Matsuzoe, *Kyushu J. Math.* **55**, 107 (2001).
9. J. Zhang, *Annals of the Institute of Statistical Mathematics* **59**, 161 (2007).
10. J. Zhang and H. Matsuzoe, Dualistic differential geometry associated with a convex function, in *Advances in Applied Mathematics and Global Optimization (in Honor of Gilbert Strang), Advances in Mechanics and Mathematics, 17*, eds. D. Gao and H. Sherali, Springer, 2009.

ASYMPTOTICS OF GENERALIZED VALUE DISTRIBUTION FOR HERGLOTZ FUNCTIONS

Y. T. CHRISTODOULIDES

Department of Mathematics and Statistics, University of Cyprus, P.O. Box 20537, 1678 Nicosia, Cyprus
E-mail: ychris@ucy.ac.cy

We consider a family of Herglotz functions F_y ($y \in \mathbb{R}$) generated from a Herglotz function F, and obtain their integral representations. We review the theory of value distribution for boundary values of Herglotz functions, and present a result for limiting generalized value distribution for translated Herglotz functions.

Keywords: Herglotz functions; Generalized value distribution.

1. Introduction

Given a Lebesgue measurable function $\mathbb{R} \to \mathbb{R}$, the distribution of values of f can be described by a mapping $\mathcal{D} : (A, S) \to \mathbb{R}$, defined for arbitrary Borel sets A, S by

$$\mathcal{D}(A, S; f) = |A \cap f_+^{-1}(S)|.$$

Here $|.|$ denotes Lebesgue measure, and $\mathcal{D}(A, S; f)$ is the Lebesgue measure of the set of points $\lambda \in A$ for which $f(\lambda) \in S$.

A case of particular interest is when $f(\lambda)$ is the real boundary value of the Weyl-Titchmarsh m-function [1,2] associated with the Schrödinger equation

$$-\frac{d^2 f(x,z)}{dx^2} + V(x) f(x,z) = z f(x,z), \tag{1}$$

where $z \in \mathbb{C}_+$ is a complex spectral parameter. Let u, v be solutions of eq. (1) which satisfy, at $x = 0$, the boundary conditions

$$\begin{aligned} u(0,z) &= 1, \qquad v(0,z) = 0, \\ u'(0,z) &= 0, \qquad v'(0,z) = 1. \end{aligned} \tag{2}$$

The m-function $m(z)$ is defined by the condition

$$u(.,z) + m(z)v(.,z) \in L^2(0,\infty). \tag{3}$$

It is analytic and has positive imaginary part in the upper half-plane $\mathbb{C}_+ = \{z \in \mathbb{C} : \operatorname{Im} z > 0\}$. Functions with these properties are called Herglotz functions. The spectral analysis of the Sturm-Liouville differential operator $T = -\frac{d^2}{dx^2} + V$ associated with eq. (1) can be carried out through an analysis of the boundary behaviour of the m-function. See for example [3], and [4], where the large x asymptotic behaviour of the solution v defined in (2) has been studied.

One of the main theoretical tools in [4] was the determination of limiting value distribution for a family of value distribution mappings $\mathcal{D}^\delta$ $(\delta > 0)$ defined by

$$\mathcal{D}^\delta(A, S; F) = \mathcal{D}(A, S; F^\delta),$$

where F^δ is the Herglotz function $F^\delta(z) = F(z + i\delta)$ obtained from a Herglotz function F by translation through distance δ parallel to the imaginary axis. It was proved that $\mathcal{D}^\delta$ converge, and precise bounds were obtained. We present in this paper an analogous result in the case of generalized value distribution for Herglotz functions.

The paper is organized as follows. In Section 2 we introduce Herglotz functions and their integral representation. We consider a one-parameter family of Herglotz functions F_y $(y \in \mathbb{R})$ generated from a Herglotz function F, and obtain their integral representations and spectral properties for the corresponding measures. In Section 3 we are concerned with value distribution for Herglotz functions. The main result regarding limiting generalized value distribution for translated Herglotz functions is stated in Theorem 3.1.

2. Herglotz functions and their integral representation

Let F be a Herglotz function, that is, analytic and with positive imaginary part in the upper half-plane $\mathbb{C}_+$. Then F admits the integral representation [5,6]

$$F(z) = a_F + b_F z + \int_{-\infty}^{\infty} \left\{ \frac{1}{t-z} - \frac{t}{t^2+1} \right\} d\rho(t), \tag{4}$$

where a_F, b_F are constants $(b_F > 0)$, and the function $\rho(t)$, which is determined up to an additive constant, is non-decreasing and right-continuous. For given Herglotz function F the constants a_F, b_F are determined by

$$a_F = \operatorname{Re} F(i), \qquad b_F = \lim_{s\to\infty} \frac{1}{s} \operatorname{Im} F(is).$$

The spectral function $\rho(t)$ gives rise to a measure μ through the relation $\mu((a,b]) = \rho(b) - \rho(a)$ for finite intervals $(a,b]$, and μ extends to Borel sets. Then μ is said to be the spectral measure corresponding to the Herglotz function F, and satisfies the convergence condition

$$\int_{\mathbb{R}} \frac{1}{1+t^2}\, d\mu(t) < \infty.$$

If $b_F = 0$ and the measure μ is finite, then we can absorb the term $\int_{\mathbb{R}} d\rho(t)t(t^2+1)^{-1}$ into the constant a_F, drop the linear term, and adopt the simpler representation

$$F(z) = a_F + \int_{-\infty}^{\infty} \frac{1}{t-z} d\rho(t). \tag{5}$$

In this case we have $F(is) = a_F + i\mu(\mathbb{R})/s + o(1/s)$ as $s \to \infty$.

By the Lebesgue decomposition theorem [7], the measure μ can be decomposed into an absolutely continuous and a singular part, $\mu = \mu_{ac} + \mu_s$, and further, the singular part can be decomposed into a singular but continuous and a discrete part, $\mu_s = \mu_{sc} + \mu_d$. The supports of the components of μ can be characterized through an analysis of the boundary behaviour of $F(z)$ as z approaches the real axis [8]. Thus,

$$\mu_s = \mu \restriction \{\lambda \in \mathbb{R} : \lim_{\varepsilon\to 0^+} \operatorname{Im} F(\lambda + i\varepsilon) = \infty\};$$

$$\mu_{ac} = \mu \restriction \{\lambda \in \mathbb{R} : \operatorname{Im} F(\lambda + i\varepsilon) \quad \text{is bounded for} \quad 0 < \varepsilon < 1\};$$

$$\mu_d = \mu \restriction \{\lambda \in \mathbb{R} : \lim_{\varepsilon\to 0^+} -i\varepsilon F(\lambda + i\varepsilon) = \mu\{\lambda\} \neq 0\}.$$

We denote by $F_+(\lambda)$ the boundary value of $F(z)$ as z approaches the real axis; that is,

$$F_+(\lambda) = \lim_{\varepsilon\to 0^+} F(\lambda + i\varepsilon), \tag{6}$$

for each $\lambda \in \mathbb{R}$ for which this limit exists finitely. It is known from the theory of boundary values of analytic functions that $F_+(\lambda)$ is defined Lebesgue almost everywhere [9]. The density function of μ_{ac} is given by $f(\lambda) = \operatorname{Im} F_+(\lambda)/\pi$, whereas the measure μ restricted to the set $\{\lambda \in \mathbb{R} : F_+(\lambda) \in \mathbb{R}\}$ is identically zero [8]. In the special case that F has real boundary value $F_+(\lambda)$ almost everywhere, the measure μ is purely singular, and $F_+(\lambda)$ takes every real value in any neighbourhood of a point of the essential spectrum of μ (see [10] for further details).

In spectral analysis we require not a single Herglotz function but a family of them. Given a Herglotz function F, consider the one-parameter family of Herglotz functions defined by the relation

$$F_y(z) = \frac{1}{y - F(z)}, \qquad z \in \mathbb{C}_+, \qquad y \in \mathbb{R}. \tag{7}$$

Denote by a_{F_y}, b_{F_y}, ρ_y, μ_y respectively the constants, spectral function, and measure corresponding to F_y, defined through the representation (1). Let $F(is) = A(s) + iB(s)$, and suppose that $b_F = \lim_{s\to\infty} \operatorname{Im} F(is)/s = \lim_{s\to\infty} B(s)/s > 0$. We must have in this case $B(s) \to \infty$ as $s \to \infty$, and it follows that

$$b_{F_y} = \lim_{s\to\infty} \frac{1}{s} \operatorname{Im} F_y(is) = \lim_{s\to\infty} \frac{1}{s} \left[\frac{B(s)}{(y - A(s))^2 + B^2(s)} \right] = 0$$

for all $y \in \mathbb{R}$.

The Lebesgue dominated convergence theorem implies that $A(s)/s \to 0$ as $s \to \infty$. Therefore, we have $F_y(is) = i(sb_F)^{-1} + o(1/s)$ as $s \to \infty$, from which we conclude that $\mu_y(\mathbb{R}) = \lim_{s\to\infty} s\operatorname{Im} F_y(is) = b_F^{-1}$ for all $y \in \mathbb{R}$.

If $b_F = 0$ and the measure μ is finite, then we can substitute $F(is) = a_F + i\mu(\mathbb{R})/s + o(1/s)$ into (7) in order to determine the large s asymptotic behaviour of $F_y(is)$. We find in this case that the measure μ_y is finite except for $y = a_F$.

These two cases, together with the resulting Herglotz representations for $F_y(z)$, are given under (i) and (ii) below. The remaining case (iii), in which $b_F = 0$ and the measure μ is infinite, follows from (i) and (ii) on reversing the roles of F and F_y and observing that $F(z) = y - (F_y(z))^{-1}$. In summary, we have:

(i) if $b_F \neq 0$, then for $y \in \mathbb{R}$ we have $b_{F_y} = 0$ and F_y has the representation

$$F_y(z) = \int_{-\infty}^{\infty} \frac{d\rho_y(t)}{t - z},$$

where μ_y is a finite measure and $\mu_y(\mathbb{R}) = b_F^{-1}$.

(ii) if $b_F = 0$ and $\mu(\mathbb{R}) < \infty$, then for $y \neq a_F$ we have $b_{F_y} = 0$ and

$$F_y(z) = \frac{1}{y - a_F} + \int_{-\infty}^{\infty} \frac{d\rho_y(t)}{t - z},$$

where μ_y is a finite measure and $\mu_y(\mathbb{R}) = \mu(\mathbb{R})(y - a_F)^{-2}$.

(iii) if $b_F = 0$ and $\mu(\mathbb{R}) = \infty$, then $b_{F_y} = 0$ for all $y \in \mathbb{R}$, and either $\mu_y(\mathbb{R}) < \infty$, or $\mu_y(\mathbb{R}) = \infty$, for all $y \in \mathbb{R}$.

We note that in each case $b_{F_y} > 0$ for at most one value of y. Eq. (7) is equivalent to $F_y(z) = F_x(z)/(1 + (y - x)F_x(z))$. By using this relation we can take any one of the family $\{F_y(z)\}$ and define the others in terms of that function.

The discrete spectrum of μ_y is determined by the boundary behaviour of F as follows. If λ is a discrete point of the measure μ_y, for some $y \in \mathbb{R}$, then $F_+(\lambda) = y$, and we have

$$\lim_{\varepsilon\to 0+} \frac{F(\lambda+i\varepsilon) - F_+(\lambda)}{i\varepsilon} = b_F + \int_{-\infty}^{\infty} \frac{d\rho(t)}{(t-\lambda)^2} = \frac{1}{\mu_y\{\lambda\}}. \tag{8}$$

To see this, note first that since $\lim_{\varepsilon\to 0+} \operatorname{Im} F_y(\lambda+i\varepsilon) = \infty$, it follows that

$$F_+(\lambda) = \lim_{\varepsilon\to 0+} F(\lambda+i\varepsilon) = \lim_{\varepsilon\to 0+} \left[y - \frac{1}{F_y(\lambda+i\varepsilon)} \right] = y.$$

Moreover, $\mu_y\{\lambda\} = -\lim_{\varepsilon\to 0+} i\varepsilon F_y(\lambda+i\varepsilon)$ implies that

$$\lim_{\varepsilon\to 0+} \frac{F(\lambda+i\varepsilon) - y}{i\varepsilon} = \lim_{\varepsilon\to 0+} \frac{1}{-i\varepsilon F_y(\lambda+i\varepsilon)} = \frac{1}{\mu_y\{\lambda\}}.$$

Taking the real part and using the representation (1), we see that this limit is also

$$\lim_{\varepsilon\to 0+} \frac{1}{\varepsilon} \operatorname{Im} F(\lambda+i\varepsilon) = b_F + \lim_{\varepsilon\to 0+} \int_{-\infty}^{\infty} \frac{d\rho(t)}{(t-\lambda)^2+\varepsilon^2} = b_F + \int_{-\infty}^{\infty} \frac{d\rho(t)}{(t-\lambda)^2},$$

which completes the proof of (8).

An immediate consequence is that the singular part of the measure μ_y is supported on the set $\{\lambda \in \mathbb{R} : F_+(\lambda) = y\}$, and hence, for any two distinct values of y, the singular parts of μ_y are mutually singular.

3. Value distribution for Herglotz functions

A theory of value distribution for boundary values of Herglotz functions has been developed in the last fifteen years [3,4,10]. By a value distribution mapping we mean a mapping $\mathcal{D} : (A, S) \to \mathcal{D}(A, S)$ which assigns an extended real non-negative number to pairs of Borel subsets A, S of $\mathbb{R}$ and satisfies the following properties:

(i) $A \to \mathcal{D}(A, S)$ defines a measure on Borel subsets of $\mathbb{R}$, for fixed S; $S \to \mathcal{D}(A, S)$ defines a measure on Borel subsets of $\mathbb{R}$, for fixed A;

(ii) $\mathcal{D}(A, \mathbb{R}) = |A|$; this implies that the measure $A \to \mathcal{D}(A, S)$ is absolutely continuous with respect to Lebesgue measure;

(iii) the measure $S \to \mathcal{D}(A, S)$ is absolutely continuous with respect to Lebesgue measure.

The value distribution mapping associated with a Herglotz function F is given by [10]

$$\mathcal{D}(A, S; F) = \int_S \mu_y(A) dy, \tag{9}$$

where $\{\mu_y\}$ ($y \in \mathbb{R}$) are the measures corresponding to a family of Herglotz functions F_y generated from F. If F has real boundary value almost everywhere, then the measure μ corresponding to F, and the measures μ_y corresponding to F_y, are purely singular, and we have [10]

$$\mathcal{D}(A, S; F) = \int_S \mu_y(A) dy = |A \cap F_+^{-1}(S)|. \tag{10}$$

Thus, in this case, $\mathcal{D}(A, S; F)$ is the Lebesgue measure of the set of points $\lambda \in A$ for which $F_+(\lambda) \in S$.

An alternative characterization of the value distribution mapping associated with a Herglotz function F uses the angle $\theta(z, S)$ subtended at the point $z \in \mathbb{C}_+$ by a Borel set S on the real axis, defined by [4]

$$\theta(z, S) = \int_S \operatorname{Im}\left[\frac{1}{t - z}\right] dt, \qquad z \in \mathbb{C}_+. \tag{11}$$

Define $\omega(z, S; F) = \theta(F(z), S)/\pi$, for $z \in \mathbb{C}_+$, and $\omega(\lambda, S; F) = \lim_{\varepsilon \to 0^+} \theta(F(\lambda + i\varepsilon), S)/\pi$, for $\lambda \in \mathbb{R}$. If it is clear from the context, we shall write $\omega(z, S)$, and $\omega(\lambda, S)$, respectively. We then have

$$\omega(\lambda, S) = \begin{cases} 1 & F_+(\lambda) \in \mathbb{R} \text{ and } F_+(\lambda) \in S, \\ 0 & F_+(\lambda) \in \mathbb{R} \text{ and } F_+(\lambda) \notin S, \\ \theta(F_+(\lambda), S)/\pi & \operatorname{Im} F_+(\lambda) > 0. \end{cases}$$

In particular, if F has real boundary value almost everywhere, then $\omega(\lambda, S)$ is a.e. the characteristic function of the set $F_+^{-1}(S)$, and we have [4]

$$\mathcal{D}(A, S; F) = \int_A \omega(\lambda, S) d\lambda = |A \cap F_+^{-1}(S)|. \tag{12}$$

The integral in eq. (12) is also a natural value distribution mapping associated with the Herglotz function F.

In [4], the determination of the asymptotic value distribution of the Herglotz function $-v'/v$, where v is the solution of the Schrödinger equation (1) which was defined in (2), was based on an important estimate about value distribution for translated Herglotz functions $F^\delta(z) = F(z + i\delta)$. If we denote $\omega^\delta(z, S; F) = \omega(z, S; F^\delta)$, for $z \in \mathbb{C}_+$, and similarly for points $\lambda \in \mathbb{R}$, it was shown that

$$\left| \int_A \omega^\delta(\lambda, S; F) d\lambda - \int_A \omega(\lambda, S; F) d\lambda \right| \leq E_A(\delta), \tag{13}$$

where A, S are arbitrary Borel sets ($|A| < \infty$), and $E_A(\delta)$ is a non-decreasing function of δ, with $\lim_{\delta \to 0^+} E_A(\delta) = 0$. Surprisingly, it was found that the bound in (13) holds uniformly for all Herglotz functions F and Borel sets S.

The results can be generalized by allowing a description of value distribution in terms of measures other than Lebesgue measure. The generalized value distribution [11] associated with a Herglotz function F is described by a measure ν_S defined by

$$\nu_S(A) = \int_S \mu_y(A) d\sigma(y). \tag{14}$$

In (14), A, S are arbitrary Borel sets, μ_y are the measures appearing in (9), and $d\sigma$ is a Herglotz measure corresponding to a Herglotz function ϕ having representation

$$\phi(z) = a_\phi + b_\phi z + \int_{-\infty}^{\infty} \left\{ \frac{1}{t-z} - \frac{t}{t^2+1} \right\} d\sigma(t). \tag{15}$$

The measure ν_S satisfies the condition $\int_{\mathbb{R}} 1/(1+t^2) d\nu_S(t) < \infty$, and therefore it is also a Herglotz measure. It can easily be verified that if the measure $d\sigma$ is absolutely continuous, then ν_S is also absolutely continuous. If F has real boundary value almost everywhere, then the measure μ is purely singular and we have $\nu_S(A) = \nu_S(A \cap F_+^{-1}(S)) = \nu_{\mathbb{R}}(A \cap F_+^{-1}(S))$. For any Borel set B we have the relation [11]

$$\nu_S(B) = \mu_{(\phi_S \circ F)}(B) - b_\phi \mu(B), \tag{16}$$

where μ is the measure corresponding to the Herglotz function F, b_ϕ is the constant appearing in the representation (15) of ϕ, and $\mu_{(\phi_S \circ F)}$ is the measure corresponding to the composed Herglotz function $\phi_S \circ F$, where ϕ_S has the same representation as that of ϕ except that integration takes place over the Borel set S instead of $\mathbb{R}$. The integral representation of composed Herglotz functions, and their boundary values, have been studied in [12].

We present a result for generalized value distribution which is analogous to the one in (13). Define first, for any Borel set S of $\mathbb{R}$ and $z \in \mathbb{C}_+$, a generalized angle subtended θ_σ by

$$\theta_\sigma(z, S) = \int_S \mathrm{Im} \left[\frac{1}{t-z} \right] d\sigma(t).$$

Thus $\theta_\sigma(z, S)$ may be thought of as an angle subtended at the point $z \in \mathbb{C}_+$ by the set S, weighted on $\mathbb{R}$ by the measure σ. We define $\omega_\sigma(z, S; F) = \theta_\sigma(F(z), S)/\pi$ for $z \in \mathbb{C}_+$, and $\omega_\sigma(\lambda, S; F) = \lim_{\delta \to 0^+} \omega_\sigma(\lambda + i\delta, S; F)$ for points $\lambda \in \mathbb{R}$.

In the remainder of the section we shall assume that the measure $d\sigma$ is absolutely continuous, with density function h_σ.

We have the relation

$$\int_S \mu_y^\delta(A)d\sigma(y) = \frac{1}{\pi}\int_{\mathbb{R}} \theta(t+i\delta, A)d\nu_S(t), \tag{17}$$

where the measures μ_y^δ correspond to the translated Herglotz functions $F_y^\delta(z) = F_y(z+i\delta)$. A detailed proof of (17) will be published elsewhere.

For any Borel set S and positive constant C, let the sets S_0, S_1 be defined by

$$S_0 = \{y \in S : h_\sigma(y) \leq C\}, \qquad S_1 = \{y \in S : h_\sigma(y) > C\}.$$

We define Herglotz functions ϕ_0, ϕ_1 having the integral representations

$$\phi_0(z) = a_\phi + b_\phi z + \int_{S_0} \left\{\frac{1}{t-z} - \frac{t}{t^2+1}\right\} d\sigma(t), \tag{18}$$

$$\phi_1(z) = \int_{S_1} \left\{\frac{1}{t-z} - \frac{t}{t^2+1}\right\} d\sigma(t), \tag{19}$$

and let ν_0, ν_1 be the measures corresponding to the composed Herglotz functions $\phi_0 \circ F$, $\phi_1 \circ F$ respectively. Since $(\phi_S \circ F)(z) = (\phi_0 \circ F)(z) + (\phi_1 \circ F)(z)$ for all $z \in \mathbb{C}_+$, we have $\mu_{(\phi_S \circ F)}(B) = \nu_0(B) + \nu_1(B)$ for any Borel set B, and from eq. (16) we obtain $\nu_S(B) = \nu_0(B) + \nu_1(B) - b_\phi \mu(B)$.

The measure $(\nu_0 - b_\phi \mu)$ is absolutely continuous with respect to Lebesgue measure, with density function bounded by C. For finite intervals $(a, b]$ whose endpoints are not discrete points of the measure $(\nu_0 - b_\phi \mu)$, this follows from the standard result [8] about Herglotz measures of intervals, by an application of the Lebesgue dominated convergence theorem. We have

$$\begin{aligned}(\nu_0 - b_\phi\mu)((a,b]) &= \lim_{\varepsilon \to 0^+} \frac{1}{\pi} \int_a^b \left\{ \int_{S_0} \operatorname{Im}\left[\frac{1}{t - F(\lambda + i\varepsilon)}\right] d\sigma(t) \right\} d\lambda \\ &\leq C(b-a),\end{aligned}$$

since $d\sigma$ is bounded by Lebesgue measure on the set S_0, and for any fixed value $\varepsilon > 0$ we have $\int_{\mathbb{R}} \operatorname{Im}[t - F(\lambda + i\varepsilon)]^{-1} dt = \pi$.

The result extends to arbitrary open intervals by considering sequences of intervals whose endpoints are not discrete points of $(\nu_0 - b_\phi\mu)$, and hence to open Borel sets.

Now fix $C > 0$. If B is any bounded Borel set, given $\varepsilon > 0$ there is an open set G containing B such that $|G| < |B| + \varepsilon/C$. Then, we have

$$(\nu_0 - b_\phi\mu)(B) \leq (\nu_0 - b_\phi\mu)(G) \leq C|G| \leq C|B| + \varepsilon,$$

and since ε was arbitrary we can infer $(\nu_0 - b_\phi\mu)(B) \le C|B|$.

Finally, the result generalizes to arbitrary Borel sets through the relation $(\nu_0 - b_\phi\mu)(B) = (\nu_0 - b_\phi\mu)(\cup_N B \cap [-N, N])$.

Let $[a,b]$ be a finite closed interval. For any Borel set $A \subseteq [a,b]$ we have, therefore,

$$\begin{aligned} &\frac{1}{\pi}\int_A \theta(t+i\delta, A^c)d(\nu_0 - b_\phi\mu)(t) \le CE_A(\delta), \\ &\frac{1}{\pi}\int_{A^c} \theta(t+i\delta, A)d(\nu_0 - b_\phi\mu)(t) \le CE_A(\delta), \end{aligned} \tag{20}$$

where

$$E_A(\delta) = \frac{1}{\pi}\int_{A^c} \theta(t+i\delta, A)dt. \tag{21}$$

Note that

$$\begin{aligned} E_A(\delta) &= \frac{1}{\pi}\int_{\mathbb{R}} \theta(t+i\delta, A)dt - \frac{1}{\pi}\int_A \{\pi - \theta(t+i\delta, A^c)\}dt \\ &= \frac{1}{\pi}\int_A \theta(t+i\delta, A^c)dt. \end{aligned}$$

It has been shown [4] that $E_A(\delta)$ is a non-decreasing function of δ, with $\lim_{\delta\to 0^+} E_A(\delta) = 0$.

We also have the following result: let $\varepsilon > 0$ be given, and take δ with $0 < \delta < 1$. Then, a constant $N = N(\varepsilon)$ can be found such that

$$\begin{aligned} &\frac{1}{\pi}\int_A \theta(t+i\delta, A^c)d\nu_1(t) < \varepsilon, \\ &\frac{1}{\pi}\int_{A^c} \theta(t+i\delta, A)d\nu_1(t) < \varepsilon, \end{aligned} \tag{22}$$

hold for all $C > N(\varepsilon)$ and for all Borel sets $A \subseteq [a,b]$, where C is the constant appearing in the definition of the sets S_0, S_1. The proof of (22) will be published elsewhere.

From eq. (17) and the definition of the measure ν_S we have

$$\begin{aligned} &\left|\int_S \mu_y^\delta(A)d\sigma(y) - \int_S \mu_y(A)d\sigma(y)\right| = \left|\frac{1}{\pi}\int_{\mathbb{R}} \theta(t+i\delta, A)d\nu_S(t) - \nu_S(A)\right| \\ &= \left|\int_A \{\frac{1}{\pi}\theta(t+i\delta, A) - 1\}d\nu_S(t) + \frac{1}{\pi}\int_{A^c} \theta(t+i\delta, A)d\nu_S(t)\right|. \end{aligned} \tag{23}$$

In (23), the integral on the left is negative, and the integral on the right is positive. Hence an upper bound for this expression is

$$\sup\left\{\frac{1}{\pi}\int_A \theta(t+i\delta, A^c)d\nu_S(t),\ \frac{1}{\pi}\int_{A^c} \theta(t+i\delta, A)d\nu_S(t)\right\}.$$

Since $\nu_S(B) = (\nu_0 - b_\phi\mu)(B) - \nu_1(B)$ for any Borel set B, combining (20) and (22) we have

$$\left| \int_S \mu_y^\delta(A) d\sigma(y) - \int_S \mu_y(A) d\sigma(y) \right| \leq C E_A(\delta) + \varepsilon, \tag{24}$$

for all $C > N(\varepsilon)$ and for all Borel sets $A \subseteq [a, b]$.

The measures μ_y^δ are absolutely continuous, with density functions $\operatorname{Im}(\lim_{\varepsilon\to 0^+} F_y^\delta(\lambda + i\varepsilon))/\pi = \operatorname{Im} F_y(\lambda + i\delta)/\pi$. Thus, we have

$$\begin{aligned}
\int_S \mu_y^\delta(A) d\sigma(y) &= \int_A \left\{ \frac{1}{\pi} \int_S \operatorname{Im}\left[\frac{1}{y - F(\lambda + i\delta)}\right] d\sigma(y) \right\} d\lambda \\
&= \int_A \frac{1}{\pi} \theta_\sigma(F(\lambda + i\delta), S) d\lambda = \int_A \omega_\sigma^\delta(\lambda, S; F) d\lambda,
\end{aligned}$$

by a change in the order of integration. The absolute continuity of $d\sigma$ implies the absolute continuity of ν_S, which has been associated with the Herglotz function $(\phi_S \circ F) - b_\phi F$. Thus, the density function h_{ν_S} of ν_S is given by $h_{\nu_S}(\lambda) = \lim_{\varepsilon\to 0^+} \operatorname{Im}(\phi_S \circ F - b_\phi F)(\lambda + i\varepsilon)/\pi$, and from the representation of ϕ_S we have

$$\begin{aligned}
h_{\nu_S}(\lambda) &= \lim_{\varepsilon\to 0^+} \frac{1}{\pi} \int_S \operatorname{Im}\left[\frac{1}{t - F(\lambda + i\varepsilon)}\right] d\sigma(t) \\
&= \lim_{\varepsilon\to 0^+} \frac{1}{\pi} \theta_\sigma(F(\lambda + i\varepsilon), S) = \omega_\sigma(\lambda, S).
\end{aligned}$$

Hence,

$$\nu_S(A) = \int_S \mu_y(A) d\sigma(y) = \int_A \omega_\sigma(\lambda, S) d\lambda.$$

We have therefore obtained the following Theorem.

Theorem 3.1. *Let $[a, b]$ be a finite closed interval, K a compact subset of $\mathbb{C}_+$, and S any Borel set. Take δ with $0 < \delta < 1$, let $E_A(\delta)$ be defined by (21), and suppose that the measure $d\sigma$ is absolutely continuous. Then, given $\varepsilon > 0$, a constant $N = N(\varepsilon)$ can be found such that*

$$\left| \int_A \omega_\sigma^\delta(\lambda, S; F) d\lambda - \int_A \omega_\sigma(\lambda, S; F) d\lambda \right| \leq C E_A(\delta) + \varepsilon, \tag{25}$$

for all $C > N(\varepsilon)$, $A \subseteq [a, b]$, and for any Herglotz function F such that $F(i) \in K$.

We remark that the condition $F(i) \in K$ is required in order to prove the inequalities (22).

The application of Theorem 3.1 in the spectral analysis of differential operators is the aim of research in progress.

Acknowledgement

The author is grateful to the Cyprus Research Promotion Foundation for support through the grant FILONE/0506/03.

References

1. E.A. Coddington and N. Levinson, *Theory of Ordinary Differential Equations*, McGraw-Hill, New York, 1955.
2. M.S.P. Eastham and H. Kalf, *Schrödinger-type Operators with Continuous Spectra*, Pitman, London, 1982.
3. S.V. Breimesser, J.D.E. Grant and D.B. Pearson, *J. Comp. Appl. Math.* **148**, 307 (2002).
4. S.V. Breimesser and D.B. Pearson, *Math. Phys., Anal. and Geom.* **3**, 385 (2000).
5. G. Herglotz, *Sächs acad. Wiss. Leipzig* **63**, 501 (1911).
6. N.I. Akhiezer and I.M. Glazman, *Theory of Linear Operators in Hilbert space I*, Pitman, London, 1981.
7. M.E. Munroe, *Measure and Integration*, Addison-Wesley, Massachusetts, 1971.
8. D.B. Pearson, *Quantum Scattering and Spectral Theory*, Academic Press, London, 1988.
9. C. Caratheodory, *Theory of Functions of a complex variable II*, Chelsea, New York, 1954.
10. D.B. Pearson, *Proc. Lond. Math. Soc.* **68**(3), 127 (1994).
11. Y.T. Christodoulides and D.B. Pearson, *Math. Phys., Anal. and Geom.* **7**, 309 (2004).
12. Y.T. Christodoulides and D.B. Pearson, *Math. Phys., Anal. and Geom.* **7**, 333 (2004).

CYCLIC HYPER-SCALAR SYSTEMS

STANCHO DIMIEV

Institute of Mathematics and Informatics,
of the Bulgarian Academy of Sciences,
Sofia 1113, Bulgaria
E-mail: sdimiev@math.bas.bg

MARIN S. MARINOV

Faculty of Applied Mathematics and Informatics,
Technical University,
Sofia 1000, P.O. Box 384, Bulgaria
E-mail: msm@tu-sofia.bg

ZHIVKO ZHELEV

Faculty of Economics, Trakia University,
Stara Zagora 6003, Bulgaria
E-mail: zhelev@uni-sz.bg

In the paper we consider cyclic hyper-number systems which are naturally related to the partial differential equations over the paracomplex algebra with zero divisors. This research is somehow parallel to the notion of the classical hypercomplex systems. At the end we shortly describe the Hankel geometry as a special cyclic geometry. Due to some page restrictions in this volume, we shall develop the analytic and the geometric approaches to this specific kind of cyclic geometry in another article of ours.

Keywords: Cyclic 4×4 matrices; Paracomplex variables; Paracomplex differentiability; Hyperbolic partial differential equations; Hankel geometry; Hankel transformations.

Let $\mathbf{K}$ denotes the field of complex numbers or the field of real numbers, i. e. $\mathbf{K} = \mathbf{R}$ or $\mathbf{K} = \mathbf{C}$. By definition $\mathbf{K}(j_0, j_1, \dots, j_n) := \{a_0 j_0 + a_1 j_1 + \cdots + a_n j_n : a_k \in \mathbf{K}, k = 0, 1, 2, \dots, n\}$, $\dim \mathbf{K}(j_0, j_1, \dots, j_n) = n + 1$, is called an $(n+1)$-dimensional hyper-$\mathbf{K}$-system with units j_k.

Addition in such systems is defined naturally by the addition of the corresponding coefficients, and the multiplication with the help of the units

(c_r^{lk}), $r = 0, 1, \ldots, n$:

$$j_k j_l = c_0^{kl} j_0 + c_1^{kl} j_1 + \cdots + c_n^{kl} j_n, \qquad j_l j_k = c_0^{lk} j_0 + c_1^{lk} j_1 + \cdots + c_n^{lk} j_n.$$

In general we have that $j_k j_l \neq j_l j_k$, $k, l = 0, 1, 2, \ldots, n$, i.e a non-commutative multiplication. We see that $\mathbf{K} \subset \mathbf{K}(j_0, j_1, j_2, \ldots, j_n)$. We set $j_0 = 1$ and therefore $j_0 j_0 = j_0$, $j_0 j_k = j_k j_0 = j_k$.

1. Hyper-real systems: K = R

The ordinary complex numbers $x + iy$ can be considered as linear combinations of the kind $x \cdot 1 + y \cdot i$ constructed from the units 1 and i with the help of the (real) parameters x and y.

Felix Klein, The Elementary Mathematics from the Point of View of the Higher.

The simplest hyper-real system is the field of complex numbers, namely

$$\mathbf{R}(1, i) := \{x + iy : x, y \in \mathbf{R},\ i^2 = -1\},$$

where i is the unique unit j_1. In fact, there are two other 2-dimensional hyper-real systems:

$$\mathbf{R}(1, \kappa) = \{x + \kappa y : x, y \in \mathbf{R},\ \kappa^2 = 1\}$$

(double-real numbers, or para-complex numbers, or hyperbolic numbers), and

$$R(1, \omega) = \{x + \omega y : x, y \in \mathbf{R},\ \omega^2 = 0\},$$

known as E. Study's numbers. The unit in the case of double-real numbers can be interpreted using the matrix representation of the hyper-real system $\mathbf{R}(1, \kappa)$. The elements of this system have the following matrix representation as 2×2 real cyclic matrices:

$$Q = \begin{pmatrix} x & y \\ y & x \end{pmatrix} = x \begin{pmatrix} 1 & 0 \\ 0 & 1 \end{pmatrix} + y \begin{pmatrix} 0 & 1 \\ 1 & 0 \end{pmatrix}, \qquad \begin{pmatrix} 0 & 1 \\ 1 & 0 \end{pmatrix}^2 = 1,$$

and we set $\kappa := \begin{pmatrix} 0 & 1 \\ 1 & 0 \end{pmatrix}$. Let us recall that the matrix representation of the real numbers can be written as follows: if $r \in \mathbf{R}$, then $r \to \begin{pmatrix} r & 0 \\ 0 & r \end{pmatrix}$.

Therefore we have $\kappa = \begin{pmatrix} 0 & 1 \\ 1 & 0 \end{pmatrix} \notin \mathbf{R}$.

The hyper-real systems above are two-dimensional systems. The following hyper-real systems are real four-dimensional. They are closely related to the non-commutative division quaternion algebras (Hamilton, 1843):

$$\mathbf{R}(1,i,j,k) = \{x_0 + ix_1 + jx_2 + kx_3 : x_0, x_1, x_2, x_3 \in \mathbf{R},$$
$$i^2 = j^2 = -1,\ k = ij\}$$

The bicomplex numbers (Segre, 1894) are: $R(1,i,j,e) = \{x_0 + ix_1 + jx_2 + ex_3 : x_0, x_1, x_2, x_3 \in \mathbf{R}, i^2 = j^2 = -1, e^2 = +1\}$. The corresponding algebra is a commutative algebra with zero divisors.

2. Hyper-complex systems

Some of the above presented hyper-real systems admit a complexification. We shall note this for the complex and bicomplex numbers [1]:

$$\mathbf{R}(1,i) = \mathbf{C}(1), \qquad \mathbf{R}(1,i,j,e) = \mathbf{C}(1,j), \qquad j^2 = -1.$$

These are true up to an isomorphism.

3. Real double-variables or para-complex variables

The real double-number system $\mathbf{R}(1,\kappa)$, $\kappa^2 = 1$, is an algebra and we denote its elements by $z = x + \kappa y$, $x, y \in \mathbf{R}$. They are also called para-complex numbers. By definition $z^* := x - \kappa y$ is called the conjugate of z. This algebra has zero divisors. We have that $dz = dx + \kappa dy$, $dz^* = dx - \kappa dy$.

One can consider differentiable real double-number functions

$$f : \mathbf{R}(1,\kappa) \to \mathbf{R}(1,\kappa), \quad \text{i.e. } f(z) = f_0(x,y) + \kappa f_1(x,y).$$

From here on we consider only continuously differentiable functions, i.e. functions with continuous partial derivatives.

The following representation for the differentials is valid:

$$df = \frac{\partial f}{\partial x}dx + \frac{\partial f}{\partial y}dy = \frac{\partial f}{\partial z}dz + \frac{\partial f}{\partial z^*}dz^*,$$

where

$$\frac{\partial}{\partial z} = 1/2\left(\frac{\partial}{\partial x} + \kappa\frac{\partial}{\partial y}\right), \quad \frac{\partial}{\partial z^*} = 1/2\left(\frac{\partial}{\partial x} - \kappa\frac{\partial}{\partial y}\right), \quad 1/\kappa = \kappa.$$

We say that $f(z)$ is holomorphic real double-number or holomorphic paracomplex function if

$$\frac{\partial f}{\partial z^*} = 0, \qquad \text{or} \qquad \frac{\partial f}{\partial x} - \kappa\frac{\partial f}{\partial y} = 0. \tag{1}$$

The condition above implies

$$\frac{\partial f_0(x,y)}{\partial x} + \kappa \frac{\partial f_1(x,y)}{\partial x} - \kappa \left(\frac{\partial f_0(x,y)}{\partial y} + \kappa \frac{\partial f_1(x,y)}{\partial y} \right) = 0,$$

and since $\kappa^2 = 1$, we get

$$\frac{\partial f_0(x,y)}{\partial x} - \frac{\partial f_1(x,y)}{\partial y} + \kappa \left(\frac{\partial f_1(x,y)}{\partial x} - \frac{\partial f_0(x,y)}{\partial y} \right) = 0.$$

Finally, we get the following system of two unknown functions f_0 and f_1:

$$\begin{aligned} \frac{\partial f_0(x,y)}{\partial x} - \frac{\partial f_1(x,y)}{\partial y} &= 0, & \frac{\partial f_1(x,y)}{\partial x} - \frac{\partial f_0(x,y)}{\partial y} &= 0, \\ \frac{\partial f_0(x,y)}{\partial x} &= \frac{\partial f_1(x,y)}{\partial y}, & \frac{\partial f_1(x,y)}{\partial x} &= \frac{\partial f_0(x,y)}{\partial y}. \end{aligned} \tag{2}$$

From (2) we get the following second order equations:

$$\frac{\partial^2 f_0(x,y)}{\partial x^2} - \frac{\partial^2 f_0(x,y)}{\partial y^2} = 0, \qquad \frac{\partial^2 f_1(x,y)}{\partial x^2} - \frac{\partial^2 f_1(x,y)}{\partial y^2} = 0. \tag{3}$$

Remark 3.1. The system (2) is a hyperbolic system of two unknown functions, and the second order equations are hyperbolic equations respectively.

Example 3.1. Let $f_0(x,y) = x$, $f_1(x,y) = y$, i.e. $f(x+\kappa y) = x + \kappa y$. The system (2) and the second order equations (3) are satisfied.

Example 3.2. Let $f_0(x,y) = x$, $f_1(x,y) = -y$, i.e. $f(x - \kappa y) = x - \kappa y$. The system (2) and the second order equations are not satisfied.

4. Four real variables, $\mathbf{K} = \mathbf{R}$

Let us consider the cyclic matrices $\begin{pmatrix} a & b & c & d \\ d & a & b & c \\ c & d & a & b \\ b & c & d & a \end{pmatrix}$. Then any real number can be represented in the form $\begin{pmatrix} r & 0 & 0 & 0 \\ 0 & r & 0 & 0 \\ 0 & 0 & r & 0 \\ 0 & 0 & 0 & r \end{pmatrix}$, which is a cyclic matrix. We denote the set of all real (4×4)-matrices by $M_4(\mathbf{R})$. This set is a non-commutative real algebra. For $r = 1$ the matrix above is the unity element in the algebra $M_4(\mathbf{R})$.

We see that the set of all cyclic matrices is a subalgebra of $M_4(\mathbf{R})$, which include the reals and the unity of $M_4(\mathbf{R})$. It is denoted by $\mathbf{R}(1, \tau, \tau^2, \tau^3)$.

In the previous paragraph we considered, in fact, cyclic (2×2)-matrices $(\mathbf{R}(1,\kappa) \subset M_2(\mathbf{R}))$.

Let us define

$$\tau := \begin{pmatrix} 0\,1\,0\,0 \\ 0\,0\,1\,0 \\ 0\,0\,0\,1 \\ 1\,0\,0\,0 \end{pmatrix}, \qquad \tau^2 := \begin{pmatrix} 0\,0\,1\,0 \\ 0\,0\,0\,1 \\ 1\,0\,0\,0 \\ 0\,1\,0\,0 \end{pmatrix}, \qquad \tau^3 := \begin{pmatrix} 0\,0\,0\,1 \\ 1\,0\,0\,0 \\ 0\,1\,0\,0 \\ 0\,0\,1\,0 \end{pmatrix},$$

and τ^4 coincides with the unity matrix I in the algebra $M_4(\mathbf{R})$. It is easy to see that $\tau^k\tau^q = \tau^{k+q}$, $k, q = 1, 2, 3, 4$ and $\tau^4 = I$.

Let (a, b, c, d) denotes the cyclic (4×4)-matrix defined by these four real numbers.

Proposition 4.1. *Each cyclic matrix (a, b, c, d) can be represented in the form:*

$$(a, b, c, d) = aI + b\tau + c\tau^2 + d\tau^3, \quad aI = a.$$

Proof. The proof is obtained by straightforward computations. □

Proposition 4.2. *Each four-real number can be represented as a sum of two double-real numbers.*

Proof. It is obvious that $a + b\tau + c\tau^2 + d\tau^3 = a + c\tau^2 + \tau(b + d\tau^2)$. Let us set $a + c\tau^2 = \alpha$ and $b + d\tau^2 = \beta$. Since $(\tau^2)^2 = 1$, it follows that α and β are double-real numbers. So we obtain that

$$a + b\tau + c\tau^2 + d\tau^3 = \alpha + \beta\tau, \quad \text{with} \quad \tau^4 = 1. \qquad \square$$

The above is often called *a doubling* of the initial real variables.

5. Four-real functions of four real variables

The set of four-real variables is denoted by $\mathbf{R}(1, \tau, \tau^2, \tau^3) \subset M_4(\mathbf{R})$. It is a commutative algebra with zero divisors which is a subalgebra of $M_4(\mathbf{R})$, or 4-dimensional commutative hyper real system. If $\xi \in \mathbf{R}(1, \tau, \tau^2, \tau^3)$, we have that

$$\xi = x + y\tau + u\tau^2 + v\tau^3 \quad \text{with} \quad x, y, u, v \in \mathbf{R}.$$

A function $\phi : \mathbf{R}(1, \tau, \tau^2, \tau^3) \to \mathbf{R}(1, \tau, \tau^2, \tau^3)$ is represented as follows:

$$\phi(x + y\tau + u\tau^2 + v\tau^3)$$
$$= \phi_0(x, y, u, v) + \phi_1(x, y, u, v)\tau + \phi_2(x, y, u, v)\tau^2 + \phi_4(x, y, u, v)\tau^3,$$

where ϕ_j, $j = 0, 1, 2, 3$ are function of four real variables.

On the other hand the function ϕ can be considered as a function of two double-real variables $\phi : [\mathbf{R}(1, \tau^2)](1, \tau) \to [\mathbf{R}(1, \tau^2)](1, \tau)$ if it can be represented as follows:

$$\phi(\alpha + \tau\beta) = \phi_0(\alpha, \beta) + \tau\phi_1(\alpha, \beta), \quad \alpha, \beta \in \mathbf{R}(1, \tau^2).$$

This means that ϕ is differentiable as a function of double-real variables if there are two functions φ_0 and φ_1 of the double-real variables α and β such that the equality above is true.

Let dx, dy, du, dv be the ordinary differentials. Then the differentials of the double-real variables α and β are

$$d\alpha = dx + \tau^2 du, \quad d\beta = dy + \tau^2 dv.$$

6. Partial derivatives with respect to the double-real variables α and β

Let $f(\xi)$ be a four-real function of four-real variables

$$\xi = x + \tau y + \tau^2 u + \tau^3 v = \begin{pmatrix} x\ y\ u\ v \\ y\ u\ v\ x \\ u\ v\ x\ y \\ v\ x\ y\ u \end{pmatrix}.$$

We have that

$$df = \frac{\partial f}{\partial x} dx + \frac{\partial f}{\partial y} dy + \frac{\partial f}{\partial u} du + \frac{\partial f}{\partial v} dv,$$

where

$$\frac{\partial f}{\partial x} = \frac{\partial f_0}{\partial x} + \tau \frac{\partial f_1}{\partial x} + \tau^2 \frac{\partial f_2}{\partial x} + \tau^3 \frac{\partial f_3}{\partial x},$$

etc. We recall that

$$\alpha = x + \tau^2 u, \qquad \beta = y + \tau^2 v, \qquad \alpha^* = x - \tau^2 u, \qquad \beta^* = y - \tau^2 v.$$

Then

$$d\alpha = dx + \tau^2 du, \quad d\beta = dy + \tau^2 v, \quad d\alpha^* = dx - \tau^2 du, \quad d\beta^* = dy - \tau^2 dv,$$

and it follows that

$$2dx = d\alpha + d\alpha^*, \qquad 2dy = d\beta + d\beta^*, \qquad 2du = \tau^2(d\alpha - d\alpha^*),$$

and $2dv = \tau^2(d\beta - d\beta^*)$. We introduce the following differential operators:

$$\frac{\partial f}{\partial \alpha} := \frac{1}{2}\left(\frac{\partial f}{\partial x} + \tau^2 \frac{\partial f}{\partial u}\right), \qquad \frac{\partial f}{\partial \beta} := \frac{1}{2}\left(\frac{\partial f}{\partial y} + \tau^2 \frac{\partial f}{\partial v}\right),$$
$$\frac{\partial f}{\partial \alpha^*} := \frac{1}{2}\left(\frac{\partial f}{\partial x} - \tau^2 \frac{\partial f}{\partial u}\right), \qquad \frac{\partial f}{\partial \beta^*} := \frac{1}{2}\left(\frac{\partial f}{\partial y} - \tau^2 \frac{\partial f}{\partial v}\right).$$

Then the following equalities are true:

$$\frac{\partial f}{\partial x} = \left(\frac{\partial f}{\partial \alpha} + \frac{\partial f}{\partial \alpha^*}\right), \qquad \frac{\partial f}{\partial y} = \left(\frac{\partial f}{\partial \beta} + \frac{\partial f}{\partial \beta^*}\right),$$
$$\tau^2\frac{\partial f}{\partial u} = \left(\frac{\partial f}{\partial \alpha} - \frac{\partial f}{\partial \alpha^*}\right), \qquad \frac{\partial f}{\partial v} = \left(\frac{\partial f}{\partial \beta} - \frac{\partial f}{\partial \beta^*}\right).$$

Proposition 6.1. *The following equalities hold:*

$$\frac{\partial f}{\partial x}dx + \frac{\partial f}{\partial u}du = \frac{\partial f}{\partial \alpha}d\alpha + \frac{\partial f}{\partial \alpha^*}d\alpha^*, \quad \frac{\partial f}{\partial y}dy + \frac{\partial f}{\partial v}dv = \frac{\partial f}{\partial \beta}d\beta + \frac{\partial f}{\partial \beta}d\beta^*. \tag{4}$$

Proof. We have that

$$\begin{aligned}
&\frac{\partial f}{\partial x}dx + \frac{\partial f}{\partial u}du \\
&= \left(\frac{\partial f}{\partial \alpha} + \frac{\partial f}{\partial \alpha^*}\right)\left(\frac{1}{2}(d\alpha + d\alpha^*)\right) + \tau^2\left(\frac{\partial f}{\partial \alpha} - \frac{\partial f}{\partial \alpha^*}\right)\left(\frac{\tau^2}{2}(d\alpha - d\alpha^*)\right) \\
&= \frac{\partial f}{\partial \alpha}d\alpha + \frac{\partial f}{\partial \alpha^*}d\alpha^*.
\end{aligned}$$

The second equality can be proved in exactly the same way. □

If we put $f(\xi) = f(\alpha + \tau\beta)$ where $\alpha = x + \tau^2 u$ and $\beta = y + \tau^2 v$, we obtain the following

Corollary 6.1. $df = \dfrac{\partial f}{\partial \alpha}d\alpha + \dfrac{\partial f}{\partial \beta}d\beta + \dfrac{\partial f}{\partial \alpha^*}d\alpha^* + \dfrac{\partial f}{\partial \beta^*}d\beta^*.$

7. Para-complex differentiability

Let f be a four-real function of four-real variables.

Definition 7.1. We say that a function f is a para-complex differentiable if

$$\frac{\partial f}{\partial \alpha^*}d\alpha^* + \frac{\partial f}{\partial \beta^*}d\beta^* = 0. \tag{5}$$

It follows that

$$\begin{aligned}
&\frac{\partial f}{\partial \alpha^*}d\alpha^* + \frac{\partial f}{\partial \beta^*}d\beta^* \\
&= \frac{1}{2}\left(\frac{\partial f}{\partial x} - \tau^2\frac{\partial f}{\partial u}\right)(dx - \tau^2 du) + \frac{1}{2}\left(\frac{\partial f}{\partial y} - \tau^2\frac{\partial f}{\partial v}\right)(dy - \tau^2 dv) \\
&= \frac{1}{2}\left(\frac{\partial f}{\partial x} - \tau^2\frac{\partial f}{\partial u}\right)dx + \frac{1}{2}\left(\frac{\partial f}{\partial y} - \tau^2\frac{\partial f}{\partial v}\right)dv \\
&\quad + \frac{1}{2}\left(\frac{\partial f}{\partial u} - \tau^2\frac{\partial f}{\partial x}\right)du + \frac{1}{2}\left(\frac{\partial f}{\partial v} - \tau^2\frac{\partial f}{\partial y}\right)dv.
\end{aligned}$$

It turns out that the differentiability condition (5) implies a new condition:

$$\frac{\partial f}{\partial x} = \tau^2\frac{\partial f}{\partial u}, \qquad \frac{\partial f}{\partial y} = \tau^2\frac{\partial f}{\partial v}. \tag{6}$$

In terms of components f_0, f_1, f_2, f_3, we have

$$\begin{aligned}
\frac{\partial f_0}{\partial x} + \tau\frac{\partial f_1}{\partial x} + \tau^2\frac{\partial f_2}{\partial x} + \tau^3\frac{\partial f_3}{\partial x} &= \tau^2\left(\frac{\partial f_0}{\partial u} + \tau\frac{\partial f_1}{\partial u} + \tau^2\frac{\partial f_2}{\partial u} + \tau^3\frac{\partial f_3}{\partial u}\right); \\
\frac{\partial f_0}{\partial y} + \tau\frac{\partial f_1}{\partial y} + \tau^2\frac{\partial f_2}{\partial y} + \tau^3\frac{\partial f_3}{\partial y} &= \tau^2\left(\frac{\partial f_0}{\partial v} + \tau\frac{\partial f_1}{\partial v} + \tau^2\frac{\partial f_2}{\partial v} + \tau^3\frac{\partial f_3}{\partial v}\right).
\end{aligned}$$

Comparing coefficients before $1, \tau, \tau^2, \tau^3$, one arrives at the next

Theorem 7.1. *The components f_j of the differentiable four-real function $f = f_0 + \tau f_1 + \tau^2 f_2 + \tau^3 f_3$ satisfy the following system of 8 first order partial differential equations*

$$\begin{aligned}
&\frac{\partial f_0}{\partial x} = \frac{\partial f_2}{\partial u}, \quad \frac{\partial f_0}{\partial u} = \frac{\partial f_2}{\partial x}, \quad \frac{\partial f_1}{\partial x} = \frac{\partial f_3}{\partial u}, \quad \frac{\partial f_1}{\partial u} = \frac{\partial f_3}{\partial x}, \\
&\frac{\partial f_0}{\partial y} = \frac{\partial f_2}{\partial v}, \quad \frac{\partial f_0}{\partial v} = \frac{\partial f_2}{\partial y}, \quad \frac{\partial f_1}{\partial y} = \frac{\partial f_3}{\partial v}, \quad \frac{\partial f_1}{\partial v} = \frac{\partial f_3}{\partial y}.
\end{aligned}$$

For the 4 second order partial differential equations of hyperbolic type, we have respectively

$$\begin{aligned}
&\frac{\partial^2 f_0}{\partial x^2} - \frac{\partial^2 f_0}{\partial u^2} = 0, \qquad \frac{\partial^2 f_1}{\partial x^2} - \frac{\partial^2 f_1}{\partial u^2} = 0, \\
&\frac{\partial^2 f_0}{\partial y^2} - \frac{\partial^2 f_0}{\partial v^2} = 0, \qquad \frac{\partial^2 f_1}{\partial y^2} - \frac{\partial^2 f_1}{\partial v^2} = 0.
\end{aligned}$$

8. Hyper four-dimensional real scalar systems and the Hankel geometry [2–4]

Here we recall another pure geometric aspect of the cyclic 4×4 matrices, namely the decomposition

$$\begin{pmatrix} a & b & c & d \\ b & c & d & a \\ c & d & a & b \\ d & a & b & c \end{pmatrix} = a\mathcal{H}_1 + b\mathcal{H}_2 + c\mathcal{H}_3 + d\mathcal{H}_4,$$

where

$$\mathcal{H}_1 = \begin{pmatrix} 1&0&0&0 \\ 0&0&0&1 \\ 0&0&1&0 \\ 0&1&0&0 \end{pmatrix}, \qquad \mathcal{H}_2 = \begin{pmatrix} 0&1&0&0 \\ 1&0&0&0 \\ 0&0&0&1 \\ 0&0&1&0 \end{pmatrix},$$

$$\mathcal{H}_3 = \begin{pmatrix} 0&0&1&0 \\ 0&1&0&0 \\ 1&0&0&0 \\ 0&0&0&1 \end{pmatrix}, \qquad \mathcal{H}_4 = \begin{pmatrix} 0&0&0&1 \\ 0&0&1&0 \\ 0&1&0&0 \\ 1&0&0&0 \end{pmatrix},$$

and $a+b+c+d \neq 0$, $a+c-b-d \neq 0$, $a \neq c$, and $b \neq d$. The last are sufficient for the matrices to be non-singular. The set of all such matrices is denoted by $\mathbf{R}(\mathcal{H}_1, \mathcal{H}_2, \mathcal{H}_3, \mathcal{H}_4)$. The following properties of the above considered matrices are valid:

$$\begin{aligned} &\mathcal{H}_1^2 = \mathcal{H}_2^2 = \mathcal{H}_3^2 = \mathcal{H}_4^2; \\ &\mathcal{H}_1\mathcal{H}_2 = \mathcal{H}_2\mathcal{H}_3 = \mathcal{H}_3\mathcal{H}_4; \\ &\mathcal{H}_1\mathcal{H}_3 = \mathcal{H}_2\mathcal{H}_4. \end{aligned}$$

The considered matrices are called Hankel matrices and the geometry induced by them is called a Hankel geometry.

Since the product of two Hankel matrices is not a Hankel matrix, $\mathbf{R}(\mathcal{H}_1, \mathcal{H}_2, \mathcal{H}_3, \mathcal{H}_4)$ is not a ring with the usual matrix multiplication, but it is a linear space. The inverse matrix of a Hankel matrix is a Hankel matrix too. Sometimes the inverse of a non-singular Hankel matrix is called *a Bézout matrix.* Hence, it is easy to formulate the following assertion:

Any product of two Hankel matrices is a linear combination of the matrices $\mathcal{H}_1\mathcal{H}_1 = \mathcal{I}$, $\mathcal{H}_1\mathcal{H}_2$, $\mathcal{H}_1\mathcal{H}_3$ and $\mathcal{H}_1\mathcal{H}_4$ defined already above.

Let now

$$\mathcal{M} = \begin{pmatrix} x & y & z & u \\ y & z & u & x \\ z & u & x & y \\ u & x & y & z \end{pmatrix} \in \mathbf{R}(\mathcal{H}_1, \mathcal{H}_2, \mathcal{H}_3, \mathcal{H}_4).$$

One can define the following transformations:

$$\begin{aligned} X &:= ax + by + cz + du, & Y &:= ay + bz + cu + dx, \\ Z &:= az + bu + cx + dy, & U &:= au + bx + cy + dz. \end{aligned} \tag{7}$$

Points (vectors) $v(x, y, z, u)$ are the points (vectors) in Hankel geometry, and (7) are the *Hankel transformations.* One can study some specific invariants under these transformations.

Let $v_1(x_1, y_1, z_1, u_1)$, $v_2(x_2, y_2, z_2, u_2)$ be two vectors in Hankel geometry. Then we define a scalar product as follows:

$$v_1 v_2 = x_1 x_2 + y_1 y_2 + z_1 z_2 + u_1 u_2.$$

It is proved that scalar product in Hankel geometry is an absolute invariant iff $a^2 + b^2 + c^2 + d^2 = 1$, $ab + bc + cd + ud = 0$ and $ac + bd = 0$ i.e. the considered cyclic matrix is an orthogonal matrix.

Analogously, it is possible to define a pseudo-scalar product of two vectors and a cross-product of three vectors. In particular, a pseudo-scalar product of two vectors v_1 and v_2 is defined as follows:

$$v_1 v_2 = x_1 x_2 + y_1 y_2 + K(z_1 z_2 + u_1 u_2). \tag{8}$$

Then the following remarkable fact is true:

Pseudo-scalar product (8) is invariant under the Hankel transformations iff $K = -1$ and these transformations are of the form

$$\begin{aligned} X &= \cosh(t)x + \sinh(t)z, & Y &= \cosh(t)y + \sinh(t)u, \\ Z &= \sinh(t)x + \cosh(t)z, & U &= \sinh(t)y + \cosh(t)u, \end{aligned}$$

i.e. iff the Hankel transformations are extensions to the Lorentz transformations and $K = -1$.

Acknowledgments

It is a great pleasure for us to acknowledge both the helpful discussion with Dr. L. Apostolova from the Institute of Mathematics & Informatics, BAS concerning these and related matters, and her invaluable help in preparing this manuscript.

References

1. S. Dimiev, Double-complex analytic functions, *Applied Complex and Quaternionic Approximation*, Ed. by R.K. Kovacheva, J. Lawrynowicz, S. Marchiafava, World Scientific, Singapore 2009, pp. 58–75.
2. Gr. Stanilov, Four-dimensional Hankel geometry, *Proceedings of the International Conference "Computer methods in Science and Education"*, 12–14 Sept. 2008, Varna, Bulgaria, (to appear).
3. M. Fiedler, Remarks on eigenvalues of Hankel matrices, IMA Preprint Series # 903, 1991, pp. 4.
4. F.J. Lander, The Bezoutian and inversion of Hankel and Toeplitz matrices, Mat. Issled. **9** (1974), 69–87. (In Russian).

PLANE CURVES ASSOCIATED WITH INTEGRABLE DYNAMICAL SYSTEMS OF THE FRENET-SERRET TYPE

P. A. DJONDJOROV* and V. M. VASSILEV

Institute of Mechanics – Bulgarian Academy of Sciences
Acad. G. Bonchev Street, Block 4, Sofia, 1113, Bulgaria
**E-mail: padjon@imbm.bas.bg*

I. M. MLADENOV

Institute of Biophysics – Bulgarian Academy of Sciences
Acad. G. Bonchev Street, Block 21, Sofia, 1113, Bulgaria
E-mail: mladenov@obzor.bio21.bas.bg

This paper deals with plane curves which being integral curves of a certain dynamical system of two degrees of freedom can be thought of as trajectories of a particle of unit mass in a potential field. Two special cases of this system that are integrable by quadratures are identified in another paper of this volume. Typical examples of curves, corresponding to each of these two cases are the so-called Lévy's elasticae and the profile curves of the Delaunay surfaces, respectively. Here their parameterizations are derived in explicit form and illustrated by a number of figures.

Keywords: Dynamical system; Integrability; Lévy's elastica; Delaunay surfaces.

1. Dynamical systems of the Frenet-Serret type

The present contribution is concerned with the integral curves of systems consisting of two second-order ordinary differential equations of form

$$\ddot{x} + \mathcal{K}(x,z)\dot{z} = 0, \qquad \ddot{z} - \mathcal{K}(x,z)\dot{x} = 0 \tag{1}$$

where (x, z) are Cartesian co-ordinates in the Euclidean plane $\mathbb{R}^2$ and dots denote derivatives with respect to the archlength s of the respective curve. Systems like (1) may be referred to as systems of Frenet-Serret type since they originate from the prominent Frenet-Serret formulas as is shown in another paper in this volume (see Ref. 1). Each system of the form (1) can be also regarded as a dynamical system of two degrees of freedom determining the motion (trajectories) of a particle of unit mass, s playing the role of time. Given a system of the aforementioned form, the function $\mathcal{K}(x,z)$ is easily identified with the curvature of its integral curves.

Two cases of dynamical systems of Frenet-Serret type that are integrable by quadratures are identified in Ref. 1. Examples of such systems within each of these cases are given below, and explicit parametric equations of the respective curves are derived for all of them.

2. Lévy's elasticae

The first example is the so-called Lévy's elasticae [3] — curves of curvature

$$\mathcal{K}(x,z) = \mathcal{K}(r) = ar^2 + c, \qquad a, c \in \mathbb{R}, \quad a \neq 0 \tag{2}$$

where

$$r = \sqrt{x^2 + z^2}.$$

It is shown in Ref. 1 that in this case, integrability in quadratures of the respective system of the form (1) can be accomplished in polar coordinates

$$x = r\cos\theta, \qquad y = r\sin\theta \tag{3}$$

in which it reduces to

$$\dot{\vartheta} = \frac{1}{r^2}\left(rU(r) - \varepsilon\right), \qquad \dot{r} = \pm\frac{1}{r}\sqrt{r^2 - \left(rU\left(r\right) - \varepsilon\right)^2} \tag{4}$$

where

$$U\left(r\right) = \frac{1}{r}\int r\,\mathcal{K}(r)\,\mathrm{d}\,r = \frac{1}{4}\left(ar^3 + 2cr\right)$$

and ε is a real number.

Instead of integrating directly for $r(s)$ and $\vartheta(r)$, we chose another path on the way to determine the explicit parametric equations of the curves with curvature specified in (2). Namely, we first differentiate the expressions (3) with respect to s, then substitute the so obtained derivatives of x and z in $\dot{x} = \cos\varphi$, $\dot{z} = \sin\varphi$, where φ is the slope angle ($\dot{\varphi} = \kappa(s) = \mathcal{K}\left(x(s), z(s)\right)$) and finally solve for $\cos\vartheta$ and $\sin\vartheta$. Substitution of the so obtained expressions in (3) yields

$$x = \frac{r\dot{r}}{r^2\dot{\vartheta}^2 + \dot{r}^2}\cos\varphi + \frac{r^2\dot{\vartheta}}{r^2\dot{\vartheta}^2 + \dot{r}^2}\sin\varphi,$$
$$z = \frac{r\dot{r}}{r^2\dot{\vartheta}^2 + \dot{r}^2}\sin\varphi - \frac{r^2\dot{\vartheta}}{r^2\dot{\vartheta}^2 + \dot{r}^2}\cos\varphi.$$

Substituting here equations (4) and taking into account the relation

$$r^2 = \frac{1}{a}\kappa(s) - c$$

that is inherent to Lévy's elasticae (see Ref. 3 as well as formula (2)), one ends with the expressions

$$\begin{aligned} x(s) &= \frac{1}{2a}\frac{\mathrm{d}\kappa(s)}{\mathrm{d}s}\cos\varphi(s) + \frac{1}{4a}(\kappa^2(s) - c^2 - 4a\varepsilon)\sin\varphi(s), \\ z(s) &= \frac{1}{2a}\frac{\mathrm{d}\kappa(s)}{\mathrm{d}s}\sin\varphi(s) - \frac{1}{4a}(\kappa^2(s) - c^2 - 4a\varepsilon)\cos\varphi(s) \end{aligned} \tag{5}$$

which produce the position vector components in terms of the curvature $\kappa(s)$ and the slope angle $\varphi(s)$, obtained by other means in Ref. 2. Thus, the problem for determination of explicit parametric equations of the curves constituting Lévy's elasticae reduces to the problem for determination of explicit expressions for the curvature $\kappa(s)$ and the slope angle $\varphi(s)$.

After the above substitutions, the equation for $\dot{r}$ can be written in the form

$$\left(\frac{\mathrm{d}\kappa(s)}{\mathrm{d}s}\right)^2 = P(\kappa(s)), \qquad P(\kappa) = 2E - \frac{1}{4}\kappa^4 + \frac{1}{2}\mu\kappa^2 + \sigma\kappa \tag{6}$$

where

$$E = -\frac{1}{8}c^4 - 2a^2\varepsilon^2 - c^2 a\varepsilon - 2ca, \qquad \mu = c^2 + 4a\varepsilon, \qquad \sigma = 4a.$$

The very same equation has been studied recently by the present authors in the context of the equilibrium shapes of lipid bilayer membranes [2] and it has been proved that depending on the values of the parameters μ, σ and E, two cases for the intrinsic equation of the curve and the corresponding slope angle $\varphi(s)$ have to be considered.

Case 1. The polynomial $P(\kappa)$ has two real roots $\alpha < \beta$ and two complex conjugate roots γ and $\delta = \bar{\gamma}$. In this case there exist both periodic and non-periodic solutions of the equation (6) for the curvature $\kappa(s)$.

If $(3\alpha + \beta)(\alpha + 3\beta) \neq 0$ and $\eta = (\gamma - \delta)/(2i) \neq 0$, then there exist periodic solutions which are given below by the following formulas

$$\kappa_1(s) = \frac{(A\beta + B\alpha) - (A\beta - B\alpha)\,\mathrm{cn}(us, k)}{(A + B) - (A - B)\,\mathrm{cn}(us, k)}, \tag{7}$$

$$\begin{aligned} \varphi_1(s) = {} & \frac{A\beta - B\alpha}{A - B}s + \frac{(A + B)(\alpha - \beta)}{2u(A - B)}\Pi\left(-\frac{(A - B)^2}{4AB}, \mathrm{am}(us, k), k\right) \\ & + \frac{\alpha - \beta}{2u\sqrt{k^2 + \frac{(A-B)^2}{4AB}}}\arctan\left(\sqrt{k^2 + \frac{(A - B)^2}{4AB}}\,\frac{\mathrm{sn}(us, k)}{\mathrm{dn}(us, k)}\right) \end{aligned} \tag{8}$$

where

$$A=\sqrt{4\eta^2+(3\alpha+\beta)^2},\qquad B=\sqrt{4\eta^2+(\alpha+3\beta)^2},\qquad u=\frac{1}{4}\sqrt{AB},$$

$$k=\frac{1}{\sqrt{2}}\sqrt{1-\frac{4\eta^2+(3\alpha+\beta)(\alpha+3\beta)}{\sqrt{[4\eta^2+(3\alpha+\beta)(\alpha+3\beta)]^2+16\eta^2(\beta-\alpha)^2}}}.$$

Non-periodic solutions are obtained in the cases in which

$$(3\alpha+\beta)(\alpha+3\beta)=\eta=0,$$

and they are of the form

$$\kappa_2(s)=\zeta-\frac{4\zeta}{1+\zeta^2 s^2},\qquad \varphi_2(s)=\zeta s-4\arctan(\zeta s),$$

where $\zeta=\alpha$ if $3\alpha+\beta=0$ and $\zeta=\beta$ if $\alpha+3\beta=0$.

Case 2. The polynomial $P(\xi)$ has four real roots $\alpha<\beta<\gamma<\delta$. Then, two periodic solutions of the equation (6) exist, i.e.,

$$\begin{aligned}\kappa_3(s)&=\delta-\frac{(\delta-\alpha)(\delta-\beta)}{(\delta-\beta)+(\beta-\alpha)\operatorname{sn}^2(us,k)},\\ \varphi_3(s)&=\delta s-\frac{\delta-\alpha}{u}\Pi\left(\frac{\beta-\alpha}{\beta-\delta},\operatorname{am}(us,k),k\right)\end{aligned}\tag{9}$$

and

$$\begin{aligned}\kappa_4(s)&=\beta+\frac{(\gamma-\beta)(\delta-\beta)}{(\delta-\beta)-(\delta-\gamma)\operatorname{sn}^2(us,k)},\\ \varphi_4(s)&=\beta s-\frac{\beta-\gamma}{u}\Pi\left(\frac{\delta-\gamma}{\delta-\beta},\operatorname{am}(us,k),k\right)\end{aligned}\tag{10}$$

where in both cases

$$u=\frac{1}{4}\sqrt{(\gamma-\alpha)(\delta-\beta)},\qquad k=\sqrt{\frac{(\beta-\alpha)(\delta-\gamma)}{(\gamma-\alpha)(\delta-\beta)}}.$$

Depending on the chosen values of the constants μ, σ and E, the curves constituting the Lévy's elasticae can be open or closed, the latter being simple or self-intersecting. It is shown in Ref. 2 that the respective simple curves are generated by the periodic solutions specified in the first case while the two solutions in the second case produce only self-intersecting curves.

Typical examples of curves, corresponding to curvatures of the form (7) are seen in Figure 1. Curves of this kind with only one axis of symmetry are necessarily self-intersecting, whereas curves of more axes of symmetry could be simple as it is illustrated in that Figure.

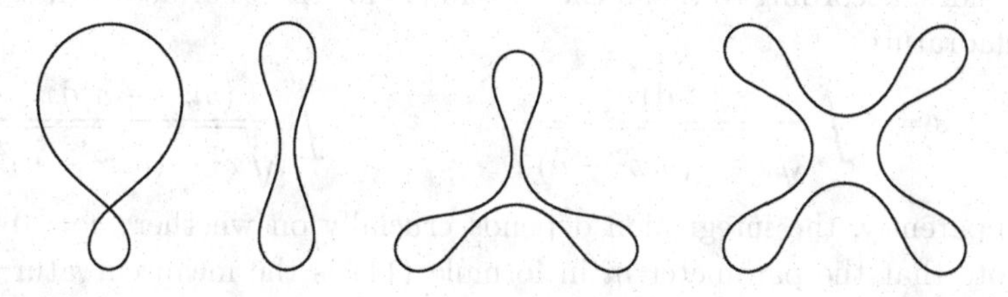

Fig. 1. Examples of closed curves of curvature (7).

Figure 2 represents typical examples of curves whose curvatures are of the forms (9) or (10). It is already pointed out that these curves are necessarily [2] self-intersecting ones, but in Figure 2 one can see that the protuberances, which are shaped due to the self-intersection, are directed inward for curvatures of the form (9) and outward for curvatures of the form (10).

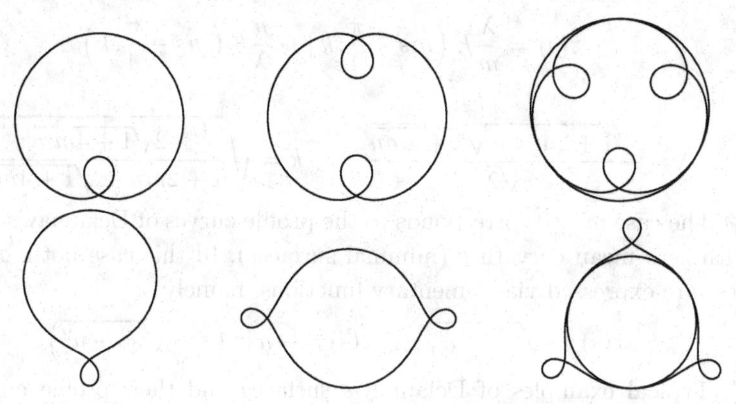

Fig. 2. Closed self-intersecting curves generated via (9) (top) and (10) (bottom).

3. Profile curves of Delaunay's surfaces

As another quite interesting example, let us consider the profile curves of classic Delaunay's surfaces [4]. Elsewhere [5] it is shown that their curvatures are specified by the general formula

$$\kappa(x) = m + \frac{n}{x^2}, \qquad x > 0, \qquad m, n \in \mathbb{R}. \tag{11}$$

Here, for typographical reasons we have replaced the canonical variables q, p in formulae (39) and (40) presented in Ref. 1 with x, respectively z. Besides, according to them, the system (1) for these surfaces reduces to the quadratures

$$s = \pm \int \frac{x\mathrm{d}x}{\sqrt{x^2 - (mx^2 - n)^2}}, \qquad z = \pm \int \frac{(mx^2 - n)\mathrm{d}x}{\sqrt{x^2 - (mx^2 - n)^2}}.$$

Apparently, the integration depends crucially on whether $m = 0$ or not. Note that the parameter m in formula (11) is the mean curvature of the Delaunay's surfaces as can be traced in more details in Ref. 5.

In the case of profile curves of Delaunay's surfaces with nonzero mean curvature ($m \neq 0$), and omitting the signs in the above integrals, $x(s)$ is expressed by the elementary function

$$x(s) = \frac{\sqrt{1 + 2mn + \sqrt{1 + 4mn}\sin(2ms)}}{\sqrt{2}m} \tag{12}$$

whereas $z(s)$ is expressed in terms of the first and the second kind elliptic integrals

$$z(s) = \frac{\lambda}{m} E\left(ms - \frac{\pi}{4}, k\right) - \frac{n}{\lambda} F\left(ms - \frac{\pi}{4}, k\right) \tag{13}$$

with

$$\lambda = \frac{\sqrt{1 + 2mn + \sqrt{1 + 4mn}}}{\sqrt{2}}, \qquad k = \sqrt{\frac{2\sqrt{1 + 4mn}}{1 + 2mn + \sqrt{1 + 4mn}}}.$$

The case $m = 0$ corresponds to the profile curves of Delaunay's surfaces with zero mean curvature (minimal surfaces). In this case both, $x(s)$ and $z(s)$ are expressed via elementary functions, namely

$$x(s) = \sqrt{s^2 + n^2}, \qquad z(s) = -n \ln\left(s + \sqrt{s^2 + n^2}\right). \tag{14}$$

Typical examples of Delaunay's surfaces and their profile curves are given in Figure 3. The most interesting profile curves — *nodary, undulary* and *catenary* are shown on the top and the corresponding surfaces obtained after their revolution about the symmetry axes are depicted at the bottom. All of them are produced by making use of the solutions of the system (1) given in (12)–(14). The nodoids and unduloids are obtained when the parameters m and n in (12)–(13) have the same, respectively different signs. The solutions to the system (1) of the form (14) correspond to catenoids. Expressions (14) imply also that since the catenoids are infinite in both directions surfaces of revolution, their shapes are independant of the sign of the parameter n.

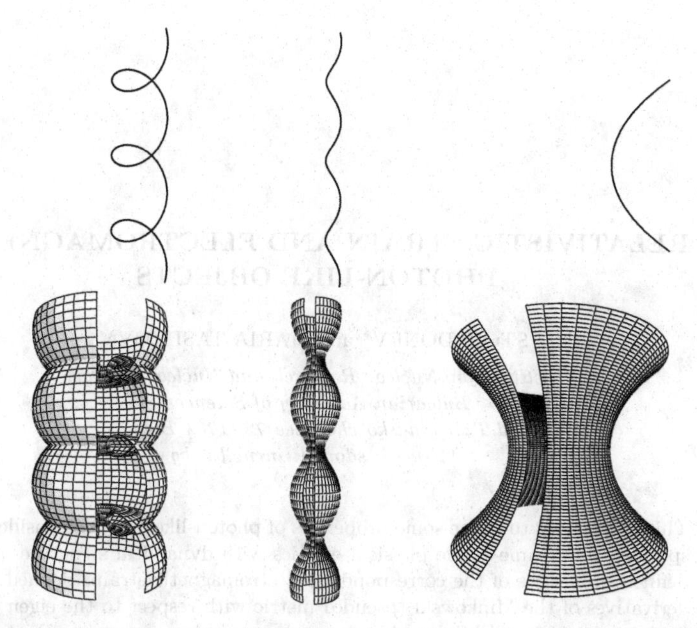

Fig. 3. Left to right: the nodary, undulary and catenary (top), and the nodoid, unduloid and catenoid (bottom).

Acknowledgments

This research is partially supported by the Bulgarian National Science Foundation under the grant B-1531/2005 and the Inter Academy contract # 23/2006 and # 35/2009 between the Bulgarian and the Polish Academies of Sciences. The authors are grateful also to Mrs. Mariana Hadzilazova for her help with the graphics files.

References

1. V. Vassilev, P. Djondjorov and I. Mladenov, Integrable Dynamical Systems of the Frenet-Serret Type, This volume pp 261–271.
2. V. Vassilev, P. Djondjorov and I. Mladenov, Cylindrical Equilibrium Shapes of Fluid Membranes, *J. Phys. A: Math. and Theor.* **41**, 435201 (2008).
3. M. Lévy, Mémoire sur un nouveau cas intégrable du problème de l'élastique et l'une de ses applications, *Journal de Mathematiques Pures et Appliquees*, Sér. 3 **X**, 5–42 (1884).
4. C. Delaunay, Sur la surface de révolution dont la courbure moyenne est constante, *Journal de Mathematiques Pures et Appliquees*, Sér. 1 **VI**, 309–315 (1841).
5. I. Mladenov, M. Hadzhilazova, P. Djondjorov and V. Vassilev, On the Intrinsic Equation Behind the Delaunay Surfaces, AIP Conference Proceedings vol. 1079, Melville-New York, 2008, pp. 274–280.

RELATIVISTIC STRAIN AND ELECTROMAGNETIC PHOTON-LIKE OBJECTS

STOIL DONEV* and MARIA TASHKOVA

Institute for Nuclear Research and Nuclear Energy,
Bulgarian Academy of Sciences,
blvd. Tzarigradsko chaussee 72, 1784 Bulgaria
**E-mail: sdonev@inrne.bas.bg*

This paper aims to relate some properties of photon-like objects, considered as spatially finite time-stable physical entities with dynamical structure, to well defined properties of the corresponding electromagnetic strains defined as Lie derivatives of the Minkowski (pseudo) metric with respect to the eigen vector fields of the Maxwell-Minkowski stress-energy-momentum tensor. First we recall the geometric sense of the concept of strain, then we introduce and discuss the notion for PhLO. We compute then the strains and establish important correspondences with the divergence terms of the energy tensor and the terms determining some internal energy-momentum exchange between the two components F and $*F$ of a vacuum electromagnetic field. The role of appropriately defined Frobenius curvature is also discussed and emphasized. Finally, equations of motion and interesting PhLO-solutions are given.

Keywords: Strain; Frobenius curvature; Photon-like object.

1. Strain

The concept of *strain* is introduced in studying elastic materials subject to external forces of different nature: mechanical, electromagnetic, etc. The classical strain describes mainly the abilities of the material to bear force-action from outside through deformation, i.e. through changing its shape, or, configuration.

The mathematical counterparts of the allowed (including reversible) deformations are the diffeomorphisms φ of a riemannian manifold (M, g), and every $\varphi(M)$ represents a possible configuration of the material considered. The essential diffeomorphisms φ must transform the metric g to some new metric $\varphi^* g$, such that $g \neq \varphi^* g$. The naturally arising tensor field $e = (\varphi^* g - g) \neq 0$ appears as a measure of the physical abilities of the material to withstand external force actions.

The continuity of the process of deformation leads to consider 1-parameter group $\varphi_t, t \in [a,b] \subset \mathbb{R}$ of deformations, i.e. of local diffeomorphisms. Let the vector field X generate φ_t, then the quantity

$$\frac{1}{2} L_X g := \frac{1}{2} \lim_{t \to 0} \frac{\varphi_t^* g - g}{t},$$

i.e. one half of the *Lie derivative* of g along X, is called (infinitesimal) *strain tensor*, or *deformation tensor*.

Remark. Further in the paper we shall work with $L_X\, g$, i.e. the factor $1/2$ will be omitted.

A consecutive relativistic approach to elasticity may be found in [3] (see also the corresponding quotations therein). In our further study we shall carry the above given Lie derivative definition of strain tensor to the case of Minkowski space-time, so we shall call $L_X\,\eta$, where $g = \eta$ is the Minkowski (pseudo)metric and X is any local nonisometry of η, just *strain tensor*.

Clearly, the term "material" does not seem to be appropriate for photon-like objects (PhLO) because, as we suggest in the next section, these objects admit NO static situations, they are of **entirely dynamical nature**, so the corresponding *relativistic strain* tensors must take care of this.

2. The Concept of photon-like object(s)

We introduce the following notion of Photon-like object(s) (we shall use the abbreviation "PhLO" for "Photon-like object(s)"):

> **PhLO are real massless time-stable physical objects with a consistent translational-rotational dynamical structure.**

We give now some explanatory comments, beginning with the term *real*. **First**, we emphasize that this term means that we consider PhLO as *really* existing *physical* objects, not as appropriate and helpful but imaginary (theoretical) entities. Accordingly, PhLO **necessarily carry energy-momentum**, otherwise, they could hardly be detected. **Second**, PhLO can undoubtedly be *created* and *destroyed*, so, no point-like and infinite models are reasonable: point-like objects are assumed to have no structure, so they cannot be destroyed since there is no available structure to be destroyed; creation of infinite physical objects (e.g. plane waves) requires infinite quantity of energy to be transformed from one kind to another for finite time-period, which seems also unreasonable. Accordingly, PhLO are *spatially finite* and have to be modeled like such ones, which is the only possibility to be consistent with their "created-destroyed" nature. It seems hardly reasonable to believe that PhLO can not be created and destroyed, and that spatially infinite and indestructible physical objects may exist at all. **Third**, "spatially finite" implies that PhLO may carry only *finite values* of physical (conservative or non-conservative) quantities. In particular, the most universal physical quantity seems to be the energy-momentum, so the model must allow finite integral values of energy-momentum to be carried by the corresponding solutions. **Fourth**, "spatially finite" means also that PhLO *propagate*, i.e. they do not "move" like classical particles along trajectories, therefore, partial differential equations should be used to describe their evolution in time.

The term "**massless**" characterizes physically the way of propagation in terms of appropriate dynamical quantities: the *integral* 4-momentum p of a PhLO should satisfy the relation $p_\mu p^\mu = 0$, meaning that its integral energy-momentum vector *must be isotropic*, i.e. to have zero module with respect to Lorentz-Minkowski (pseudo)metric in $\mathbb{R}^4$. If the object considered has spatial and time-stable structure, so that the translational velocity of every point where the corresponding field functions are different from zero must be equal to c, we have in fact null direction in the space-time *intrinsically determined* by a PhLO. Such a direction is formally defined by a null vector field $\bar{\zeta}, \bar{\zeta}^2 = 0$. The integral trajectories of this vector field are isotropic (or null) *straight lines* as is traditionally assumed in physics. It follows that with every PhLO a null straight line direction is *necessarily* associated, so, canonical coordinates $(x^1, x^2, x^3, x^4) = (x, y, z, \xi = ct)$ on $\mathbb{R}^4$ may be chosen such that in the corresponding coordinate frame ζ to have only two non-zero components of magnitude 1: $\bar{\zeta}^\mu = (0, 0, -\varepsilon, 1)$, where $\varepsilon = \pm 1$ accounts for the two directions along the coordinate z (further such a coordinate system will be called $\bar{\zeta}$-adapted and will be of main usage). Our PhLO propagates *as a whole* along the $\bar{\zeta}$-direction, so the corresponding energy-momentum tensor $T_{\mu\nu}$ of the model must satisfy the corresponding *local isotropy (null) condition*, namely, $T_{\mu\nu}T^{\mu\nu} = 0$ (summation over the repeated indices is throughout used).

The term "**translational-rotational**" means that besides translational component along $\bar{\zeta}$, the propagation necessarily demonstrates some rotational (in the general sense of this concept) component in such a way that *both components exist simultaneously and consistently.* It seems reasonable to expect that such kind of behavior should be consistent only with some distinguished spatial shapes. Moreover, if the Planck relation $E = h\nu$ must be respected throughout the evolution, the rotational component of propagation should have *time-periodical* nature with time period

$T = \nu^{-1} = h/E = \text{const}$, and one of the two possible, *left* or *right*, orientations. It seems reasonable also to expect periodicity in the spatial shape of PhLO, which somehow to be related to the time periodicity.

The term "**dynamical structure**" means that the propagation is supposed to be necessarily accompanied by an *internal energy-momentum redistribution*, which may be considered in the model as energy-momentum exchange between (or among) some appropriately defined subsystems. It could also mean that PhLO live in a dynamical harmony with the outside world, i.e. *any outside directed energy-momentum flow should be accompanied by a parallel inside directed energy-momentum flow.*

3. The Electromagnetic strain tensors

From formal point of view the relativistic formulation of classical electrodynamics in vacuum (zero charge density) is based on the following assumptions. The configuration space is the Minkowski space-time $M = (\mathbb{R}^4, \eta)$ where η is the pseudometric with $sign(\eta) = (-,-,-,+)$ with the corresponding volume 4-form $\omega_o = dx \wedge dy \wedge dz \wedge d\xi$ and Hodge star $*$ defined by $\alpha \wedge \beta = \eta(*\alpha, \beta)\omega_o$, α and β are p-form and $(4-p)$-form respectively. The electromagnetic filed is described by two closed 2-forms $(F, *F) : \mathbf{d}F = 0,\ \mathbf{d} * F = 0$. The physical characteristics of the field are deduced from the following stress-energy-momentum tensor field

$$T_\mu{}^\nu(F, *F) = -\frac{1}{2}\left[F_{\mu\sigma}F^{\nu\sigma} + (*F)_{\mu\sigma}(*F^{\nu\sigma})\right]. \tag{1}$$

In the non-vacuum case the allowed energy-momentum exchange with other physical systems is given in general by the divergence

$$\nabla_\nu T_\mu^\nu = \frac{1}{2}\Big[F^{\alpha\beta}(\mathbf{d}F)_{\alpha\beta\mu} + (*F)^{\alpha\beta}(\mathbf{d}*F)_{\alpha\beta\mu}\Big] = F_{\mu\nu}(\delta F)^\nu + (*F)_{\mu\nu}(\delta * F)^\nu, \tag{2}$$

where $\delta = *\mathbf{d}*$ is the co-derivative. Since $\mathbf{d}F = 0, \mathbf{d} * F = 0$, this divergence is obviously equal to zero on the vacuum solutions: its both terms are zero. Therefore, energy-momentum exchange between the two component-fields F and $*F$, which should be expressed by the terms $(*F)^{\alpha\beta}(\mathbf{d}F)_{\alpha\beta\mu}$, and $F^{\alpha\beta}(\mathbf{d} * F)_{\alpha\beta\mu}$, summation over $\alpha < \beta$, is NOT allowed on the solutions of $\mathbf{d}F = 0, \mathbf{d} * F = 0$. This shows that the widely used 4-potential approach (even if two 4-potentials U, U^* are introduced so that $\mathbf{d}U = F$, $\mathbf{d}U^* = *F$ locally) to these equations excludes any possibility to individualize two energy-momentum exchanging time-stable subsystems of the field that are mathematically represented by F and $*F$.

According to the above introduced and discussed concept of PhLO we have to impose the condition $T_{\mu\nu}T^{\mu\nu} = 0$ on the energy tensor (1). As is well known [3] this is equivalent to zero eigenvalues of T_μ^ν, which implies zero invariants $I_1 = \frac{1}{2}F_{\mu\nu}F^{\mu\nu} = \frac{1}{2}F_{\mu\nu}(*F)^{\mu\nu} = 0$, and that T_μ^ν admits just one null eigen direction defined by the vector field $\bar{\zeta}$, $\bar{\zeta}^2 = 0$, determining a null straight-line direction along which the energy density propagates translationally. So, the spatial-like electric and magnetic 1-form components of the field, further denoted correspondingly by A and A^*, in this case are mutually orthogonal and with equal magnitudes: $\eta(A, A^*) = 0$, $A^2 = (A^*)^2$, and orthogonal to the η-corresponding to $\bar{\zeta}$ 1-form ζ: $\eta(A,\zeta) = \eta(A^*,\zeta) = 0$. Under these conditions it is possible to find, and convenient to use, a canonical coordinate system $(x, y, z, \xi = ct)$ such that $\zeta = \varepsilon dz + d\xi$, $A = u\,dx + p\,dy$, $A^* = \varepsilon\,p\,dx - \varepsilon u\,dy$, $F = A \wedge \zeta$, $*F = A^* \wedge \zeta$, where (u, p) are two functions, and $\varepsilon = \pm 1$ carries information about the direction of propagation along the coordinate z of our PhLO.

Remark 3.1. All further identifications of tangent and cotangent objects will be made by η, and if A is 1-form we shall denote by $\bar{A}$ the corresponding vector field.

Since we have three generic vector fields $(\bar{\zeta}, \bar{A}, \bar{A}^*)$, we compute the corresponding three strain tensors.

Proposition 3.1. *The following relations hold:*

$$L_{\bar{\zeta}}\,\eta = 0, \qquad (L_{\bar{A}}\,\eta)_{\mu\nu} \equiv D_{\mu\nu} = \begin{Vmatrix} 2u_x & u_y + p_x & u_z & u_\xi \\ u_y + p_x & 2p_y & p_z & p_\xi \\ u_z & p_z & 0 & 0 \\ u_\xi & p_\xi & 0 & 0 \end{Vmatrix},$$

$$(L_{\bar{A}^*}\,\eta)_{\mu\nu} \equiv D^*_{\mu\nu} = \begin{Vmatrix} -2\varepsilon p_x & -\varepsilon(p_y + u_x) & -\varepsilon p_z & -\varepsilon p_\xi \\ -\varepsilon(p_y + u_x) & 2\varepsilon u_y & \varepsilon u_z & \varepsilon u_\xi \\ -\varepsilon p_z & \varepsilon u_z & 0 & 0 \\ -\varepsilon p_\xi & \varepsilon u_\xi & 0 & 0 \end{Vmatrix}.$$

Proof. Immediately verified. □

Let $[X, Y]$ denote the Lie bracket of vector fields on M, and $\phi^2 = u^2 + p^2$, $\psi = \mathrm{arctg}(p/u)$, i.e. $u = \phi\cos\psi$, $p = \phi\sin\psi$. We note that the local conservation law $\nabla_\nu T_\mu^\nu = 0$ reduces in our case to $L_{\bar{\zeta}}\phi^2 = 0$, so (u, p) would be running waves if $L_{\bar{\zeta}}\psi = 0$ too.

We give now some important from our viewpoint relations.

$$D(\bar\zeta,\bar\zeta)=D^*(\bar\zeta,\bar\zeta)=0,$$

$$D(\bar\zeta)\equiv D(\bar\zeta)_\mu dx^\mu\equiv D_{\mu\nu}\bar\zeta^\nu dx^\mu=(u_\xi-\varepsilon u_z)dx+(p_\xi-\varepsilon p_z)dy,$$

$$D(\bar\zeta)^\mu\frac{\partial}{\partial x^\mu}\equiv D^\mu_\nu\bar\zeta^\nu\frac{\partial}{\partial x^\mu}=-(u_\xi-\varepsilon u_z)\frac{\partial}{\partial x}-(p_\xi-\varepsilon p_z)\frac{\partial}{\partial y}=-[\bar A,\bar\zeta],$$

$$D_{\mu\nu}\bar A^\mu\bar\zeta^\nu=-\frac{1}{2}\Big[(u^2+p^2)_\xi-\varepsilon(u^2+p^2)_z\Big]=-\frac{1}{2}L_{\bar\zeta}\phi^2, \tag{3}$$

$$D_{\mu\nu}\bar A^{*\mu}\bar\zeta^\nu=-\varepsilon\Big[u(p_\xi-\varepsilon p_z)-p(u_\xi-\varepsilon u_z)\Big]=-\varepsilon\mathbf{R}=-\varepsilon\phi^2\,L_{\bar\zeta}\psi, \tag{4}$$

where $\mathbf{R}\equiv u(p_\xi-\varepsilon p_z)-p(u_\xi-\varepsilon u_z)$. We also have:

$$D^*(\bar\zeta)=\varepsilon\Big[-(p_\xi-\varepsilon p_z)dx+(u_\xi-\varepsilon u_z)dy\Big],$$

$$D^*(\bar\zeta)^\mu\frac{\partial}{\partial x^\mu}\equiv(D^*)^\mu_\nu\bar\zeta^\nu\frac{\partial}{\partial x^\mu}=-\varepsilon(p_\xi-\varepsilon p_z)\frac{\partial}{\partial x}+(u_\xi-\varepsilon u_z)\frac{\partial}{\partial y}=[\bar A^*,\bar\zeta],$$

$$D^*_{\mu\nu}\bar A^{*\mu}\bar\zeta^\nu=-\frac{1}{2}\Big[(u^2+p^2)_\xi-\varepsilon(u^2+p^2)_z\Big]=-\frac{1}{2}L_{\bar\zeta}\phi^2, \tag{5}$$

$$D^*_{\mu\nu}\bar A^\mu\bar\zeta^\nu=\varepsilon\Big[u(p_\xi-\varepsilon p_z)-p(u_\xi-\varepsilon u_z)\Big]=\varepsilon\mathbf{R}=\varepsilon\phi^2\,L_{\bar\zeta}\psi. \tag{6}$$

Clearly, $D(\bar\zeta)$ and $D^*(\bar\zeta)$ are linearly independent in general:

$$D(\bar\zeta)\wedge D^*(\bar\zeta)=\varepsilon\Big[(u_\xi-\varepsilon u_z)^2+(p_\xi-\varepsilon p_z)^2\Big]dx\wedge dy$$
$$=\varepsilon\phi^2(\psi_\xi-\varepsilon\psi_z)^2\,dx\wedge dy\neq 0.$$

Recall now that every 2-form F defines a linear map $\tilde F$ from 1-forms to 3-forms through the exterior product: $\tilde F(\alpha):=\alpha\wedge F$, where $\alpha\in\Lambda^1(M)$. Moreover, the Hodge $*$-operator, composed now with $\tilde F$, gets $\tilde F(\alpha)$ back to $*\tilde F(\alpha)\in\Lambda^1(M)$. We readily obtain now

$$\begin{aligned}
D(\bar\zeta)\wedge F&=D^*(\bar\zeta)\wedge(*F)=D(\bar\zeta)\wedge A\wedge\zeta=D^*(\bar\zeta)\wedge A^*\wedge\zeta\\
&=-\varepsilon\Big[u(p_\xi-\varepsilon p_z)-p(u_\xi-\varepsilon u_z)\Big]dx\wedge dy\wedge dz\\
&\quad-\Big[u(p_\xi-\varepsilon p_z)-p(u_\xi-\varepsilon u_z)\Big]dx\wedge dy\wedge d\xi\\
&=-\phi^2\,L_{\bar\zeta}\psi\,(\varepsilon\,dx\wedge dy\wedge dz+dx\wedge dy\wedge d\xi)\\
&=-\mathbf{R}\,(\varepsilon\,dx\wedge dy\wedge dz+dx\wedge dy\wedge d\xi),\\
D(\bar\zeta)\wedge(*F)&=-D^*(\bar\zeta)\wedge F=D(\bar\zeta)\wedge A^*\wedge\zeta=-D^*(\bar\zeta)\wedge A\wedge\zeta\\
&=\frac{1}{2}\Big[(u^2+p^2)_\xi-\varepsilon(u^2+p^2)_z\Big](dx\wedge dy\wedge dz+\varepsilon\,dx\wedge dy\wedge d\xi).
\end{aligned}$$

Thus we get

$$*\left[D(\bar{\zeta}) \wedge A \wedge \zeta\right] = *\left[D^*(\bar{\zeta}) \wedge A^* \wedge \zeta\right]$$
$$= -\varepsilon \mathbf{R}\, \zeta = -i(\overline{*F})\mathbf{d}F = i(\bar{F})\mathbf{d}(*F), \qquad (7)$$
$$*\left[D(\bar{\zeta}) \wedge A^* \wedge \zeta\right] = -*\left[D^*(\bar{\zeta}) \wedge A \wedge \zeta\right]$$
$$= \frac{1}{2} L_{\bar{\zeta}}\phi^2\, \zeta = i(\bar{F})\mathbf{d}F = i(\overline{*F})\mathbf{d}(*F), \qquad (8)$$

where, if (F, G) are respectively a 2-form and a 3-form, we have denoted

$$i(\bar{F})G = F^{\mu\nu} G_{\mu\nu\sigma} dx^\sigma, \ \mu < \nu.$$

4. Discussion

The above relations show various dynamical aspects of the energy-momentum redistribution during evolution of our PhLO in terms of electromagnetic strain. Let's look closer to some of them.

Relation (8) shows how the local conservation law $\nabla_\nu T^\nu_\mu = 0$, being equivalent to $L_{\bar{\zeta}}\phi^2 = 0$, and meaning that the energy density ϕ^2 propagates just translationally along $\bar{\zeta}$, can be expressed in terms of the strain components. On the other hand the phase function ψ would satisfy $L_{\bar{\zeta}}\psi = 0$ ONLY if the introduced quantity $\mathbf{R}$ is equal to zero: $\mathbf{R} = 0$. Now, what is the sense of $\mathbf{R}$ reveals the following

Proposition 4.1. *The following relations hold:*

$$\mathbf{d}A \wedge A \wedge A^* = 0; \qquad \mathbf{d}A^* \wedge A^* \wedge A = 0;$$
$$\mathbf{d}A \wedge A \wedge \zeta = \varepsilon\left[u(p_\xi - \varepsilon p_z) - p(u_\xi - \varepsilon u_z)\right]\omega_o;$$
$$\mathbf{d}A^* \wedge A^* \wedge \zeta = \varepsilon\left[u(p_\xi - \varepsilon p_z) - p(u_\xi - \varepsilon u_z)\right]\omega_o.$$

Proof. Immediately verified. □

These relations say that the 2-dimensional Pfaff system (A, A^*) is completely integrable for any choice of the two functions (u, p), while the two 2-dimensional Pfaff systems (A, ζ) and (A^*, ζ) are NOT completely integrable in general, and the same curvature factor

$$\mathbf{R} = u(p_\xi - \varepsilon p_z) - p(u_\xi - \varepsilon u_z)$$

determines their nonintegrability. Hence, the condition $\mathbf{R} \neq 0$ allows in principle rotational component of propagation, i.e. u and p NOT to be plane waves, so consistency between the translational and rotational components of propagation is possible in principle.

As we mentioned above it is possible the translational and rotational components of the energy-momentum redistribution to be represented in form depending on the $\bar{\zeta}$-directed strains $D(\bar{\zeta})$ and $D^*(\bar{\zeta})$. So, the local translational changes of the energy-momentum carried by the two components F and $*F$ of our PhLO are given by the two 1-forms $*\big[D(\bar{\zeta})\wedge A^*\wedge\zeta\big]$ and $*\big[D^*(\bar{\zeta})\wedge A\wedge\zeta\big]$), and the local rotational ones are given by the 1-forms $*\big[D(\bar{\zeta})\wedge A\wedge\zeta\big]$ and $*\big[D^*(\bar{\zeta})\wedge A^*\wedge\zeta\big]$. In fact, the 1-form $*\big[D(\bar{\zeta})\wedge A\wedge\zeta\big]$ determines the strain that tends to "leave" the 2-plane defined by (A,ζ) and the 1-form $*\big[D^*(\bar{\zeta})\wedge A^*\wedge\zeta\big]$ determines the strain that tends to "leave" the 2-plane defined by (A^*,ζ). So, the particular kind of nonintegrability of (A,ζ) and (A^*,ζ) mathematically guarantees some physical intercommunication through establishing local dynamical equilibrium (eq.(7))) between two subsystems of the field, that are mathematically represented by F and $*F$.

Since the PhLO is free, i.e. no energy-momentum is lost or gained, this means that the two (null-field) components F and $*F$ exchange locally *equal* energy-momentum quantities:

$$*\Big[D(\bar{\zeta})\wedge A\wedge\zeta\Big]=*\Big[D^*(\bar{\zeta})\wedge A^*\wedge\zeta\Big].$$

Now, the local energy-momentum conservation law

$$\nabla_\nu\big[F_{\mu\sigma}F^{\nu\sigma}+(*F)_{\mu\sigma}(*F)^{\nu\sigma}\big]=0,$$

i.e. $L_{\bar{\zeta}}\phi^2=0$, requires the corresponding strain-fluxes $*\big[D^*(\bar{\zeta})\wedge A\wedge\zeta\big]$ and $*\big[D(\bar{\zeta})\wedge A^*\wedge\zeta\big]$ to become zero. It seems important to note that, only dynamical relation between the energy-momentum change and strain fluxes exists, so NO analog of the assumed in elasticity theory generalized Hooke law, i.e. linear relation between the stress tensor and the strain tensor [1], seems to exist here. This clearly goes along with the fully dynamical nature of PhLO, i.e. linear relations exist between the divergence terms of our stress tensor $\frac{1}{2}\big[-F_{\mu\sigma}F^{\nu\sigma}-(*F)_{\mu\sigma}(*F)^{\nu\sigma}\big]$ and the $\bar{\zeta}$-directed strain fluxes as given by equations (2), (7), (8).

The constancy of the translational propagation and the required translational-rotational compatibility suggest also constancy of of the rotational component of propagation, i.e. $L_{\bar{\zeta}}\psi=\text{const}$, so we consider the equations

$$L_{\bar{\zeta}}\phi^2=0,\qquad L_{\bar{\zeta}}\psi=\frac{\kappa}{l_o},$$

where $\kappa = \pm 1$ and l_o is a parameter of dimension of length. These equations admit the following solutions:

$$u(x, y, z, \xi) = \phi(x, y, \xi + \varepsilon z) \cos(-\varepsilon\kappa\frac{z}{l_o} + \text{const }),$$

$$p(x, y, z, \xi) = \phi(x, y, \xi + \varepsilon z) \sin(-\varepsilon\kappa\frac{z}{l_o} + \text{const }),$$

where $\phi(x, y, \xi + \varepsilon z)$ is an arbitrary smooth finite function, so it can be chosen to have compact support inside an appropriate 3-region such, that at every moment our PhLO to be localized inside an one-step long helical cylinder wrapped up around the z-axis (the figures bellow).

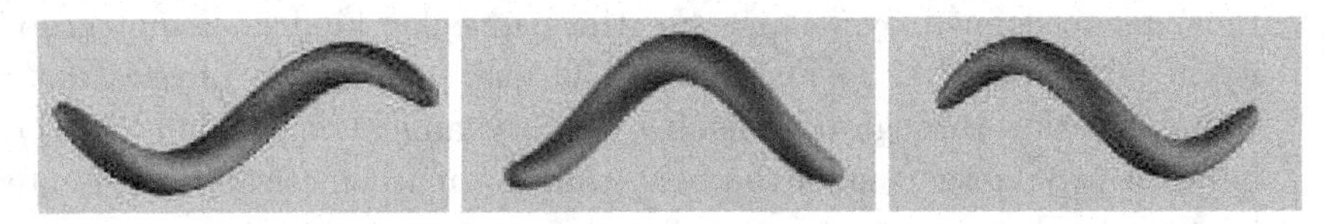

Fig. 1. Theoretical example with $\kappa = -1$. The translational propagation is directed left-to-right.

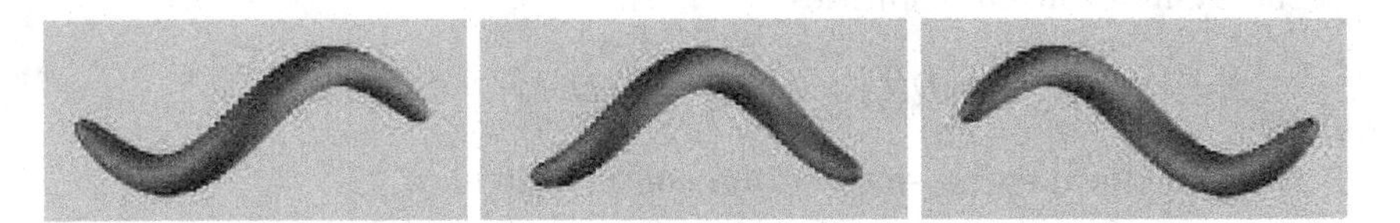

Fig. 2. Theoretical example with $\kappa = 1$. The translational propagation is directed left-to-right.

5. Conclusion

We introduced and discussed a concept of photon-like object(s) (PhLO) as real, massless time stable physical object(s) with an intrinsically compatible translational-rotational dynamical structure. So, PhLO are spatially finite, with every PhLO a straight-line null direction is necessarily associated, and the corresponding stress-energy-momentum tensor must be isotropic: $T_{\mu\nu}T^{\mu\nu} = 0$. We showed that in the frame of the relativistic formulation of vacuum electrodynamics two essential strain tensors D and D^* can be introduced as corresponding Lie derivatives of the flat Minkowski metric. In terms of the components of these strain tensors can be defined important characteristics of the internal dynamics of PhLO, in particular, internal energy-momentum exchange between the F-component and $*F$-component of the free field was explicitly obtained. Definite relations of the projections of D and D^* along the null direction of translational propagation to the Lie brackets of the electric and magnetic components were also given. We defined amplitude ϕ and phase ψ of PhLO and showed that the plane wave character of ϕ is admissible and corresponds to the local energy conservation. The non plane wave character of ψ guarantees the availability of rotational component of propagation, which property turned out to be well defined in terms of corresponding Frobenius curvature. The physical understanding of the Frobenius nonintegrability of some subcodistributions in this context should read: there is internal energy-momentum exchange between two individualized subsystems of PhLO mathematically represented in our case by F and $*F$.

Assuming constant nature of the rotation we come to an extension of the Riemann-Clifford-Einstein idea for linear relation between energy-density and Riemannian curvature to linear relation between energy-density and Frobenius curvature as a measure of nonintegrability. This helps also to define linear equations of motion, which equations admit spatially finite solutions of helical structure along the corresponding spatial direction of propagation.

Finally we note that integrating any of the two 4-forms $l_o/c\,\mathbf{d}A\wedge A\wedge\zeta$ and $l_o/c\ \mathbf{d}A^*\wedge A^*\wedge\zeta$ over the 4-volume $\mathbb{R}^3\times 4l_o$ gives $\pm E.T$, where $T=\nu^{-1}=4l_o/c$ and E is the integral energy of the solution. This is naturally interpreted as *action for one period*, and represents an analog of the famous Planck formula $E.T=h$.

As for the question *what is the "material" that PhLO are made of* we leave the corresponding reasoning for the future.

Acknowledgment

This study was partially supported by Contract Φ−1515 with the Bulgarian Science foundation.

References

1. M. Wernig-Pichler, *Relativistic Elastodynamics*, Thesis, arXiv:gr-qc/0605025 (2006).
2. J. Synge, *Relativity: The Special Theory*, North Holland, Amsterdam, 1958.
3. J. Marsden, T. Hughes, *Mathematical foundations of Elasticity*, Prentice Hall 1983; Reprinted by Dover Publications, 1994.

A CONSTRUCTION OF MINIMAL SURFACES IN FLAT TORI BY SWELLING

NORIO EJIRI

Department of Mathematics, Meijo University,
Tempaku, Nagoya 466-8555, Japan
E-mail: ejiri@ccmfs.meijo-u.ac.jp

1. Introduction

Micallef [6] proved that a full, compact, orientable, stable, hyperelliptic minimal surface in a Riemannian flat torus is holomorphic with respect to a suitable orthogonal complex structure on the torus. Hence it is homologically area-minimizing by the Wirtinger inequality. On the other hand, there are many full, compact, orientable, stable, non-hyperelliptic minimal surfaces that are not holomorphic with respect to all orthogonal complex structure in Riemannian flat tori [2,4]. However, since there are at least two, full, compact, orientable, homotopically area-minimizing, non-hyperelliptic minimal surfaces of genus 4 that are not holomprphic with respect to all orthogonal complex structures in some Riemannian flat torus of dimension 8 [4, Corollary 12.5], we consider whether the stableness of minimal surfaces implies the area-minimizingness. Let M_g denote the moduli of Riemann surfaces of genus $g \geq 4$. Let n be a positive integer and $M_g(n)$ the subset of of $M_g(g \geq 4)$ such that the Riemann surface corresponding to each point of $M_g(n)$ admits a stable minimal immersion into a torus of dimension 7 which contains at least n homotopic, compact, orientable, stable, minimal surfaces of genus g with different areas.

Theorem 1.1. *$M_g(n)$ is dense in M_g.*

Our idea is as follows: Put a compact, orientable, immersed minimal surface of genus g with only trivial Jacobi fields, which are caused by parallel translations of the torus, in a torus of dimension $< 2g$ and swell the torus to the higher dimensional (indefinite) torus. Then we find that many homotopic, minimal surfaces of genus g with different areas appear in the swelled thin torus.

Arezzo and Micallef [2] have given stable nonholomorphic minimal immersion of Σ_γ into $T^{2\gamma-2k}$, for $k = 1, 2$ or 3 and sufficently large γ (if $\gamma \geq 7, 9, 10$, then $k = 1, 2, 3$, respectively). They expect that such examples also exit for any $k \leq \gamma - 4$. From our Theorem 1.1, we can infer the result that their expectation is true.

2. Preliminary

We reviw some results [4]. Let M be a Riemann surface of genus g, $\{A_i, B_i\}$ a canonical homology basis and $\{\psi_i\}$ the associated 1 forms such that $\int_{A_i} \psi_j = \delta_{ij}$ holds. Then the matrix $\tau = (\tau_{ij})$ of size g defined by $\tau_{ij} = \int_{B_j} \psi_i$ is called the Riemann matrix, which provides an element of the Siegel upper half space H_g (the space of complex symmetric matrices τ of size g such that $\operatorname{Im}\tau$ is positive (see, for example, [5])). Let RM be the space of Riemann matrices in H_g. Let $RM_{\textit{non-hyper}}$ and $RM_{\textit{hyper}}$ be the space of Riemann matrices of non-hyperelliptic Riemann surfaces and hyperelliptic Riemann surfaces, respectively. Then $RM_{\textit{non-hyper}}$ and $RM_{\textit{hyper}}$ are the $(3g-3)$ and $(2g-1)$- dimensional complex submanifold, respectively, $RM = RM_{\textit{non-hyper}} \cup RM_{\textit{hyper}}$ and $RM_{\textit{hyper}}$ is the singularity of RM (see, for example, [1]).

Let $\mathbf{R}^N$ denote the real vector space of dimension N given by vectors of ${}^t(x_1, \ldots, x_N)$, where $x_1, \ldots, x_N$ are real numbers. Then the indefinite innnerproduct of type (p, q) is defined by $(x_1, \ldots, x_N) I_{p,q} {}^t(y_1, \ldots, y_N)$, where $p + q = N$,

$$I_{p,q} = \begin{pmatrix} -I_p & 0 \\ 0 & I_q \end{pmatrix}$$

and I_p is an identity matrix of size p. Let Γ be a lattice group of $\mathbf{R}^N$ and T^N the indefinite flat torus $\mathbf{R}^N/\Gamma$ of dimension N with the innerproduct of type (p, q).

Let Ψ denote the $\mathbf{R}^{2g}$-valued 1-form given by

$${}^t(\operatorname{Re}\psi_1, \ldots, \operatorname{Re}\psi_g, \operatorname{Im}\psi_1, \ldots, \operatorname{Im}\psi_g).$$

Let F be a harmonic map of M into T^N. Then, since the differential dF of F is the $\mathbf{R}^N$-valued harmonic 1-form, there exists the $(N, 2g)$-matrix Q such that $dF = Q\Psi$. We define the periods of F by

$$\int_{A_j} dF = \ell_j, \quad \int_{B_j} dF = \ell_{g+j}, \quad j = 1, \ldots, g$$

and set $L = (\ell_1, \ldots, \ell_{2g})$. Let T_τ denote the matrix

$$\begin{pmatrix} I & \mathrm{Re}\,\tau \\ 0 & \mathrm{Im}\,\tau \end{pmatrix}.$$

Then $QT_\tau = L$ holds and hence, we obtain $dF = LT_\tau^{-1}\Psi$. We call L the period matrix of F with respect to $\{A_i, B_i\}$.

In this paper, we consider the case that the periods of F generate Γ.

If $L = T_\tau$, then F is called the Albanese map of M into the Jacobian variety whose lattice group is generated by column vectors of I_g and τ.

Theorem 2.1 ([4]). *The energy of F is given by $\frac{1}{2}\mathrm{tr}(P(\tau)\widetilde{L})$, where*

$$L_{11} = {}^tL_1 I_{p,q} L_1,\ L_{12} = {}^tL_1 I_{p,q} L_2,\ L_{21} = {}^tL_2 I_{p,q} L_1,\ L_{22} = {}^tL_2 I_{p,q} L_2,$$

$$\widetilde{L} = \begin{pmatrix} L_{11} & L_{12} \\ L_{21} & L_{22} \end{pmatrix}$$

and

$$P(\tau) = \begin{pmatrix} (\mathrm{Im}\,\tau) + (\mathrm{Re}\,\tau)(\mathrm{Im}\,\tau)^{-1}(\mathrm{Re}\,\tau) & -(\mathrm{Re}\,\tau)(\mathrm{Im}\,\tau)^{-1} \\ -(\mathrm{Im}\,\tau)^{-1}(\mathrm{Re}\,\tau) & (\mathrm{Im}\,\tau)^{-1} \end{pmatrix}.$$

We consider a harmonic map of a Riemann surface of genus g into T^N whose period matrix is only $L = (L_1, L_2)$ for a suitable canonical homology basis and select such harmonic map for all Riemann surfaces. Note that these harmonic maps are homotopic. Then we obtain the function on RM, which can be extended to the function $E_{\widetilde{L}}(\tau) = \frac{1}{2}tr(P(\tau)\widetilde{L})$ on H_g. We call it the enegy function. We relate the critical points of the energy function to minimal surfaces.

Theorem 2.2 ([4]). *Let τ be an element of $RM_{non\text{-}hyper}$. Then the restriction of $E_{\widetilde{L}}$ to $RM_{non\text{-}hyper}$ is critical at τ if and only if there exist the unique Riemann surface M and the two canonical homology basis $\{A_i, B_i\}$, $\{-A_i, -B_i\}$ up to automorphisms of M such that $dF = \pm LT_\tau^{-1}\Psi$ is weakly conformal. Similarly, let τ be an element of RM_{hyper}. Then the restriction of $E_{\widetilde{L}}$ to RM_{hyper} is critical at τ if and only if there exist the unique Riemann surface M and the unique canonical homology basis $\{A_i, B_i\}$ up to automorphisms of M such that $dS = LT_\tau^{-1}\Psi$ is weakly conformal.*

If the ambient torus is a Riemannian flat torus, then we can define index_A and $\mathrm{nullity}_A$ by the index and the nullity of Jacobi operators for minimal surfaces, respectively.

Let $\mathrm{index}_{\widetilde{L}}$ and $\mathrm{nullity}_{\widetilde{L}}$ be the index and the nullity of the hessian of $E_{\widetilde{L}}$. Then we related $\mathrm{index}_{\widetilde{L}}$ and $\mathrm{nullity}_{\widetilde{L}}$ to index_A and $\mathrm{nullity}_A$.

In this paper, we need the following [4].

Fact 1. *For a critical point* $\tau \in RM_{non\text{-}hyper}$, $\text{index}_{E_{\widetilde{L}}} = \text{index}_A$ *holds. If the map is immersed, then* $\text{nullity}_{\widetilde{L}} = \text{nullity}_A - N$.

Fact 2. *For a critical point* $\tau \in RM_{hyper}$ *of the restriction of* $E_{\widetilde{L}}$, *there is a non-negative integer* $\alpha \leq g-2$ *such that* $\text{index}_{\widetilde{L}} + \alpha = \text{index}_A$ holds. *If the map is immersed, then* $\text{nullity}_{\widetilde{L}} + 2(g-2) - 2\alpha = \text{nullity}_A - N$.

3. Minimal surfaces in Riemannian flat tori

Let Γ_0 be a lattice group of $\mathbf{R}^N$ and T^N the Riemannian flat torus $\mathbf{R}^N/\Gamma_0$. Let M be a compact, orientable, minimal surface of genus g in T^N with $N < 2g$ and F the branched minimal immersion of M into T^N. Since the periods of F generate Γ_0, $F_* : H_1(M, \mathbf{Z}) \to H_1(T^N, \mathbf{Z})$ is surjective.

Lemma 3.1. *There exists a canonical homology basis* $\{A_i, B_i\}$ *satisfying* $F_*(A_1) = 0$.

Proof. Let $\operatorname{Ker} F_*$ be the kernel of F_*. Then $\operatorname{Ker} F_*$ is of rank $2g - N$ of $H_1(M, \mathbf{Z})$ and there exists a basis $e_1, \ldots, e_{2g}$ of $H_1(M, \mathbf{Z})$ and integers $k_1, \ldots, k_{2g-N}$ such that $k_1 e_1, \ldots, k_{2g-N} e_{2g-N}$ is a basis of $\operatorname{Ker} F_*$. Thus $k_1 = \cdots = k_{2g-N} = 1$ and $e_1, \ldots, e_{2g-N}$ is a basis of $\operatorname{Ker} F_*$. Let $a \cdot b$ denote the intersection number of a and b for $a, b \in H_1(M, \mathbf{Z})$. Putting $A_1 = e_1$, we can construct B_1 such that $A_1 \cdot B_1 = 1$ as follows: Integers $e_1 \cdot e_j$ are relatively prime, because the determinant of the matrix $(e_i \cdot e_j)$ is ± 1. Thus we have integers α_k such that $B_1 = \sum \alpha_k e_k$ and $A_1 \cdot B_1 = 1$. Since $\{X \in H_1(M, \mathbf{Z}) \mid A_1 \cdot (X) = B_1 \cdot (X) = 0\}$ is of dimension $2g - 2$, we have a basis $e_1, \ldots, e_{2g-2}$ of $\{X \in H_1(M, \mathbf{Z}) \mid A_1 \cdot (X) = B_1 \cdot (X) = 0\}$. We set $A_2 = e_1$ and can construct B_2 by the same method as above. Continuing this process, we can obtain a canonical homology basis seeked in Lemma 3.1. □

Note that the canonical homology basis $\{A_i, B_i\}$ given in Lemma 3.1 is not unique.

Lemma 3.2. *Let* m *be an integer and* $\{\widetilde{A_i}, \widetilde{B_i}\}$ *the canonical homology bassis defined by* $\widetilde{A_i} = A_i$, $i = 1, \ldots, g$, $\widetilde{B_1} = B_1 + mA_1$, $\widetilde{B_i} = B_i$, $i = 2, \ldots, g$. *Then the associated holomorphic* 1 *forms* $\widetilde{\psi_i}$ *with respect to* $\{\widetilde{A_i}, \widetilde{B_i}\}$ *are* ψ_i *and* $F_*(\widetilde{A_1}) = 0$ *holds. Furthermore we denote by* τ_0 *and* $\tau_0(m)$ *the associated Riemann matrices with respect to* $\{A_i, B_i\}$ *and* $\{\widetilde{A_i}, \widetilde{B_i}\}$, *respectively. Then*

$$\tau_0(m) = \tau_0 + \begin{pmatrix} m & 0 & \cdots & 0 \\ 0 & 0 & \cdots & 0 \\ \vdots & \vdots & \ddots & \vdots \\ 0 & 0 & \cdots & 0 \end{pmatrix}.$$

Proof. It is easy to see that $\{\widetilde{A_i}, \widetilde{B_i}\}$ is a canonical homology basis, $\widetilde{\psi_i} = \psi_i$ and $F_*\widetilde{A_1} = 0$. Then we see

$$\int_{\widetilde{B_i}} \widetilde{\psi_j} = \int_{B_i + mA_i\delta_{1i}} \psi_j = \tau_{0ij} + m\delta_{ij}\delta_{1i}. \qquad \square$$

Let L_0 be the period matrix of F with respect to $\{A_i, B_i\}$ and $\widetilde{L_0} = {}^tL_0L_0$. Then τ_0 is a critical point of $E_{\widetilde{L_0}}$ restricted to $RM_{non\text{-}hyper}$ or RM_{hyper}.

Lemma 3.3. *The period matrix of F with respect to $\{\widetilde{A_i}, \widetilde{B_i}\}$ is also given by L_0. So $\tau_0(m)$ is also a critical point of $E_{\widetilde{L_0}}$ restricted to $RM_{non\text{-}hyper}$ or RM_{hyper}.*

Proof. By $F_*(A_1) = 0$, $F_*(\widetilde{A_1}) = 0$, $F_*(\widetilde{A_i}) = F_*(A_i)$, $F_*(\widetilde{B_1}) = F_*(B_1 + mA_1) = F_*(B_1)$ and $F_*(\widetilde{B_i}) = F_*(B_i)$ for $i = 2, \ldots, g$. $\square$

Note $L_0 = (0\,\ell_2 \cdots \ell_{2g})$. We define the $(N+1, 2g)$-matrix L_s $(s > 0)$ by

$$\begin{pmatrix} \sqrt{s} & 0 & \cdots & 0 \\ 0 & \ell_2 & \cdots & \ell_{2g} \end{pmatrix}.$$

It is easy to see the following.

Lemma 3.4. *The column vectors of L_s generate a lattice group of $\mathbf{R}^{N+1}$.*

Remark 3.1. The obtained torus is $S^1(\frac{\sqrt{s}}{2\pi}) \times T^N$. We may consider $\{$*one point*$\} \times T^N$ as T^N and say that T^N is swelled to $S^1(\frac{\sqrt{s}}{2\pi}) \times T^N$.

Lemma 3.5. *Let $\widetilde{L_s}$ be ${}^tL_sI_{p,q}L_s$, where $p = 0$ or 1. Then*

$$\widetilde{L_s} = \widetilde{L_0} + \begin{pmatrix} \pm s & 0 \\ 0 & 0 \end{pmatrix}.$$

So we obtain the following.

Lemma 3.6. $E_{\widetilde{L_s}}(\tau) = \frac{1}{2}\mathrm{tr}(P(\tau)\widetilde{L_0}) + \frac{1}{2}\mathrm{tr}(P(\tau)\begin{pmatrix} \pm s & 0 \\ 0 & 0 \end{pmatrix})$.

Lemma 3.7. *$E_{\widetilde{L_s}}(\tau_0(m)) = \frac{\pm s}{2}Q(m) + E_{\widetilde{L_0}}(\tau_0)$, where*

$$Q(m) = (\mathrm{Im}\,\tau_0)^{-1}_{11}m^2 + 2m\sum_i(\mathrm{Im}\,\tau_0)^{-1}_{1i}\mathrm{Re}\,(\tau_0)_{i1} + \mathrm{Im}(\tau_0)_{11}$$
$$+\sum_{ijk}\mathrm{Re}(\tau_0)_{ij}(\mathrm{Im}\,\tau_0^{-1})_{jk}\mathrm{Re}(\tau_0)_{ki}.$$

$Q(m)$ is a polynomial of degree 2 with respect to m, because $\mathrm{Im}\,\tau_0$ is positive.

From now, If M is non-hyperelliptic, then, we assume that M is immersed and has only trivial Jacobi fields ($\mathrm{nullity}_A = N$) and hence $\mathrm{nullity}_{\widetilde{L_0}} = 0$ by Fact 1. If M is hyperelliptic, then, we assume that $2(g-2) - 2\alpha = \mathrm{nullity}_A - N$ holds. By Fact 2, again $\mathrm{nullity}_{\widetilde{L_0}} = 0$ holds. So the hessian of $E_{\widetilde{L_0}}$ is non-degenerate at $\tau_0(m)$.

For non-degenerate critical points of a function with parameteres, we obtain the following (see, for example, [7]).

Lemma 3.8. *There is a positive integer ε_m and a smooth curve $\tau_s(m)$, $0 \le s \le \varepsilon_m$ starting at $\tau_0(m)$ such that $\tau_s(m)$ is a non-degenerate critical point of $E_{\widetilde{L_s}}$.*

By the Taylor expansion,

$$E_{\widetilde{L_s}}(\tau_s(m)) = E_{\widetilde{L_0}}(\tau_0(m)) + \left[\frac{d}{ds}\bigg|_{s=0} E_{\widetilde{L_s}}(\tau_0(m))\right]s$$
$$+\left[\frac{d}{ds}\bigg|_{s=0} E_{\widetilde{L_0}}(\tau_s(m))\right]s + O(s^2).$$

Since $\tau_0(m)$ is a critical point of $E_{\widetilde{L_0}}$,

$$\frac{d}{ds}\bigg|_{s=0} E_{\widetilde{L_0}}(\tau_s(m)) = 0.$$

By Lemma 3.7,

$$\frac{d}{ds}\bigg|_{s=0} E_{\widetilde{L_s}}(\tau_0(m)) = \frac{\pm 1}{2}Q(m).$$

Thus we obtain

Lemma 3.9. $E_{\widetilde{L_s}}(\tau_s(m)) = E_{\widetilde{L_0}}(\tau_0(m)) + \frac{\pm s}{2}Q(m) + O(s^2)$.

Hence for any positive integer n, there are $m_1, \ldots, m_n$ such that $Q(m_i)$ are different for each other. So $E_{\widetilde{L_s}}(\tau_s(m_i))$ are all different for small $s(\le \varepsilon_{m_i})$ by Lemma 3.9. The minimal surfaces corresponding to the critical points $\tau_s(m_i)$, $i = 1, \ldots, n$ have different area. Namely, we obtain n different homotopic minimal surfaces in a torus of dimension $N+1$ corresponding to $\widetilde{L_s}$.

Theorem 3.1. *Let M be a compact, orientable, immersed minimal surface of genus g with only trivial Jacobi fields in a Riemannian flat torus T^N of dimension $N < 2g$. Then, by swelling of T^N for a natural number n, we obtain n homotopic minimal surfaces with different areas in a swelled indefinite torus of dimension $N+1$. If M is non-hyperelliptic and hyperelliptic, then obtained minimal surfaces are non-hyperelliptic and hyperelliptic, respectively. When the ambient space is a Riemannian flat torus, these minimal surfaces have the same index and nullity as M. In particular, if M is stable, then obtained minimal surfaces are stable.*

Corollary 3.1. *Let M be a holomorphic curve of genus g with only trivial infinitesimal holomorphic deformations in a complex torus T^N of complex dimension $N < g$. Let n be a natural number. Then, by swelling of T^N we obtain n homotopic stable minimal surfaces with different areas in a swelled Riemannian flat torus of real dimension $2N+1$.*

Proof. A holomorphic curve of genus g with only trivial infinitesimal holomorphic deformations in a complex torus T^N is immersed and has only trivial Jacobi fields. □

Colombo and Pirola [3] proved the density of the subset of M_g for $g \geq 4$ such that the Riemann surface corresponding to each element of the subset admits a holomorphic map with only infinitesimal holomorphic deformation in complex tori of dimension 3. So we obtain Theorem 1.1 by Corollary 3.1.

Remark 3.2. We may deform a complex torus of dimension 3 to a real Riemannian flat torus of dimension 6 which contains a stable minimal surface. However, the obtained minimal surface is again holomorphic [2].

Remark 3.3. We may consider a holomorphic curve in a complex torus of dimension 2. However, the holomorphic curve has non-trivial infinitesimal holomorphic deformations. So, we would like to know what occurs when we swell the torus to tori of higher dimension.

We do not know whether one of stable minimal surfaces obtained in Corollary 3.1. is homotopically area-minimizing.

Let $\widetilde{L_1} = {}^tT_{\tau_0}T_{\tau_0}$. We set $\widetilde{L_s} = (1-s)\widetilde{L_0} + s\widetilde{L_1}$. Since τ_0 is a critical point of $E_{\widetilde{L_0}}$ and $E_{\widetilde{L_1}}$, τ_0 is a critical point of $E_{\widetilde{L_s}}$ by Theorem 7.1 [4]. Since

$\widetilde{L_s}, 1 \geq s > 0$ is positive-definite, the square root of $L_s = \sqrt{\widetilde{L_s}}$ exists and the column vectors generate a lattice group of $\mathbf{R}^{2g}$ which gives a torus T_s^{2g} of dimension $2g$. This gives a swelling of T^N to the Jacobi variety and a minimal surface M with the fixed conformal structure in T_s^{2g}. We see that if M is a holomorphic curve in a complex torus of dimension 3, then T_s^{2g} is isomorphic to the Jacobi variety and M is again holomorphic in T_s^{2g}. We can prove that our argument is useful in this case. So we get the following.

Corollary 3.2. *The dense set given by Colombo and Pirola of M_g, $(g \geq 4)$ has the property as follows: The Jacobi variety of the Riemann surface corresponding to each point of the dense set admits a Riemannian flat metric depending on each positive integer n which contains at least n different compact, orientable, non area-minimizing, stable minimal surfaces of genus g homotopic to the Albanese map, which is the only one area-minimizing minimal surface.*

We consider another application of Theorem 3.1. Ross [8] proved that Schwarz' P and D surfaces have $\text{index}_A = 1$ and only trivial Jacobi fields. So we get the following.

Corollary 3.3. *When we swell the torus containing Schwarz' P and D surfaces to a Riemannian flat torus of dimension 4, many different hyperelliptic minimal surfaces of genus 3 appear in the Riemannian flat torus of dimension 4.*

Corollary 3.4. *When we swell the torus containing Schwarz' P and D surfaces to a Minkowski flat torus of dimension 4, many different hyperelliptic minimal surfaces of genus 3 appear in the Minkowski flat torus of dimension 4.*

For Schwarz' P surface, one of obtained minimal surfaces may be isometric to Schwarz' P surface [4].

References

1. L.V. Ahlfors, *The complex analytic structure of the space of closed Riemann surfaces, analytic functions (Nevanlinna et al., eds)*, Princeton Univ. Press, 1960, 45–66
2. C. Arezzo and M.J. Micallef, *Minimal surfaces in flat tori*, Geom. Funct. Anal. 10 (2000), 679–701.
3. E. Colombo and G.P. Pirola, *Some density results for curves with non-simple jacobians*, Math. Ann. 288 (1990), 161–178.

4. N. Ejiri, *A Differential-Geometric Schottoky Problem, and Minimal Surfaces in Tori*, Contemporary Mathematics 308 (2002), 101–144.
5. H. Farkas and H. Kra, *Riemann Surfaces*, Springer, 1980.
6. M.J. Micallef, *Stable minimal surfaces in flat tori*, Contemporary Math. 49 (1986), 73–78.
7. T. Poston and I. Stewart, *Catastrophe Theory and its applications*, Pitman Publishing Limited, 1978.
8. M. Ross, *Schwarz' P and D surfaces are stable*, Diff. Geom. App. 2 (1992), 179–195.

ON NLS EQUATIONS ON BD.I SYMMETRIC SPACES WITH CONSTANT BOUNDARY CONDITIONS

V. S. GERDJIKOV and N. A. KOSTOV

Institute for Nuclear Research and Nuclear Energy
Bulgarian Academy of Sciences
1784 Sofia, Bulgaria
E-mail: gerjikov@inrne.bas.bg, nakostov@inrne.bas.bg

We analyze the algebraic aspects of solving the multicomponent nonlinear Schrödinger (MNLS) equations related to the symmetric spaces of **BD.I**-type. This analysis includes: i) the spectral properties of the MNLS equations under the nonvanishing (constant) boundary conditions; ii) the construction of new equations of MNLS type imposing additional reductions; iii) the involutivity of their integrals of motion proven by using the method of the classical R-matrix; iv) brief but explicit description of their hierarchies of Hamiltonian structures.

Keywords: Multicomponent NLS eqs.; Constant boundary conditions; Classical R-matrix.

1. Introduction

We analyze the algebraic aspects of solving the multicomponent nonlinear Schrödinger (MNLS) equations [1–5] equations related to the symmetric spaces.

In Section 2 we formulate the Lax representation for the MNLS on **BD.I** type symmetric spaces [1,5] and give examples of new reductions. In Sections 3 and 4 we formulate the MNLS equations and outline the spectral properties of the Lax operator for the class of potentials with constant boundary conditions (CBC). In Section 5 we derive the regularized Hamiltonian for the MNLS with CBC and prove the involutivity of their integrals of motion by using the classical R-matrix method.

2. MNLS eqs on BD.I-symmetric spaces

These symmetric spaces are $SO(n+2)/SO(n) \times SO(2)$. The Lax pairs for the corresponding MNLS eqs. with vanishing boundary conditions take the form:

$$\begin{aligned} L\psi &\equiv \left(i\frac{d\psi}{dx} + U(x,t,\lambda) \right) \psi(x,t,\lambda) = 0, \\ M\psi &\equiv \left(i\frac{d\psi}{dt} + V_0(x,t) + \lambda q(x,t) - \lambda^2 J) \right) \psi(x,t,\lambda) = -\psi\lambda^2 J, \\ U(x,\lambda) &= q(x,t) - \lambda J, \qquad V_0(x,t) = i\mathrm{ad}_J^{-1}\frac{dq}{dx} + \frac{1}{2}\left[\mathrm{ad}_J^{-1} q, q(x,t) \right], \\ q &= \begin{pmatrix} 0 & \vec{q}^T & 0 \\ \vec{p} & 0 & s_0\vec{q} \\ 0 & \vec{p}^T s_0 & 0 \end{pmatrix}, \qquad J = \begin{pmatrix} 1 & 0 & 0 \\ 0 & 0 & 0 \\ 0 & 0 & -1 \end{pmatrix}. \end{aligned} \tag{1}$$

Here s_0 is the matrix used to define the orthogonality condition. For $n = 2r+1$ we have $s_0 = \sum_{k=1}^{n} (-1)^k E_{k,\bar{k}}$ with $\bar{k} = n+1-k$ and $(E_{kl})_{sp} = \delta_{ks}\delta_{lp}$. The corresponding MNLS have the form:

$$\begin{aligned} i\partial_t \vec{q} + \partial^2_{xx}\vec{q} + 2(\vec{p},\vec{q})\vec{q} - (\vec{q}s_0\vec{q})\, s_0\vec{p} = 0, \\ i\partial_t \vec{p} - \partial^2_{xx}\vec{p} - 2(\vec{p},\vec{q})\vec{p} + (\vec{p}s_0\vec{p})\, s_0\vec{q} = 0. \end{aligned} \tag{2}$$

The reductions of these equations are obtained following the method of Mikhailov [6] using automorphisms of the corresponding Lie algebra of finite order, see also [7]. The typical reduction with an automorphism K_1 from the Cartan subgroup with $K_1 = \mathbb{1}$ gives rise to $\vec{p} = \vec{q}^*$ and:

$$i\partial_t \vec{q} + \partial^2_{xx}\vec{q} + 2(\vec{q}^\dagger,\vec{q})\vec{q} - (\vec{q}^T s_0\vec{q})\, s_0\vec{q}^* = 0. \tag{3}$$

For $n = 3$ with $K_1 = \mathrm{diag}\,(\epsilon_1, \epsilon_2, 1, \epsilon_2, \epsilon_1)$, $\epsilon_{1,2}^2 = 1$ we get $p_2 = \epsilon_1\epsilon_2 q_2^*$, $p_3 = \epsilon_1 q_3^*$, $p_4 = \epsilon_1\epsilon_2 q_4^*$, which gives a 3-component system of NLS equation

$$\begin{aligned} &iq_{2,t} + q_{2,xx} + 2\epsilon_1(\epsilon_2|q_2|^2 + |q_3|^2)q_2 + \epsilon_1\epsilon_2 q_3^2 q_4^* = 0, \\ &iq_{3,t} + q_{3,xx} + 2\epsilon_1 q_2 q_4 q_3^* + \epsilon_1(2\epsilon_2|q_2|^2 + 2\epsilon_2|q_4|^2 + |q_3|^2)q_3 = 0, \\ &iq_{4,t} + q_{4,xx} + 2\epsilon_1(\epsilon_2|q_4|^2 + |q_3|^2)q_4 + \epsilon_1\epsilon_2 q_3^2 q_2^* = 0. \end{aligned} \tag{4}$$

A second type of reductions are produced via Weyl reflections. Taking $n = 2r+1$ and K_2 to correspond to $S_{e_2}S_{e_3}\cdots S_{e_r}$ leads to $\vec{p} = s_0\vec{q}^*$ and the MNLS eq. (3) becomes:

$$i\partial_t \vec{q} + \partial^2_{xx}\vec{q} + 2(\vec{q}^\dagger s_0\vec{q})\vec{q} - (\vec{q}^T s_0\vec{q})\, \vec{q}^* = 0. \tag{5}$$

For $r = 2$ we get $p_2 = q_4^*, p_3 = -q_3^*, p_4 = q_2^*$ and another inequivalent system of 3 NLS equations:

$$\begin{aligned} &iq_{2,t} + q_{2,xx} + 2(q_2 q_4^* - |q_3|^2)q_2 + q_3^2 q_2^* = 0, \\ &iq_{3,t} + q_{3,xx} - 2q_2 q_4 q_3^* + (2q_2 q_4^* + 2q_4 q_2^* - |q_3|^2)q_3 = 0, \\ &iq_{4,t} + q_{4,xx} + 2(q_4 q_2^* - |q_3|^2)q_4 + q_3^2 q_4^* = 0. \end{aligned} \tag{6}$$

3. MNLS with Constant Boundary Conditions

Below we fix up the boundary conditions on q in such a way that the two asymptotic operators $L_\pm = id/dx + U_\pm(\lambda)$ with

$$U(x,t,\lambda) = q(x,t) - \lambda J, \qquad U_\pm(\lambda) \equiv \lim_{x\to\pm\infty} U(x,t,\lambda) = q_\pm - \lambda J \quad (7)$$

have the same continuous spectrum. Here $q_\pm = \lim_{x\to\pm\infty} q(x,t)$ and $\vec{q}_\pm = \lim_{x\to\pm\infty} \vec{q}(x,t)$. Besides we consider only solutions which have time-independent limits for $x \to \infty$. This can be done by modifying the M-operators as follows:

$$\begin{aligned} M\psi &\equiv \left(i\frac{d\psi}{dt} + V_0(x,t) - V_{00} + \lambda q(x,t) - \lambda^2 J) \right) \psi(x,t,\lambda) \\ &= -\psi(x,t,\lambda)\lambda^2 J, \qquad V_{00} = \lim_{x\to\infty} V_0(x,t) = \lim_{x\to-\infty} V_0(x,t). \end{aligned} \quad (8)$$

The last condition in eq. (8) is satisfied if and only if $\vec{q}_- = e^{i\theta}\vec{q}_+$, where the overall phase θ is an integral of motion. In other words

$$q_+ = u_\theta^{-1} q_- u_\theta, \qquad u_\theta = e^{-i\theta J/2} \quad (9)$$

so $U_+(\lambda)$ and $U_-(\lambda)$ are related by a similarity transformation and therefore have the same sets of eigenvalues.

The corresponding regularized MNLS with $\vec{p} = -\vec{q}^*$ have the form:

$$i\vec{q}_t + \vec{q}_{xx} - (2(\vec{q}^\dagger, \vec{q}) - \rho^2)\vec{q} + (\vec{q}_\pm^\dagger, \vec{q})\vec{q}_\pm + (\vec{q}^T s_0 \vec{q}) s_0 \vec{q}^* - (\vec{q}_\pm^T, s_0 \vec{q}) s_0 \vec{q}_\pm^* = 0. \quad (10)$$

Note that due to eq. (9) the coefficients in eq. (10) are the same for both choices of the subscript $+$ and $-$.

4. Spectral properties of BD.I-type MNLS with CBC

Using the typical reduction $\vec{p} = -\vec{q}^\dagger$, the eigenvalues of $U_\pm(\lambda)$ are the roots of the characteristic polynomial:

$$\begin{aligned} &\mu^{n-2}(\mu^4 - \mu^2(2a + \lambda^2) + a^2 - b) = 0, \\ &a = -(\vec{q}_\pm^\dagger, \vec{q}_\pm) = -\rho^2, \qquad b = |(\vec{q}_\pm^T s_0 \vec{q}_\pm)|^2. \end{aligned} \quad (11)$$

They are given by:

$$\mu_{1,n+2}^2(\lambda) = \frac{\lambda^2}{2} + a + \sqrt{\frac{\lambda^4}{4} + a\lambda^2 + b}, \; \mu_{2,n+1}^2(\lambda) = \frac{\lambda^2}{2} + a - \sqrt{\frac{\lambda^4}{4} + a\lambda^2 + b},$$

and $\mu_{3,4,\ldots,n} = 0$. The Jost solutions are determined by their asymptotics for $x \to \pm\infty$:

$$\begin{aligned} &\lim_{x\to\infty} \psi(x,\lambda) e^{i\mu(\lambda)x} = u_{0,+}, \qquad \lim_{x\to-\infty} \phi(x,\lambda) e^{i\mu(\lambda)x} = u_{0,-}, \\ &q_\pm - \lambda J = -u_{0,\pm}\mu(\lambda)\hat{u}_{0,\pm}, \qquad \mu(\lambda) = -\mathrm{diag}\,(\mu_1(\lambda), \ldots, \mu_{n+2}(\lambda)), \end{aligned} \quad (12)$$

and the scattering matrix is defined as $T(t,\lambda) = \psi^{-1}\phi(x,t,\lambda)$. The continuous spectrum of $L_{\rm as}$ lies on those lines in the complex λ-plane on which $\operatorname{Im}\mu_j(\lambda) = 0$. With the typical reduction $\vec{p} = -\vec{q}^*$ we get that the spectrum fills in the two semiaxis $|\operatorname{Re}\lambda| > \rho_0$, where $\rho_0 = \sqrt{2\rho^2 + 2\sqrt{\rho^4 - b}}$, see the figure.

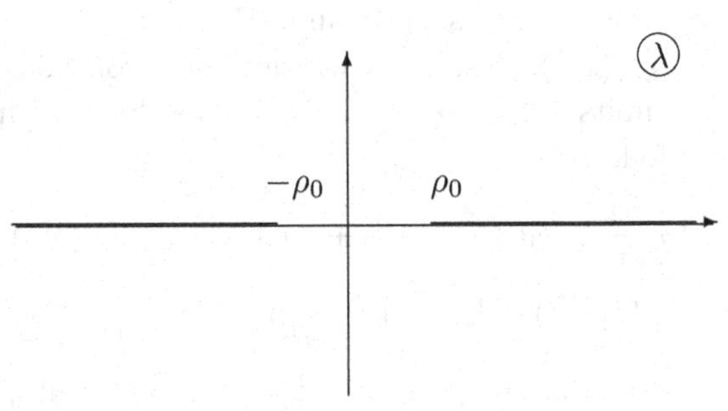

Fig. 1. The continuous spectrum of $L_\pm$ for **BD.I**-type MNLS with typical reduction $\vec{p} = -\vec{q}^\dagger$ and $\rho^4 > b$.

The characteristic equation determines a Riemannian surface $\mathfrak{R}$ which is characterized by the cuts on the figure. On each of the leaves of $\mathfrak{R}$ we can introduce a fundamental analytic solution (FAS). Let us denote by $\mathfrak{R}^+$ (resp. $\mathfrak{R}^-$) the leaf on which

$$\begin{aligned} &\operatorname{Im}\mu_1(\lambda) > \operatorname{Im}\mu_2(\lambda) > 0, \qquad \text{for} \quad \lambda \in \mathfrak{R}^+, \\ &\operatorname{Im}\mu_1(\lambda) < \operatorname{Im}\mu_2(\lambda) < 0, \qquad \text{for} \quad \lambda \in \mathfrak{R}^-. \end{aligned} \tag{13}$$

Then the FAS $\chi^\pm(x,t,\lambda)$ that are analytic on $\mathfrak{R}^\pm$ are constructed by using the generalized Gauss decomposition of the scattering matrix [8]:

$$\begin{aligned} &\chi^\pm(x,t,\lambda) = \phi(x,t,\lambda)S^\pm(t,\lambda) = \psi(x,t,\lambda)T^\mp(t,\lambda)D^\pm(\lambda), \\ &T(t,\lambda) = T^-(t,\lambda)D^+(\lambda)\hat{S}^+(t,\lambda) = T^+(t,\lambda)D^-(\lambda)\hat{S}^-(t,\lambda), \end{aligned} \tag{14}$$

where $T^+(t,\lambda)$, $S^+(t,\lambda)$ (resp. $T^-(t,\lambda)$, $S^-(t,\lambda)$ are block-upper-triangular (resp. block-lower-triangular) whose diagonal elements are equal to 1 and whose block-matrix structure is compatible with $\mu(\lambda)$. The block-diagonal factors $D^\pm(\lambda) = \operatorname{bdiag}\left((m_1^\pm)^{\pm1}, (m_2^\pm)^{\pm1}, \boldsymbol{\tilde{m}}_3^\pm, (m_2^\pm)^{\mp1}, (m_1^\pm)^{\mp1}\right)$ have the same block structure with block dimensions $1, 1, n-2, 1, 1$. Note that $D^\pm(\lambda)$ are analytic functions for $\lambda \in \mathfrak{R}^\pm$ and are the generating functionals of the integrals of motion. In the particular case of $n = 3$ $T^+(t,\lambda)$, $S^+(t,\lambda)$ (resp. $T^-(t,\lambda)$, $S^-(t,\lambda)$ are simply upper-triangular (resp. lower-triangular) whose diagonal elements are equal to 1.

Hamiltonian properties

The phase space $\mathcal{M}$ for non-vanishing boundary conditions is a nonlinear space: $\mathcal{M} \equiv \cup_\theta \mathcal{M}_\theta$,

$$\mathcal{M}_\theta = \left\{ q(x,t) = \Pi_{0J} q(x,t), \quad \lim_{x\to\pm\infty} q(x,t) = q_\pm, \; q_+ = u_\theta^{-1} q_- u_\theta \right\}.$$

Here $\Pi_{0J} = \mathrm{ad}_J^{-1} \mathrm{ad}_J$ and u_θ is an integral of motion which must be fixed up in order to have non-degenerate symplectic structures.

The invariants of the transfer matrix $T(\lambda)$ such as, e.g. its principal minors generate integrals of motion, i.e. if all $\mathrm{Im}\,\mu_j$ are different we have only r independent series of conserved quantities. The methods of deriving of these integrals as functionals of the potential q is based on the Wronskian relation [7]:

$$\begin{aligned}
\mathrm{tr}\left[i(\chi^+)^{-1} \frac{d\chi^+}{d\lambda} C - \frac{d\mu(\lambda)}{d\lambda} xC \right]\Bigg|_{x=-\infty}^{\infty} \\
= i \sum_{k=1}^{r} \frac{d\delta_k^+}{d\lambda} \mathrm{tr}\,(H_k C) + i \mathrm{tr}\left(\hat{\psi}_0(\lambda) \frac{d\psi_0}{d\lambda} C - \hat{\phi}_0(\lambda) \frac{d\phi_0}{d\lambda} C \right) \\
= \int_{-\infty}^{\infty} dx\, \mathrm{tr}\, \left[J R_C(x,t,\lambda) - \lambda \mu^{-1}(\lambda) C \right], \quad (15)
\end{aligned}$$

where C is a constant element of $\mathfrak{h}$ and $R_C(x,t,\lambda) = \chi^+ C (\chi^+)^{-1}(x,t,\lambda)$ is a natural generalization of the diagonal of the resolvent of L. It satisfies the equation:

$$i \frac{dR_C}{dx} + [q(x,t) - \lambda J, R_C] = 0, \quad \lim_{x\to\infty} R_C(x,t,\lambda) = \psi_0(\lambda) C \psi_0^{-1}(\lambda). \quad (16)$$

Eq. (16) allows one to derive the recurrent relations for evaluating the expansion coefficients

$$R_C(x,t,\lambda) = C_0 + \sum_{k=1}^{\infty} R_k \lambda^{-k}(x,t), \quad \psi_0 C \psi_0^{-1}(\lambda) = C_0 + \sum_{k=1}^{\infty} C_k \lambda^{-k}. \quad (17)$$

In what follows we fix up $C = J$.

The trace identities for the MNLS type equations with CBC can be derived by inserting the asymptotic expansions of $R_J(x,\lambda)$ and $\boldsymbol{m}_k^+(\lambda)$:

$$\ln \boldsymbol{m}_k^+(\lambda) = \sum_{p=1}^{\infty} I_p^{(k)} \lambda^{-p}, \qquad k = 1, \ldots, r, \quad (18)$$

in both sides of (15) and equating the corresponding coefficients of λ^{-p}. Here $\boldsymbol{m}_k^+(\lambda)$ are the upper principal minors of order k of $T(\lambda)$, which are all analytic on the sheet $\mathfrak{R}^+$. The matrix elements of $D^+(\lambda)$ are expressed by them as $m_1^+ = \boldsymbol{m}_1^+(\lambda)$, $m_2^+ = \boldsymbol{m}_2^+(\lambda)/\boldsymbol{m}_1^+(\lambda)$, etc.

The first three of the local integrals of motion coming from the principal series with $C = J$, see [9]:

$$
\begin{aligned}
I_k^{(1)} &= \int_{-\infty}^{\infty} dx \,\mathrm{tr}\, \left(JR_{k+1}(x,t) - C_{k+1}\right), \\
I_1^{(1)} &= \frac{i}{2}\int_{-\infty}^{\infty} dx\, \left(\langle q, q(x,t)\rangle + 2\rho^2\right) = i\int_{-\infty}^{\infty} dx\, \left((\vec{q}^{\,\dagger}, \vec{q}(x,t)) - \rho^2\right), \\
I_2^{(1)} &= \frac{1}{2}\int_{-\infty}^{\infty} dx\, \langle q, \mathrm{ad}_J^{-1} q_x\rangle = \frac{1}{2}\int_{-\infty}^{\infty} dx\, \left((\vec{q}_x^{\,\dagger}, \vec{q}) - (\vec{q}^{\,\dagger}, \vec{q}_x)\right), \\
I_3^{(1)} &= \frac{i}{6}\int_{-\infty}^{\infty} dx\, \{-\langle q_x, q_x\rangle + \langle Y(x,t), Y(x,t)\rangle - \langle Y_0, Y_0\rangle\} \\
&= \frac{i}{3}\int_{-\infty}^{\infty} dx\, \left\{ (\vec{q}_x^{\,\dagger}, \vec{q}_x) + (\vec{q}^{\,\dagger}, \vec{q})^2 - \frac{1}{2}(\vec{q}^{\,\dagger} s_0 \vec{q}^{\,*})(\vec{q}^{\,T} s_0 \vec{q}) - \rho^4 + \frac{b}{2} \right\}
\end{aligned}
$$

where

$$
Y(x,t) = \frac{1}{2}\left[q, \mathrm{ad}_J^{-1} q(x,t)\right], \qquad Y_0 = \lim_{x\to\pm\infty} Y(x,t). \tag{19}
$$

The correct use of the Wronskian relation (15) allowed us to derive renormalized integrals of motion, i.e. ones that converge for $q(x,t) \in \mathcal{M}$.

However among the integrals in this series one can not find the Hamiltonian of the MNLS (10). The obvious candidate for Hamiltonian $-3iI_3^{(1)}$ has two defects: i) its gradient $\delta I_3^{(1)}/\delta q^T(x,t)$ does not vanish for $x \to \pm\infty$ and ii) it does not produce the terms in (10) depending on the boundary values $\vec{q}_\pm$. In order to deal with these difficulties we have to regularize $-3iI_3^{(1)}$ by adding additional integrals of motion. For the scalar NLS with CBC the regularization is obtained just by taking proper linear combination of $-3iI_3^{(1)}$ and $-iI_1^{(1)}$. For the MNLS related to the **BD.I**-type of symmetric spaces this still does not solve the problem. Therefore we need to consider an additional series of integrals, which generically have non-local densities. Fortunately among the simplest of them one may find local ones. For example, there exist first integrals of zeroth order with local densities:

$$
\tilde{I}_1^{(l)} = \frac{1}{2}\int_{-\infty}^{\infty} dx\, \left\langle Y(x,t) - Y_0, Y_0^{2l-1}\right\rangle, \qquad l = 1, \dots \tag{20}
$$

Note, that since $Y_0 \in so(n+2)$ then Y_0^{2l+1} belongs to $so(n+2)$ too for any positive integer l. Besides $Y_0 \in so(n+2)$ and satisfies characteristic polynomial equation of order $2r+1$, so at most r of them are independent.

In particular $\tilde{I}_1^{(l)} = -i\rho^2 I_1^{(l)} + K_1$ where

$$K_1 = \frac{1}{2}\int_{-\infty}^{\infty} dx \left\{ (\vec{q}^{\dagger}, \vec{q}_{\pm})(\vec{q}_{\pm}^{\dagger}, \vec{q}(x,t)) - \rho^4 - (\vec{q}^{\dagger} s_0 \vec{q}_{\pm}{}^{*})(\vec{q}_{\pm}^{T} s_0 \vec{q}(x,t)) + b \right\}. \tag{21}$$

Using these integrals we can regularize the other higher integrals. For example, the Hamiltonian of the MNLSE (10) is obtained by:

$$\begin{aligned}
\boldsymbol{H}_{\rm MNLS} &= -3iI_3^{(1)} - 2\tilde{I}_1^{(l)} \\
&= \frac{1}{2}\int_{-\infty}^{\infty} dx \, \{ -\langle q_x, q_x \rangle + \langle Y(x,t) - Y_0, Y(x,t) - Y_0 \rangle \} \\
&= \int_{-\infty}^{\infty} dx \left\{ (\vec{q}_x^{\dagger}, \vec{q}_x) + (\vec{q}^{\dagger}, \vec{q})^2 - \rho^2 (\vec{q}^{\dagger}, \vec{q}) - (\vec{q}^{\dagger}, \vec{q}_{\pm})(\vec{q}_{\pm}^{\dagger}, \vec{q}(x,t)) \right. \\
&\quad \left. - \frac{1}{2}(\vec{q}^{\dagger} s_0 \vec{q}^{*})(\vec{q}^{T} s_0 \vec{q}(x,t)) + (\vec{q}^{\dagger} s_0 \vec{q}_{\pm}^{*})(\vec{q}_{\pm}^{T} s_0 \vec{q}(x,t)) + \rho^4 - \frac{1}{2} b \right\}.
\end{aligned} \tag{22}$$

It is easy to check that $\boldsymbol{H}_{\rm MNLS}$ is a regular integral of motion since the gradient $\delta \boldsymbol{H}_{\rm MNLS}/\delta q^T(x,t)$ vanishes for both $x \to \infty$ and $x \to -\infty$; it also gives rise to the MNLS equation (10).

In analyzing the Hamiltonian properties of the MNLS with CBC we will make use of the classical r-matrix approach. It allows one to write down in compact form the Poisson brackets of the transfer (monodromy) matrix. Since our problem is ultra-local in the terminology of Faddeev then the definition of r is independent on the boundary conditions. Using [1] we find

$$r(\lambda,\mu) = \frac{1}{\lambda - \mu}\left(\sum_{\alpha \in \Delta^+} (E_\alpha \otimes E_{-\alpha} + E_{-\alpha} \otimes E_\alpha) + \sum_{j=1}^{r} H_j \otimes H_j \right), \tag{23}$$

where Δ^+ is the set of positive roots of simple Lie algebra $\mathfrak{g}$ and E_α and H_j are its Cartan-Weyl generators. In order to derive the Poisson brackets for the MNLS on the whole axis with CBC we need to take into account the corresponding oscillations coming from the Jost solutions.

Skipping the details we just write down the expressions for the Poisson brackets between the matrix elements of $T(\lambda)$:

$$\left\{ T(\lambda) \underset{,}{\otimes} T(\mu) \right\} = r_+(\lambda,\mu) T(\lambda) \otimes T(\mu) - T(\lambda) \otimes T(\mu) r_-(\lambda,\mu), \tag{24}$$

$$\begin{aligned}
r_\pm(\lambda,\mu) &= \lim_{x \to \pm\infty} \tau_\pm^{-1}(x,\lambda,\mu) r(\lambda,\mu) \tau_\pm(x,\lambda,\mu), \\
\tau_+(x,\lambda,\mu) &= \psi_0(\lambda) e^{-iJ(\lambda)x} \otimes \psi_0(\mu) e^{-iJ(\mu)x}, \\
\tau_-(x,\lambda,\mu) &= \phi_0(\lambda) e^{-iJ(\lambda)x} \otimes \phi_0(\mu) e^{-iJ(\mu)x},
\end{aligned}$$

where $\{T(\lambda) \underset{,}{\otimes} T(\mu)\}_{ij,kl} \equiv \{T_{ij}(\lambda), T_{kl}(\mu)\}$.

An important and difficult problem here is to take correctly into account the the threshold singularities of $T_{kl}(\lambda)$ at the end points of the continuous spectrum.

As a consequence of (24) we get the involution properties of $\boldsymbol{m}_k^+(\lambda)$:

$$\{\boldsymbol{m}_k^+(\lambda), \boldsymbol{m}_j^+(\lambda')\} = 0, \tag{25}$$

for all values of $1 \leq i, j \leq r$ and λ and λ' taking values on the continuos spectrum of L. From (25) there follows that $\{I_p^{(k)}, I_s^{(l)}\} = 0$ for all positive values of p and s, and for all $1 \leq k, l \leq r$. A consequence of eq. (25) is the involutivity of the integrals of the principal series $\{I_1^{(k)}, I_s^{(1)}\} = 0$. This is a necessary condition in proving the complete integrability of the MNLS equations with CBC.

Of course the rigorous proof of the complete integrability and the derivation of the basic properties of the MNLS equations must be based on the completeness relation of the relevant 'squared solutions' of L. For the single component NLS such relation has been proposed in [10]; for the multicomponent systems this is still open question.

5. Discussion

Recently it was discovered by Wadati [11], that the $so(5)$ and $so(7)$ MNLS with vanishing BC have important applications to BEC with spin $F = 1$ and $F = 2$ respectively. The dark solitons are also relevant for the BEC [12] and attract attention, which enhances the interest to the MNLS type models.

The soliton solutions of MNLS with non-vanishing boundary conditions have been reported in 1983 (see Ref. 2) for the $su(n+m)/s(u(n) \otimes su(m))$ case. Later in [13] it was worked out for some special choices of the boundary constants $q_+^2 = q_-^2 = \rho^2 \mathbb{1}$. Then it is possible to introduce uniformization variable and the dressing can be done as for vanishing boundary conditions. In the generic case however, uniformization variables do not exist, which makes the problem still more difficult. Deriving the dressing factors for the MNLS for the symmetric spaces of types **C.II** and **D.III** requires substantial changes even for vanishing BC (see [7,14–16]); doing the same for CBC is still a bigger challenge.

References

1. A.P. Fordy, P.P. Kulish, Commun. Math. Phys. **89** (1983) 427–443.

2. V.S. Gerdjikov, P.P. Kulish. *Multicomponent nonlinear Schrödinger equation in the case of nonzero boundary conditions*, Journal of Mathematical Sciences **30**, No 4, 2261–2269 (1985).
3. V.S. Gerdjikov. *Algebraic and Analytic Aspects of N-wave Type Equations*, Contemporary Mathematics **301**, 35–68 (2002); nlin.SI/0206014.
4. V.S. Gerdjikov, G.G. Grahovski, N.A. Kostov. *On the multi-component NLS type equations on symmetric spaces and their reductions*, Theor. Math. Phys. **144** No. 2 1147–1156 (2005).
5. C. Athorne, A. Fordy, *Generalised KdV and MKDV Equations Associated with Symmetric Spaces*, J. Phys. A: Math. Gen. **20** (1987) 1377–1386.
6. A.V. Mikhailov, The Reduction Problem and the Inverse Scattering Problem, *Physica D*, **3D** 73–117, 1981.
7. G.G. Grahovski, V.S. Gerdjikov, N.A. Kostov, V.A. Atanasov. *New integrable multi-component NLS type equations on symmetric spaces: $\mathbb{Z}_4$ and $\mathbb{Z}_6$ reductions*, In: Eds. Ivailo Mladenov, Manuel de Leon, *Geometry, Integrability and Quantization*, Softex, Sofia (2006); pp. 154–175; nlin.SI/0603066.
8. V.S. Gerdjikov, N.A. Kostov, T.I. Valchev, *Solutions of multi-component NLS models and Spinor Bose-Einstein condensates*, Physica D, doi:10.1016/j.physd.2008.06.007; ArXiv:0802.4398.
9. V.S. Gerdjikov, D.J. Kaup, N A. Kostov, T.I. Valchev, *Bose-Einstein condensates and multi-component NLS models on symmetric spaces of* ***BD.I****-type, Expansions over squared solutions*, Reported at Conference on Nonlinear Science and Complexity, Porto, Portugal, July 28–31, 2008.
10. V.V. Konotop, V.E. Vekslerchik, Direct perturbation theory for dark solitons, Phys. Rev. **49**, 2397–2407 (1994).
11. J. Ieda, T. Miyakawa and M. Wadati, *Exact Analysis of Soliton Dynamics in Spinor Bose-Einstein Condensates*, Phys. Rev Lett. **93**, (2004), 194102; *Matter-wave solitons in an $F = 1$ spinor Bose-Einstein condensate*, J. Phys. Soc. Jpn. **73** (2004) 2996.
12. M. Uchiyama, J. Ieda and M. Wadati, *Dark solitons in $F = 1$ spinor Bose–Einstein condensate*, J. Phys. Soc. Jpn. **75** (2006) 064002.
13. B. Prinari, M. Ablowitz, G. Biondini, J. Math. Phys. **47**, 063508 (2006).
14. V.S. Gerdjikov, G.G. Grahovski, N.A. Kostov, *Reductions of N-wave interactions related to low–rank simple Lie algebras I: $\mathbb{Z}_2$-reductions*, J. Phys. A: Math & Gen. **34**, 9425–9461 (2001).
15. V.S. Gerdjikov, *Selected Aspects of Soliton Theory. Constant boundary conditions*, In: Prof. G. Manev's Legacy in Contemporary Aspects of Astronomy, Gravitational and Theoretical Physics, Eds.; V. Gerdjikov, M. Tsvetkov, Heron Press Ltd, Sofia, 2005. pp. 277–290; nlin.SI/0604004.
16. V.S. Gerdjikov, D.J. Kaup, N.A. Kostov, T.I. Valchev, *On classification of soliton solutions of multicomponent nonlinear evolution equations*, J. Phys. A: Math. Theor. **41** (2008) 315213 (36pp).

ORTHOGONAL ALMOST COMPLEX STRUCTURES ON $S^2 \times \mathbf{R}^4$

HIDEYA HASHIMOTO* and MISA OHASHI

Department of Mathematics, Meijo University,
Tempaku, Nagoya 468-8502, Japan
** E-mail: hhashi@ccmfs.meijo-u.ac.jp*

1. Introduction

The purpose of this paper is to construct the family of induced almost complex structures on $S^2 \times \mathbf{R}^4$ (compatible with the induced metric) in purely imaginary octonions $\mathrm{Im}\mathfrak{C}$. It is well known that any orientable hypersurface M^6 in $\mathrm{Im}\mathfrak{C}$ can admit the almost Hermitian structure ([1,2]) $(J, \langle , \rangle)$ where $\langle , \rangle$ and J are the induced metric on M^6 and the almost complex structure compatible with the metric, respectively. Also, note that the structure group of the tangent bundle of M can reduce to $SU(3)$.

We construct the 1-parameter family of the isometric imbeddings from $S^2 \times \mathbf{R}^4$ to purely imaginary octonions. As Riemaniann manifolds, they coincide. However, the induced almost complex structures are different. We will prove this, and show that the induced almost Hermitian structures of 1-parameter family are all homogeneous.

2. Preliminaries

Let $\mathbf{H}$ be the skew field of all quaternions with canonical basis $\{1, i, j, k\}$, which satisfy

$$i^2 = j^2 = k^2 = -1, \; ij = -ji = k, \; jk = -kj = i, \; ki = -ik = j.$$

The octonions (or Cayley algebra) $\mathfrak{C}$ over $\mathbf{R}$ can be considered as a direct sum $\mathbf{H} \oplus \mathbf{H} = \mathfrak{C}$ with the following multiplication

$$(a + b\varepsilon)(c + d\varepsilon) = ac - \bar{d}b + (da + b\bar{c})\varepsilon,$$

where $\varepsilon = (0,1) \in \mathbf{H} \oplus \mathbf{H}$ and $a, b, c, d \in \mathbf{H}$, where the symbol "$^-$" denote the conjugation of the quaternion. For any $x, y \in \mathfrak{C}$, we have

$$\langle xy, xy \rangle = \langle x, x \rangle \langle y, y \rangle$$

which is called "normed algebra" in ([3]). The octonions is a non-commutative, non-associative alternative division algebra. The group of automorphisms of the octonions is the exceptional simple Lie group

$$G_2 = \{ g \in SO(8) \mid g(xy) = g(x)g(y) \text{ for any } x, y \in \mathfrak{C} \}.$$

Since $g(1) = 1$, we have $G_2 \subset SO(7) \simeq SO(\operatorname{Im}\mathfrak{C})$ where $\operatorname{Im}\mathfrak{C} = \{x \in \mathfrak{C} \mid \langle x, 1 \rangle = 0\}$ (which is called purely imaginary octonions).

3. Hypersurfaces in $\operatorname{Im}\mathfrak{C}$

In ([1,2]), it is proved that any orientable hypersurface $\varphi : M^6 \to \operatorname{Im}\mathfrak{C}$ of purely imaginary octonions admits the almost complex (Hermitian) structure defined by

$$\varphi_*(JX) = \varphi_*(X)\xi,$$

where ξ is the (oriented) unit normal vector field on M^6. The structure equations of G_2 is given by

Proposition 3.1 ([1]). *Let $\varphi : M^6 \to \operatorname{Im}\mathfrak{C}$ be an isometric immersion from an oriented 6-dimensional manifold M^6 to the purely imaginary octonions and ξ be an unit normal vector field on M^6. Then*

$$d\varphi = f\omega + \bar{f}\bar{\omega}, \tag{1}$$

$$d\xi = f(-2\sqrt{-1}\,\theta) + \bar{f}(2\sqrt{-1}\,\bar{\theta}), \tag{2}$$

$$df = \xi(-\sqrt{-1}\,{}^t\bar{\theta}) + f\kappa + \bar{f}[\theta], \tag{3}$$

and the integrability conditions imply that

$$d\omega = -\kappa \wedge \omega - [\theta] \wedge \bar{\omega}, \tag{4}$$

$$0 = -{}^t\bar{\theta} \wedge \omega + {}^t\theta \wedge \bar{\omega}, \tag{5}$$

$$d\theta = -\kappa \wedge \theta + [\bar{\theta}] \wedge \bar{\theta}, \tag{6}$$

$$d\kappa = -\kappa \wedge \kappa + 3\theta \wedge {}^t\bar{\theta} - ({}^t\theta \wedge \bar{\theta})I_3, \tag{7}$$

where $(\xi, f, \bar{f})$ is a G_2-adapted (local) frame field along φ, and θ, κ are $M_{3\times 1}(\mathbf{C})$, $M_{3\times 3}(\mathbf{C})$ valued 1-forms on M^6, respectively.

By (5), we see that there exist $M_{3\times 3}$-valued matrices $\mathfrak{A}, \mathfrak{B}$ satisfying

$$\sqrt{-1}\,\theta = {}^t\mathfrak{B}\omega + \overline{\mathfrak{A}}\bar{\omega},$$

where ${}^t\mathfrak{A} = \mathfrak{A}$. More explicitly, there exists a unitary frame $\{e_i, Je_i\}$ $(i = 1, 2, 3)$, such that

$$f_i = (1/2)(e_i - \sqrt{-1}Je_i).$$

We denote by II the second fundamental form of M^6. Then the components of second fundamental form are given by

$$\mathfrak{A}_{ij} = \langle \mathrm{II}(f_i, f_j), \xi\rangle, \quad \mathfrak{B}_{ij} = \langle \mathrm{II}(f_i, \bar{f}_j), \xi\rangle.$$

Let $\varphi, \varphi' : M^6 \to \operatorname{Im}\mathfrak{C}$ be two isometric immersions from an oriented 6-dimensional manifold M^6 to $\operatorname{Im}\mathfrak{C}$. If there exist $g \in G_2$, $a \in \operatorname{Im}\mathfrak{C}$ satisfying

$$g \circ \varphi + a = \varphi',$$

then we say that φ and φ' are G_2-congruent. We can easily see that, if φ and φ' are G_2-congruent, then the induced almost Hermitian structures coincide. However, the induced almost Hermitian structure of $\tilde{g} \circ \varphi$ are different from the one of φ in general, for $\tilde{g} \in SO(7)$.

4. On G_2 frame field on $S^2 \times \mathbf{R}^4$

In this section, we give an explicit representation of G_2-frame fields on $S^2 \times \mathbf{R}^4 \subset \operatorname{Im}\mathfrak{C}$, and the G_2-invariants. Let $q \in S^3(\subset \mathbf{H})$ be the unit quaternion. We define the map $\pi : S^3 \to S^2$ such that $\pi(q) = qi\bar{q}$, which is called the Hopf map.

Next we define the 1-parameter family of imbeddings from $S^2 \times \mathbf{R}^4$ to $\operatorname{Im}\mathfrak{C}$, as follows

$$\begin{aligned}\varphi_\alpha(qi\bar{q}, \tilde{y}) &= \cos(\alpha)qi\bar{q} + \sin(\alpha)(qi\bar{q})\varepsilon \\ &\quad + y_0\varepsilon + y_1(-\sin(\alpha)i + \cos(\alpha)i\varepsilon) \\ &\quad + y_2(-\sin(\alpha)j + \cos(\alpha)j\varepsilon) \\ &\quad + y_3(-\sin(\alpha)k + \cos(\alpha)k\varepsilon),\end{aligned} \tag{8}$$

where $qi\bar{q} \in S^2$ and $\tilde{y} = (y_0, y_1, y_2, y_3) \in \mathbf{R}^4$, for some fixed $\alpha \in [0, \pi/3]$.

Note that the imbeddings are equivariant in the following sense.

Let $\rho_{III} : Sp(1) \to G_2$ be the representation of the Lie subgroup $Sp(1)$ of G_2, which is defined by

$$\rho_{III}(q)(a + b\varepsilon) = qa\bar{q} + (qb\bar{q})\varepsilon, \tag{9}$$

where $a, b \in \mathbf{H}$ (see [5]). In fact, we may show that ρ_{III} satisfies

$$\rho_{III}(q)(a + b\varepsilon)\rho_{III}(q)(c + d\varepsilon) = \rho_{III}(q)(ac - \bar{d}b + (da + b\bar{c})\varepsilon),$$

for any $a, b, c, d \in \mathbf{H}$. From (8) and (9), it follows immediately that the imbedding φ_α is rewritten as

$$\begin{aligned}\varphi_\alpha(qi\bar{q}, \tilde{y}) = \rho_{III}(q)(&\cos(\alpha)i + \sin(\alpha)i\varepsilon) \\ &+ y_0\varepsilon + y_1(-\sin(\alpha)i + \cos(\alpha)i\varepsilon) \\ &+ y_2(-\sin(\alpha)j + \cos(\alpha)j\varepsilon) \\ &+ y_3(-\sin(\alpha)k + \cos(\alpha)k\varepsilon).\end{aligned} \tag{10}$$

Therefore, we see that the imbeddings are equivariant and the induced almost Hermitian structures are homogeneous for all $\alpha \in [0, \pi/3]$. In fact, we define the G_2-adapted frame fields by

$$\begin{aligned}\xi &= \Big\{\rho_{III}(q)(\cos(\alpha) \cdot i + \sin(\alpha)i\varepsilon)\Big\}, \\ f_1 &= \frac{1}{2}\Big\{\rho_{III}(q)(-\sin(\alpha) \cdot i + \cos(\alpha)i\varepsilon - \sqrt{-1}\,(\varepsilon))\Big\}, \\ f_2 &= \frac{1}{2}\Big\{\rho_{III}(q)(j - \sqrt{-1}\,(-\cos(\alpha)k + \sin(\alpha)k\varepsilon))\Big\}, \\ f_3 &= -\frac{1}{2}\Big\{\rho_{III}(q)(-\sin(\alpha)k - \cos(\alpha)k\varepsilon - \sqrt{-1}\,j\varepsilon)\Big\}.\end{aligned}$$

Then we may see that (f_1, f_2, f_3) is a $SU(3)$−frame field on $\varphi_\alpha(S^2 \times \mathbf{R}^4)$.

To calculate the G_2 invariants, we define the local 1-forms μ_1, μ_2 on S^2 by

$$\mu_1 = \langle d(qi\bar{q}), qj\bar{q}\rangle, \quad \mu_2 = \langle d(qi\bar{q}), qk\bar{q}\rangle.$$

Then, we obtain

$$\begin{aligned}\omega^1 &= dy_1 - \sqrt{-1}\,dy_0, \\ \omega^2 &= \cos(\alpha)\mu_1 - \sin(\alpha)dy_2 + \sqrt{-1}\,(-\cos(2\alpha)\mu_2 + \sin(2\alpha)dy_3), \\ \omega^3 &= \sin(2\alpha)\mu_2 + \cos(2\alpha)dy_3 - \sqrt{-1}\,(\sin(\alpha)\mu_1 + \cos(\alpha)dy_2),\end{aligned}$$

at $q = 1$. Since

$$\begin{aligned}d\xi &= \cos(\alpha)dqi\bar{q} + \sin(\alpha)(dqi\bar{q})\varepsilon \\ &= \cos(\alpha)(qj\bar{q} \otimes \mu_1 + qk\bar{q} \otimes \mu_2) + \sin(\alpha)\left((qj\bar{q})\varepsilon \otimes \mu_1 + (qk\bar{q})\varepsilon \otimes \mu_2\right),\end{aligned}$$

we get

$$d\xi = \cos(\alpha)(j \otimes \mu_1 + k \otimes \mu_2) + \sin(\alpha)\left(j\varepsilon \otimes \mu_1 + k\varepsilon \otimes \mu_2\right),$$

at $q=1$. Hence we have

$$\mathfrak{B} = -\frac{1}{4}\begin{pmatrix} 0 & 0 & 0 \\ 0 & \cos^2(\alpha)+\cos^2(2\alpha) & \frac{\sqrt{-1}}{2}(\sin(2\alpha)-\sin(4\alpha)) \\ 0 & -\frac{\sqrt{-1}}{2}(\sin(2\alpha)-\sin(4\alpha)) & \sin^2(\alpha)+\sin^2(2\alpha) \end{pmatrix},$$

$$\mathfrak{A} = -\frac{1}{4}\begin{pmatrix} 0 & 0 & 0 \\ 0 & \cos^2(\alpha)-\cos^2(2\alpha) & -\frac{\sqrt{-1}}{2}(\sin(2\alpha)+\sin(4\alpha)) \\ 0 & -\frac{\sqrt{-1}}{2}(\sin(2\alpha)+\sin(4\alpha)) & -\sin^2(\alpha)+\sin^2(2\alpha) \end{pmatrix}. \quad (11)$$

Thererfore from (11), we get the G_2 invariants on $S^2 \times \mathbf{R}^4$ given by

$$\mathrm{tr}({}^t\overline{\mathfrak{B}}\,\mathfrak{B}) = \frac{1}{8}(1+\cos^2(3\alpha)), \quad \mathrm{tr}({}^t\overline{\mathfrak{A}}\,\mathfrak{A}) = \frac{1}{8}(1-\cos^2(3\alpha)).$$

Summing up the above arguments, we have

Theorem 4.1. *For $\alpha \in \mathbf{R}$ $(0 \leq \alpha \leq \pi/3)$, let $(S^2 \times \mathbf{R}^4, \varphi_\alpha)$ be defined as above. The family of the imbeddings φ_α induce the 1-parameter family of the almost complex structures J_α on $S^2 \times \mathbf{R}^4$, which are not G_2-conguent to each other. Moreover the induced almost Hermitian structure $(J_\alpha, \langle,\rangle)$ is (1,2)-symplectic iff $\alpha = 0$ or $\pi/3$.*

We here note that φ_0 and $\varphi_{\pi/3}$ are G_2-congruent. The almost Hermitian manifold $(M, J, \langle,\rangle)$ is said to be $(1,2)$-*symplectic* if $(d\omega)^{(1,2)} = 0$, where $\omega = \langle J\ ,\ \rangle$ is the canonical 2-form (or Kähler form) on M. In our situation, $(d\omega)^{(1,2)} = 0$, is equivalent to $\mathfrak{A} = 0$.

5. Orbits related to G_2

It is well known that 6-dimensional, 5-dimensional spheres can be represented as follows

$$S^6 = G_2/SU(3), \quad S^5 = SU(3)/SU(2). \quad (12)$$

Let $G_2^+(\mathbf{R}^7)$ be the Grassmann manifold of oriented 2-planes in $\mathbf{R}^7$ and let $V_2(\mathbf{R}^7)$ be the Stiefel manifold of oriented orthonormal 2-frames in $\mathbf{R}^7$. Then

$$V_2(\mathbf{R}^7) = \{(u,v) \in S^6 \times S^6 \mid \langle u,v\rangle = 0\}.$$

By (12) we see that

$$V_2(\mathbf{R}^7) \simeq G_2/SU(2), \quad G_2^+(\mathbf{R}^7) \simeq G_2/U(2).$$

A 3-dimensional vector space V in $\mathrm{Im}\,\mathfrak{C}$ is called *associative* if $\mathrm{span}_{\mathbf{R}}\{u, v, uv\} = V$, where $\{u,v\}$ is an oriented orthonormal pair of V. We also note that the Grassmann manifold $G_{ass}(\mathrm{Im}\,\mathfrak{C})$ of associative 3-planes are given by

$$G_{ass}(\mathrm{Im}\,\mathfrak{C}) \simeq G_2/SO(4).$$

We note that the representation ρ from $SO(4)(\simeq Sp(1) \times Sp(1)/Z_2)$ to G_2 is given by

$$\rho(q_1, q_2)(a + b\varepsilon) = q_1 a \overline{q_1} + (q_2 a \overline{q_1})\varepsilon,$$

where $(q_1, q_2) \in Sp(1) \times Sp(1)$ and $a + b\varepsilon \in \mathrm{Im}\,\mathfrak{C}$. Let S^2 be a 2-dimensional sphere in $\mathrm{Im}\,\mathfrak{C}$. Then there exists a 3-dimensional vector space V such that $S^2 \subset V$. In general, V is not necessarily associative.

If V is an associative 3-plane, then the imbedding from $S^2 \times \mathbf{R}^4$ to $\mathrm{Im}\,\mathfrak{C}$ is G_2-congruent to $\varphi_0(S^2 \times \mathbf{R}^4)$.

Next we assume that V is not an associative 3-plane. Let $\{v_1, v_2, v_3\}$ be the oriented orthonormal base of V. From assumption

$$\mathrm{span}_{\mathbf{R}}\{v_1, v_2, v_1 v_2\} \neq V.$$

We set

$$\cos\theta = \langle v_1 v_2, v_3 \rangle,$$

then, we may choose a vector $u_1 \in V$, $(|u_1| = 1, \langle u_1, v_1 v_2 \rangle = 0)$ satisfying

$$v_3 = \cos\theta v_1 v_2 + \sin\theta u_1.$$

By algebraic properties of G_2, we can take $g \in G_2$ satisfying

$$g(i) = v_1, \quad g(j) = v_2, \quad g(k) = v_1 v_2, \quad g(\varepsilon) = u_1.$$

From which, we may assume that S^2 is included in the 3-dimensional vector space $\mathrm{span}_{\mathbf{R}}\{g(i), g(j), g(\cos\theta k + \sin\theta\varepsilon)\}$, for some $\theta \in [0, \pi/2]$. By reparametrizing 3-dimensional subspace in $\mathrm{Im}\mathfrak{C}$ suitably, we may assume that

$$S^2 \subset \mathrm{span}_{\mathbf{R}}\{\cos\alpha i + \sin\alpha i\varepsilon,\ \cos\alpha j + \sin\alpha j\varepsilon,\ \cos\alpha k + \sin\alpha k\varepsilon\},$$

for some $\alpha \in [0, \pi/3]$. Therefore we obtain

Proposition 5.1. *Let φ be any isometric imbedding from $S^2 \times \mathbf{R}^4$ to* $\mathrm{Im}\,\mathfrak{C}$. *Then there exist a $g \in G_2$ and $\alpha \in [0, \pi/3]$ such that*

$$g \circ \varphi = \varphi_\alpha.$$

Hence the moduli space (up to the action of G_2) of isometric imbedddings from $S^2 \times \mathbf{R}^4$ to $\mathrm{Im}\,\mathfrak{C}$ coincide with $\{\varphi_\alpha \mid \alpha \in [0, \pi/3]\}$.

References

1. R.L. Bryant, *Submanifolds and special structures on the octonions*, J. Diff. Geom. 17 (1982) 185–232.
2. E. Calabi, *Construction and properties of some 6-dimensional almost complex manifolds*, Trans. A. M. S. 87 (1958) 407–438.
3. R. Harvey and H.B. Lawson, *Calibrated geometries*, Acta Math. 148 (1982) 47–157.
4. H. Hashimoto, T. Koda, K. Mashimo and K. Sekigawa, *Extrinsic homogeneous almost Hermitian 6-dimensional submanifolds in the octonions*, Kodai Math. J. 30 (2007) 297–321.
5. H. Hashimoto and K. Mashimo, *On some 3-dimensional CR submanifolds in* S^6, Nagoya Math. J. 156 (1999) 171–185.
6. H. Hashimoto, *Characteristic classes of oriented 6-dimensional submanifolds in the octonians*, Kodai Math. J. 16 (1993) 65–73.
7. H. Hashimoto, *Oriented 6-dimensional submanifolds in the octonions III*, Internat. J. Math and Math. Sci. 18 (1995) 111–120.
8. S. Kobayashi and K. Nomizu, *Foundations of Differential geometry II*, Wiley-Interscience, New York, 1968.

PERSISTENCE OF SOLUTIONS FOR SOME INTEGRABLE SHALLOW WATER EQUATIONS

D. HENRY

School of Mathematical Sciences, Dublin City University,
Dublin 9, Ireland
E-mail: david.henry@dcu.ie
www.dcu.ie

In the following article we present some recently derived results describing how a class of solutions of a family of nonlinear dispersive equations evolve over time. We describe the physical relevance of certain integrable members of this family, including applications in modelling shallow water waves over a flat bed and the propagation of waves in hyperelastic rods. The results contained in this article are of relevance to the soliton solutions of these equations.

Keywords: b-equations; Camassa-Holm; Degasperis-Procesi; Hunter-Saxton; Compact support; Exponential decay.

1. Introduction

The equations of motion for a perfect (homogeneous and inviscid) fluid in two dimensions are described by Euler's equation,

$$\begin{cases} u_t + uu_x + vu_y = -P_x, \\ v_t + uv_x + vv_y = -P_y - g, \end{cases}$$

where $(u(x,y,t), v(x,y,t))$ is the velocity field, $P(x,y,t)$ is the pressure function and g is the gravitation constant of acceleration. This equation, coupled with the equation of mass conservation and equations describing boundary conditions on the surface and at the bottom of the fluid domain, describe the full governing equations for the water wave problem. The equations of motion are highly intractible, due to nonlinearity, and this difficulty is compounded by the fact that the free-surface of the fluid is also an unknown element of the problem [26]. In general the full governing equations are not well understood. As a first step in tackling the water wave problem, we can study approximations to the full governing equations. Such approximations can be derived in a variety of ways, however most follow from the introduction of small shallowness or amplitude paramters. We assume the water motion is two-dimensional over a flat bed, where the water has a typical depth h_0— this is a reasonable assumption for water flowing in a channel, or for wave motion in the ocean. In the case where the scale of the amplitude is small compared to that of the water depth then the parameter $\epsilon = a/h_0$ is tiny, and where the scale of the wavelength is long compared to that of the water depth then the shallowness parameter $\delta = h_0/\lambda$ is very small. If we perform asymptotic expansions in terms of these parameters [24,25], then we find that, to the first order, the motion is described by the well-known Korteweg-deVries equation,

$$u_t - uu_x + u_{xxx} = 0$$

and to the next asymptotic order we obtain a nonlinear dispersive partial differential

$$u_t - u_{txx} + buu_x = bu_x u_{xx} + uu_{xxx} \tag{1}$$

where, depending on how we manipulate various coefficients, we can have $b = 2, 3$ [24].

2. The b-equations

The family of equations Eq. (1), parameterised by the real coefficient $b \in \mathbb{R}$, are commonly known as the b-equations, and have been the focus of much research in their own right. However, as mentioned above, there are two special cases of great interest.

The Camassa-Holm (CH) equation ($b = 2$) arises in a variety of different contexts. It has been widely studied since 1993 when Camassa and Holm [3] proposed it as a model for the unidirectional propagation of shallow water waves over a flat bed. In the shallow water setting $u(x, t)$ represents the horizontal velocity of the fluid motion at a certain depth over a flat bed in nondimensional variables [24,25]. The equation was originally derived in 1981 as a bi-Hamiltonian equation with infinitely many conservation laws by Fokas and Fuchssteiner [19]. It also models axially symmetric waves in hyperelastic rods [12].

The Degasperis-Procesi (DP) equation ($b = 3$) was originally derived by Degasperis and Procesi [14], and it was shown to be formally integrable by Degasperis, Holm and Hone [13] when they obtained a Lax pair and a bi-Hamiltonian structure for it. Concerning the existence of solutions to the DP equation, certain classical solutions exist for all times, whereas others blow up in finite time [17,28,32] — a situation which occurs for the CH equation [4,7] but not for KdV (where all classical solutions are global).

The solitary wave solutions of the DP and CH equations are solitons [2,11,13,16] — in fact they also admit a class of weak solutions which are peaked-solitons (peakons) [2,3,13]. A property which is unique to the DP equation is the existence of discontinuous shock wave and periodic shock wave solutions [18].

Geometrically, the KdV and CH equations are re-expressions of geodesic flow on the Bott-Virasoro group and the diffeomorphism group respectively [1,4,9,27]. However, the DP equation has not currently been afforded such a geometric interpretation.

The main reason the CH and DP equations have generated such interest is that the CH and DP equations are the only members of the b-equation family with a bi-Hamiltonian structure and which are integrable [24]. In what follows we present a summary of recently derived results which describe how solutions of Eq. (1) which initially have compact support, or solutions which initially have a weighted asymptotic exponential decay rate, retain these properties as they evolve over the duration of their existence.

If $p(x) = \frac{1}{2}e^{-|x|}$, $x \in \mathbb{R}$, then $(1 - \partial_x^2)^{-1} f = p * f$ for all $f \in L^2(\mathbb{R})$ and so $p * m = u$, where $*$ denotes convolution in the spatial variable. If we define the function

$$\mathcal{F}^b(x,t) := \frac{3-b}{2} u_x^2(x,t) + \frac{b}{2} u^2(x,t),$$

which depends on the parameter $b \in \mathbb{R}$, then we can re-express equation Eq. (1) as

$$u_t + uu_x + \partial_x p * \mathcal{F}^b = 0, \qquad x \in \mathbb{R},\ t \geq 0. \tag{2}$$

The space $H^s(\mathbb{R})$, is the space of functions f for which the following sum is finite,

$$\|f\|_s = \left(\sum_{n=-\infty}^{\infty} (1+n^2)^s |\hat{f}(n)|^2 \right)^{\frac{1}{2}} < \infty,$$

where $\hat{f}(n)$ denotes the Fourier series of f. The above expression for $\|f\|_s$ defines a norm on $\mathbb{H}^s(\mathbb{R})$. An application of Kato's semigroup theory to Eq. (2) ensures, for each fixed $b \in \mathbb{R}$, the local existence of solutions to Eq. (2) for $u_0 \in \mathbb{H}^s(\mathbb{R})$, where $s > 3/2$ [21]. We will regards these strong solutions of Eq. (2) to be weak solutions of Eq. (1).

2.1. *Compactly supported solutions*

Note that we can rewrite Eq. (1) as

$$m_t + bu_x m + m_x u = 0, \tag{3}$$

where $m = u - u_{xx}$ represents the momentum density. Compactly supported solutions of Eq. (3) represent localised perturbations, and it is of interest to see how such perturbations develop over time. The first result we present tells us that, for each fixed $b \in \mathbb{R}$, initially compactly supported classical solutions of Eq. (3) remain compactly supported throughout their evolution.

Proposition 2.1 ([20]). *Assume that $u_0 \in \mathbb{H}^4(\mathbb{R})$ is such that $m_0 = u_0 - u_{0,xx}$ has compact support. If $T = T(u_0) > 0$ is the maximal existence time of the unique solution $u(x,t)$ to Eq. (3) with initial data $u_0(x)$, then for any $t \in [0,T)$ the C^1 function $x \mapsto m(x,t)$ has compact support.*

Proof. We fix $b \in \mathbb{R}$, and associate to the solution m the family $\{\varphi(\cdot,t)\}_{t\in[0,T)}$ of increasing C^2 diffeomorphisms of the line defined by

$$\varphi_t(x,t) = u(\varphi(x,t),t), \quad t \in [0,T), \tag{4}$$

with

$$\varphi(x,0) = x, \quad x \in \mathbb{R}. \tag{5}$$

We find that the following identity holds:

$$m(\varphi(x,t),t)\cdot \varphi_x^b(x,t) = m(x,0), \quad x \in \mathbb{R},\ t \in [0,T). \tag{6}$$

Note that if we are dealing with the case $b = 0$, then Eq. (6) gives

$$m(\cdot,t) = m_0 \qquad x \in \mathbb{R},\ t \in [0,T),$$

and so $m(\cdot,t)$ is automatically compactly supported. For the rest of the cases, we infer from Eq. (4)–Eq. (5) that

$$\varphi_x(x,t) = \exp\left(\int_0^t u_x(\varphi(x,s),s)\,ds\right), \quad x \in \mathbb{R},\ t \in [0,T). \tag{7}$$

It follows that if m_0 is supported in the compact interval $[a,b]$, then since $\varphi_x(x,t) > 0$ on $\mathbb{R}\times[0,T)$ from Eq. (7), we can conclude from Eq. (6) that $m(\cdot,t)$ has its support in the interval $[\varphi(a,t),\varphi(b,t)]$. □

It is worth noting that the Hunter-Saxton (HS) equation,

$$u_{xxt} + 2u_x u_{xx} + uu_{xxx} = 0, \tag{8}$$

which is an integrable equation that arises in the theoretical study of nematic liquid crystals [23], arises from the CH equation in what is called the high frequency limit — essentially we end up with $m = -u_{xx}$ in Eq. (3). Following basic computations we see that the above Proposition also holds for the HS equation, and so any solution of the HS where m which is initially compactly supported, retains this property for all future times of existence. Since $m = -u_{xx}$, it follows that if u is a solution of the HS equation Eq. (8) with $u_0 = u(x,0)$ compactly supported, then u is compactly supported throughout its evolution. In general, the $b-$equations do not share this property, as we see from the following theorem.

Theorem 2.1 ([20]). *Fix $b \in [0,3]$ in Eq. (1). Assume that the function $u_0 \in \mathbb{H}^4(\mathbb{R})$ has compact support. Let $T > 0$ be the maximal existence time of the unique solution $u(x,t)$ to Eq. (1) with initial data $u_0(x)$. If at every $t \in [0,T)$ the C^2 function $x \mapsto u(x,t)$ has compact support, then u must be identically zero.*

A consequence of this result is that any solution to Eq. (1) which is initially compactly supported, for $b \in [0,3]$, instantly loses this property — to see this, apply the statement of the Theorem to u on the interval $[0,\epsilon]$, for any $\epsilon > 0$.

2.2. *Soliton-type solutions*

We now study the evolution of solutions of Eq. (1) which admit a larger class of initial data. First, we extend our definition of solutions to allow weak solutions of Eq. (1), that is, classical solutions of Eq. (2).

We adopt the following notation: we write $|f(x)| \sim O(g(x))$ as $x \to \infty$ if $\lim_{x\to\infty} |f(x)|/g(x) = C$, where C is a constant (allowed to be zero); we write $|f(x)| \sim o(g(x))$ as $x \to \infty$ if $\lim_{x\to\infty} |f(x)|/g(x) = 0$.

The next result provides an asymptotic description of how solutions of Eq. (2) propagate, if they initially decay asymptotically at a weighted exponential rate. Such a class of solutions trivially include the functions whose initial data is compactly supported, but more importantly, since the CH and DP are integrable, it includes soliton type solutions— solutions with rapid decay outside a localised region.

The solitons of the CH equation are weak solutions of the form $u(x,t) = ce^{-|x-ct|}$, where c is both the speed and the amplitude. This travelling wave is known as a peakon (peaked soliton) and it is smooth everywhere except at the cusp (where $x - ct = 0$), and here it is not differentiable. There are also multi-peakon solutions for the CH equation [2], which consist of a finite sum of peakons. The DP equation also admits multi-peakon solutions [30] as well as solutions which display even more exotic properties [29].

The evolution of such solutions is particularly interesting in the context of CH as an equation describing the propagation of waves in a rod. This equation acts as a model for nonlinear waves in cylindrical hyperelastic rods, with the solutions representing the radial stretch relative to a pre-stressed state [12]. Solitary waves in solids are interesting from the point of view of applications, as they are easy to detect because they do not change their shapes during propagation and can be used to determine material properties and to detect flaws.

We state the following result, which was obtained in Ref. 21, and describes the persistence of decay properties for the entire family of b–equations.

Theorem 2.2 ([21]). *Let $b \in \mathbb{R}$. For $s > 3/2$, let $T > 0$ be the maximal existence time of the strong solution $u \in C([0,T), \mathbb{H}^s(\mathbb{R}))$ to equation Eq. (2) with initial data $u_0 = u(x,0)$. Suppose there is a $\theta \in (0,1)$ such that both*

$$|u_0(x)| \quad \text{and} \quad |u_{0,x}(x)| \sim O(e^{-\theta|x|}) \quad \text{as} \quad |x| \to \infty. \tag{9}$$

Then both

$$|u(x,t)| \quad \text{and} \quad |u_x(x,t)| \sim O(e^{-\theta|x|}) \quad \text{as} \quad |x| \to \infty, \tag{10}$$

uniformly in the time interval $\mathcal{I}$, where

$$\mathcal{I} = \begin{cases} [0, T-\epsilon] & \text{for any} \quad \epsilon \in (0,T), \quad \text{if} \quad T < \infty, \\ [0, \mathcal{T}] & \text{for any} \quad \mathcal{T} > 0, \quad \text{if} \quad T = \infty. \end{cases} \tag{11}$$

The condition on the interval $\mathcal{I}$ is necessary to ensure uniformity — for solutions of the CH and DP equations which breakdown then wave-breaking occurs [4,7,28,32], that is, $\|u(.,t)\|_\infty$ is bounded for all $t \in [0,T]$ but

$$\lim_{t\to T} \sup \|u_x(t)\|_\infty = \infty. \tag{12}$$

The next Theorem sharpens the result of Theorem 2.1 by prescribing an asymptotic rate of decay which no non-trivial solution of *Eq.* (2) can satisfy at any more than one instant. Since a compactly supported solutions automatically satisfies this rate, therefore any non-trivial compactly supported solution to *Eq.* (1) must instantly lose its compact support.

Theorem 2.3 ([21]). *Let $b \in [0,3]$. For $s > 3/2$, let $T > 0$ be the maximal existence time of the strong solution $u \in C([0,T), \mathbb{H}^s(\mathbb{R}))$ to equation*

Eq. (2) with initial data $u_0 = u(x,0)$. *Suppose there is a* $\delta \in (1/2, 1)$ *such that*

$$|u_0(x)| \sim o(e^{-|x|}) \quad \text{and} \quad |u_{0,x}(x)| \sim O(e^{-\delta|x|}) \quad \text{as} \quad |x| \to \infty. \tag{13}$$

If there is a $t_1 \in (0,T)$ *such that*

$$|u(x,t_1)| \sim o(e^{-|x|}) \quad \text{as} \quad |x| \to \infty, \tag{14}$$

then $u \equiv 0$.

We saw in Theorem 2.1 that, for fixed $b \in [0,3]$, every nontrivial solution of Eq. (1) which is initially compactly supported must instantly lose this property. The following result provides a more explicit description of this process.

Theorem 2.4 ([21]). *For* $b \in \mathbb{R}$, *let* u *be a nontrivial solution of equation Eq. (2) with maximal time of existence* $T > 0$ *and* $u \in C([0,T), \mathbb{H}^s(\mathbb{R}))$ *for* $s > 5/2$.

(a) If the initial data $u_0(x) = u(x,0)$ *is initially compactly supported on* $[\alpha, \beta]$ *then for* $t \in [0,T)$ *we have*

$$u(x,t) = \begin{cases} 1/2E_+(t)e^{-x} & \text{for} \quad x > \varphi(\beta,t), \\ 1/2E_-(t)e^{x} & \text{for} \quad x < \varphi(\alpha,t), \end{cases} \tag{15}$$

where E_+ *and* E_- *are continuous nonvanishing functions with* $E_+(0) = E_-(0) = 0$; *and when* $b \in [0,3]$ *then* $E_+ > 0$ *and* $E_- < 0$ *with* E_+ *strictly increasing and* E_- *strictly decreasing for* $t \in (0,T)$.

(b) Suppose for some constant $\mu > 0$ *we have*

$$u_0, u_{0,x}, u_{0,xx} \sim O(e^{-(1+\mu)|x|}) \quad \text{as} \quad |x| \to \infty, \tag{16}$$

then for $t \in [0,T)$ *we have*

$$m(x,t) \sim O(e^{-(1+\mu)|x|}) \quad \text{as} \quad |x| \to \infty, \tag{17}$$

and

$$\lim_{x \to \pm\infty} e^{\pm x} u(x,t) = 1/2E_\pm(t). \tag{18}$$

Proof. Since $u = p * m$ let us re-express u in the form

$$u(x,t) = \frac{1}{2}e^{-x} \int_{-\infty}^{x} e^{y} m(y,t)dy + \frac{1}{2}e^{x} \int_{x}^{\infty} e^{-y} m(y,t)dy. \tag{19}$$

Notice Proposition 2.1 tells us that if u_0 is initially supported on the compact interval $[\alpha, \beta]$ then, for any $t \in [0, T)$, our function $m(\cdot, t)$ is supported on the compact interval $[\varphi(\alpha, t), \varphi(\beta, t)]$. To prove (a) we define

$$E_+(t) = \int_{\varphi(\alpha,t)}^{\varphi(\beta,t)} e^y m(y,t)\mathrm{d}y \text{ and } E_-(t) = \int_{\varphi(\alpha,t)}^{\varphi(\beta,t)} e^{-y} m(y,t)\mathrm{d}y, \tag{20}$$

with

$$\begin{aligned} u(x,t) &= p(x) * m(x,t) = \frac{1}{2}e^{-x}E_+(t), \qquad & x > \varphi(\beta,t), \\ u(x,t) &= p(x) * m(x,t) = \frac{1}{2}e^{x}E_-(t), \qquad & x < \varphi(\alpha,t). \end{aligned} \tag{21}$$

It follows from the relations in Eq. (21) that

$$\begin{aligned} u(x,t) = -u_x(x,t) = u_{xx}(x,t) &= \frac{1}{2}e^{-x}E_+(t), \qquad & x > \varphi(\beta,t), \\ u(x,t) = u_x(x,t) = u_{xx}(x,t) &= \frac{1}{2}e^{x}E_-(t), \qquad & x < \varphi(\alpha,t), \end{aligned}$$

and so $E_+(0) = E_-(0) = 0$ by definition of $\{\alpha, \beta\}$. Since $m(\varphi(\alpha,t)) = m(\varphi(\beta,t)) = 0$, for fixed t we have

$$\frac{\mathrm{d}E_+(t)}{\mathrm{d}t} = \int_{\varphi(\alpha,t)}^{\varphi(\beta,t)} e^y m_t(y,t)\mathrm{d}y = \int_{-\infty}^{\infty} e^y m_t(y,t)\mathrm{d}y. \tag{22}$$

If $b \in [0,3]$, then from Eq. (3) and integration by parts, and using the fact that both u and consequently m have compact support, we get

$$\begin{aligned} \frac{\mathrm{d}E_+(t)}{\mathrm{d}t} &= -b\int_{-\infty}^{\infty} e^y u_x m \mathrm{d}y - \int_{-\infty}^{\infty} e^y u m_x \mathrm{d}y \\ &= \int_{-\infty}^{\infty} e^y \left(\frac{b}{2}u^2 + \frac{3-b}{2}u_x^2\right)\mathrm{d}y > 0. \end{aligned}$$

Thus $E_+(t)$ is initially zero and strictly increasing for all $t \in [0, T)$. In a similar manner one can show that, when $b \in [0,3]$, E_- is also initially zero but strictly decreasing for $t \in [0, T)$. This proves part (a) of the theorem.

To prove (b) we write equation Eq. (3) in the form

$$m_t(x,t) + u(x,t)m_x(x,t) = -bu_x(x,t)m(x,t), \quad x \in \mathbb{R},\ t \in [0,T), \tag{23}$$

and proceeding as in Theorem 2.2 we find that

$$\sup_{t\in\mathcal{I}} \|m(t)e^{(1+\mu)|x|}\|_\infty \leq c(\mathcal{I})\|m(0)e^{(1+\mu)|x|}\|_\infty, \tag{24}$$

where $c(\mathcal{I})$ is a constant depending on the choice of time interval $\mathcal{I}$ for $\sup_{t\in\mathcal{I}}(\|u_x(\tau)\|_\infty + \|u(\tau)\|_\infty)$. We note that for any $\theta \in (0,1)$

$$u(t), u_x(t), u_{xx}(t) \sim O(e^{-\theta|x|}) \quad \text{as} \quad |x| \to \infty, \tag{25}$$

meaning that all of the integrals in the computations above are well-defined. The property Eq. (18) follows by an approach similar to the one performed before, taking into account Eq. (19) and defining $E_{\pm}$ as in Eq. (20) but with $\alpha = -\infty$, $\beta = \infty$. □

References

1. V. Arnold, B. Khesin, "Topological Methods in Hydrodynamics", Springer-Verlag, New York, 1998.
2. R. Beals, D.H. Sattinger, and J. Szmigielski, *Inverse Problems* **15**, 1–4 (1999).
3. R. Camassa and D. Holm, *Phys. Rev. Lett.* **71**, 1661–1664 (1993).
4. A. Constantin, *Ann. Inst. Fourier (Grenoble)*, **50**, 321–362 (2000).
5. A. Constantin, *Proc. Roy. Soc. London* **457**, 953–970 (2001).
6. A. Constantin, *J. Math. Phys.* **46**, P023506 (2005).
7. A. Constantin and J. Escher, *Acta Mathematica*, **181**, 229–243 (1998).
8. A. Constantin, V. Gerdjikov and R. Ivanov, *Inverse Problems* **22**, 2197–2207 (2006).
9. A. Constantin and B. Kolev, *Comment. Math. Helv.* **78**, 787–804 (2003).
10. A. Constantin and H. P. McKean, *Comm. Pure Appl. Math.* **52**, 949–982 (1999).
11. A. Constantin and W. Strauss, *Comm. Pure Appl. Math.* **53**, 603–610 (2000).
12. H.H. Dai, *Acta Mech.* **127**, 193-207 (1998).
13. A. Degasperis, D. Holm, and A. Hone, *Theoret. Math. Phys.* **133**, 1463–1474 (2002).
14. A. Degasperis and M. Procesi, in "Symmetry and Perturbation Theory", World Scientific, Singapore, 1999.
15. H.R. Dullin, G. Gottwald, D.D. Holm, *Fluid Dynam. Res.* **33** 73–95 (2003).
16. P. Drazin and R. S. Johnson, "Solitons — an Introduction", Cambridge University Press, Cambridge, 1989.
17. J. Escher, Y. Liu and Z. Yin, *J. Funct. Anal.* **241**, 457–485 (2006).
18. J. Escher, Y. Liu and Z. Yin, *Indiana Univ. Math. J.*, **56**, 87–117 (2007).
19. A.S. Fokas and B. Fuchssteiner, *Physica D* **4**, 47–66 (1981).
20. D. Henry, *Dyn. Contin. Discrete Impuls. Syst. A* **15**, 145–150 (2008).
21. D. Henry, *Nonlinear Anal.* **70**, 1565–1573 (2008).
22. A. Himonas, G. Misiolek, G. Ponce and Y. Zhou, *Comm. Math. Phys.* **271**, 511–522 (2007).
23. J. K. Hunter and R. Saxton, *SIAM J. Appl. Math.* **51**, 1498–1521 (1991).
24. R. Ivanov, *Phil. Trans. Roy. Soc. London* **365** 2267–2280 (2007).
25. R.S. Johnson, *J. Fluid Mech.* **455**, 63–82 (2002).
26. R.S. Johnson, *A Modern Introduction to the Mathematical Theory of Water Waves*, Cambridge Univ. Press, Cambridge, 1997.
27. S. Kouranbaeva, *J. Math. Phys.* **40**, 857–868 (1999).
28. Y. Liu, Z. Yin, *Comm. Math. Phys.* **267**, 801–820 (2006).
29. H. Lundmark, *J. Nonlinear Sci.* **17**, 169–198 (2007).
30. H. Lundmark and J. Szmigielski, *Inverse Problems* **19**, 1241–1245 (2003).

31. H.P. McKean, in "Global Analysis", Lecture Notes in Mathematics, Springer-Verlag, Berlin, 1979, pp. 83–200.
32. Z. Yin, *J. Nonl. Math. Phys.* **10**, 10–15 (2003).

SOME GEOMETRIC PROPERTIES AND OBJECTS RELATED TO BÉZIER CURVES

M. J. HRISTOV

"St. Cyril and St. Methodius" University of Veliko Tarnovo,
Faculty of Education, Department of Algebra and Geometry
E-mail: m.hristov@uni-vt.bg

Classical Bézier curves (the parabolic case) with the natural barycentric description are considered. The explicit expressions for lenght, curvature, shape etc. are given. It is introduced the notion trajectory of Bézier curve and by means of the projective isotomic conjugation the corresponding envelope conic surfaces in the space are obtained.

Keywords: Barycentric coordinates and orbits; Projective isotomic conjugation; Bézier curves; Curvature; Shape; Envelope.

1. Preliminaries

1.1. *Barycentric coordinates and barycentric trajectories*

Let α be fixed euclidean plane, $\triangle ABC$ be fixed triangle in λ and $\mathbb{E}_2^*(\alpha)$ be the extended euclidean plane — the basic model of the real projective plane $\mathbb{RP}_2$. The set of points (resp. lines) in $\mathbb{E}_2^*(\alpha)$ will be denoted by Σ (resp. Λ). For $G \in \Sigma$, Λ_G will stand for the set of lines, incident ($\mathbf{z}$) with G (known as star of lines with center G), for $g \in \Lambda$, Σ_g will stand for the set of points $\mathbf{z}\, g$ and U_g denotes the infinity (unproper) point of g. The infinity line in α is the set of all infinity points and will be denoted by ω.

For any point $P \in \Sigma$ and $p \in \mathbb{R}$ there exist unique $x, y, z \in \mathbb{R}$, so that

$$\begin{cases} p \cdot \overrightarrow{OP} = x \cdot \overrightarrow{OA} + y \cdot \overrightarrow{OB} + z \cdot \overrightarrow{OC} \\ x + y + z = p. \end{cases}$$

The ordered triple (x, y, z) is said to be triple of barycentric coordinates (b-coordinates) [4–6] of P with respect to basic Möbius triangle ABC and it will be written $P(x, y, z)$. The case $p = 0$ corresponds to $P \in \omega$ and $p \neq 0$ describes finite euclidean points. From projective point of view, b-coordinates of P are the affine coordinates with respect to space coordinate system

$$\mathcal{K} = \{O(\not\in\alpha), a_1(\|\overrightarrow{OA}), a_2(\|\overrightarrow{OB}), a_3(\|\overrightarrow{OC})\},$$

of a vector $p\|\overrightarrow{OP}$ and $\alpha : x + y + z = p$. That is why we do not use the terminology "homogeneous relative (absolute) b-coordinates" [4–6].

From algebraic point of view b-coordinates x, y, z of finite $P \in \Sigma$ are the unique solution of the system of linear equations (SLE)

$$\begin{cases} \lambda y + z = 0 \\ \mu x + (1-\lambda)y = 0, \\ (1-\mu)x = p \end{cases} \tag{1}$$

where $\lambda = (BCA') \stackrel{\text{def}}{=} \dfrac{\overline{BA'}}{\overline{CA'}}$ and $\mu = (AA'P)$ are the affine ratios, always $\neq 1$ for each finite points P and $A' = PA \cap BC$ (it is assumed) and $\lambda = 1 \iff A' = U_{BC}$. We rewrite (1) in the matrix form $A.X_1 = B$, where

$$A = \begin{pmatrix} 1 & \lambda & 0 \\ 0 & 1-\lambda & \mu \\ 0 & 0 & 1-\mu \end{pmatrix}, \qquad B = \begin{pmatrix} 0 \\ 0 \\ p \end{pmatrix}, \qquad X_1 = \begin{pmatrix} z_1 \\ y_1 \\ x_1 \end{pmatrix}.$$

Thus, in transposed form,

$$X_1^T = B^T (A^{-1})^T = \left(\frac{p\lambda\mu}{(1-\lambda)(1-\mu)}; -\frac{p\mu}{(1-\lambda)(1-\mu)}; \frac{p}{1-\mu} \right).$$

The triangle matrix A is of stochastic type by columns (left(L)-stochastic-like) and generates the affine transfomation φ_A in $\mathbb{E}_2^*(\alpha)$. The reason to use "-like" is that A is not non-negative. The set $\mathcal{A}$ of all such matrixes is multiplicative, non-commutative group. We express the commutator $[A_1, A_2]$ of such matrixes in the form

$$[A_1, A_2] = \begin{vmatrix} \lambda_1 & \lambda_2 \\ \mu_1 & \mu_2 \end{vmatrix} \cdot D \tag{2}$$

where

$$D = \begin{pmatrix} 0 & 0 & 1 \\ 0 & 0 & -1 \\ 0 & 0 & 0 \end{pmatrix}, \qquad A_i = \begin{pmatrix} 1 & \lambda_i & 0 \\ 0 & 1-\lambda_i & \mu_i \\ 0 & 0 & 1-\mu_i \end{pmatrix}.$$

Moreover $\mathcal{A}$ is Lie group. We call $\mathcal{A}$ *L-stochastic-like Lie group with basic Möbius triangle ABC.*

By the consideration of the powers of $A \in \mathcal{A}$, we get the following

Definition 1.1. The orbit of the point $P(x, y, z)$ under the action of the matrix $A \in \mathcal{A}$, i.e. the set of points $\{P_n(x_n, y_n, z_n) = \varphi_A^{1-n}(P),\ n \in \mathbb{N}\}$, such that (x_n, y_n, z_n) is the unique solution of SLE

$$A^n X_n = B, \qquad X_n = \begin{pmatrix} z_n \\ y_n \\ x_n \end{pmatrix}, \quad B = \begin{pmatrix} 0 \\ 0 \\ p \end{pmatrix}$$

is said to be *barycentric trajectory (b-trajectory)* of the point $P(\equiv P_1)$ with respect to basic $\triangle ABC$.

We get the explicit form of the b-trajectory in the form

$$\begin{aligned} z_n &= p + \frac{p}{\lambda - \mu}\left[\frac{\mu}{(1-\lambda)^n} - \frac{\lambda}{(1-\mu)^n}\right], \\ y_n &= \frac{p\mu}{\lambda - \mu}\left[\frac{1}{(1-\mu)^n} - \frac{1}{(1-\lambda)^n}\right], \\ x_n &= \frac{p}{(1-\mu)^n}. \end{aligned} \tag{3}$$

Any point $P_n(x_n, y_n, z_n) \in \alpha : x + y + z = p$ and determines

$$\lambda_n = (BCA_n) = -\frac{z_n}{y_n}, \qquad \mu_n = (AA_nP_n) = -\frac{y_n + z_n}{x_n}$$

where $A_n = AP_n \cap BC$. Thus, the position of P_n is determined by

$$\overrightarrow{CP_n} = \frac{1}{1-\mu_n}\overrightarrow{CA} - \frac{\mu_n}{1-\mu_n}\frac{1}{1-\lambda_n}\overrightarrow{CB}.$$

1.2. *Barycentric characterization of the duality in the projective plane. Projective isotomic conjugation*

Further we assume $P(x, y, z) : x + y + z = 1$, i.e. $\alpha : x + y + z = 1$.

In the following theorem we resume some well known facts from the analytic geometry of $\mathbb{E}_2^*(\alpha)$.

Theorem 1.1. *Let in the plane α of a basic $\triangle ABC$ be given in b-coordinates the points $A'(a_1, a_2, a_3)$, $B'(b_1, b_2, b_3)$ and $C'(c_1, c_2, c_3)$. Let*

$$\mathcal{M}(A'B'C') = \begin{pmatrix} a_1 & a_2 & a_3 \\ b_1 & b_2 & b_3 \\ c_1 & c_2 & c_3 \end{pmatrix} \qquad \textit{and} \qquad \Delta' = \det \mathcal{M}(A'B'C')$$

be the matrix and it's determinant, formed by the ordered triple (A', B', C'). Then:

(i) *A', B', C' are incident with a line* $\iff \Delta' = 0$;

(ii) *the ratio of the areas* $\dfrac{S_{A'B'C'}}{S_{ABC}} = |\Delta'|$;

(iii) *the barycentric equation of the line* $a' : \begin{cases} \mathsf{z}\ B'(b_1, b_2, b_3) \\ \mathsf{z}\ C'(c_1, c_2, c_3) \end{cases}$ *is*

$$a' : \begin{vmatrix} x & y & z \\ b_1 & b_2 & b_3 \\ c_1 & c_2 & c_3 \end{vmatrix} = 0 \iff a' : \mathcal{A}_{23}.x + \mathcal{A}_{31}.y + \mathcal{A}_{12}.z = 0.$$

The minors $\mathcal{A}_{ij}$ are called b-coordinates of the line a' and it is written $a'[\mathcal{A}_{23}, \mathcal{A}_{31}, \mathcal{A}_{12}]$;

(iv) *the b-coordinates of the intersect point $P = m \cap n$ of the lines $m[\mathcal{M}_1, \mathcal{M}_2, \mathcal{M}_3]$ and $n[\mathcal{N}_1, \mathcal{N}_2, \mathcal{N}_3]$ are $P\left(\frac{p_{23}}{\Delta}, \frac{p_{31}}{\Delta}, \frac{p_{12}}{\Delta}\right)$, where p_{ij} are the minors, formed with $\{i, j\}$-columns of the matrix*

$$\mathcal{P}(m, n) = \begin{pmatrix} \mathcal{M}_1 & \mathcal{M}_2 & \mathcal{M}_3 \\ \mathcal{N}_1 & \mathcal{N}_2 & \mathcal{N}_3 \end{pmatrix} \qquad \textit{and} \qquad \Delta = \begin{vmatrix} 1 & 1 & 1 \\ \mathcal{M}_1 & \mathcal{M}_2 & \mathcal{M}_3 \\ \mathcal{N}_1 & \mathcal{N}_2 & \mathcal{N}_3 \end{vmatrix};$$

(v) *the b-equation of a conic $\mathcal{C}$ is $\mathcal{C} : (x, y, z).A.(x, y, z)^T = 0$, where A is 3×3 real symmetric matrix. In particular, $\mathcal{C}$ passes through the vertexes of the basic triangle iff $a_{ii} = 0$, $i = 1, 2, 3$.*

There are several kinds of conjugate maps (isogonal, isotomic) with respect to fixed triangle, defined as colineations over $\mathbb{E}_2^*$, for example by means of point multiplication [6]. The theorems of Ceva and Menelaus give rise

Definition 1.2. *Projective ($\mathbb{RP}_2$-) isotomic conjugation* with respect to basic $\triangle ABC$ is the map

$$\begin{aligned} \pi : \Sigma \backslash \{\Sigma_{AB} \cup \Sigma_{BC} \cup \Sigma_{CA}\} &\longrightarrow \Lambda \backslash \{AB \cup BC \cup CA\} \\ P = AA' \cap BB' \cap CC' &\longmapsto \pi(P) = p\ \mathsf{z}\ A^*, B^*, C^*, \end{aligned}$$

where the points I', I^* are harmonically-conjugated to the pair (J, K), i.e. $I', I^*\ \mathsf{z}\ JK$ and $(JKI') = -(JKI^*)$ for $I, J, K \in \{A, B, C\}$. The Ceva point P is called *pole*, the Menelaus line $p = \pi(P)$ is called *polar line* and the map π is known also as *polarity* w.r.t. to $\triangle ABC$.

The image of the medicenter G is $\pi(G) = \omega$.

The following theorem is easy to check by using Theorem 1.1.

Theorem 1.2.

(i) *The* $\mathbb{RP}_2$*–isotomic conjugation is involutive corelation, i.e.* $\pi \neq \mathrm{id}$ *(the identity),* $\pi^2 = \mathrm{id}$ *(or* $\pi^{-1} \equiv \pi$*) and analytically in b-coordinates*

$$\Sigma \ni P(p_1, p_2, p_3) \overset{\pi}{\longmapsto} \pi(P) = p\,[p_2p_3,\, p_3p_1,\, p_1p_2\,] \in \Lambda.$$

(ii) *The sets of stars of lines in the plane of basic* $\triangle ABC$ *and the set of conics, incident with the vertexes* A, B, C *are equivalent in the sence of the* $\mathbb{RP}_2$*-isotomic conjugation.*

2. Bézier curves and related geometric properties and objects

Bézier curves, as basic tools for Computer Aided Geometric Design [1], are intensively studied by many autors. We recall the well known

Definition 2.1. Let $\triangle A_0A_1A_2$ be fixed triangle with vertex-vector-positions $a_i = \overrightarrow{OA_i}$, $i = 0, 1, 2$. The polynomial curve of second degree

$$\mathcal{C} : r(t) = (1-t)^2 a_0 + 2t(1-t)a_1 + t^2 a_2, \quad 0 \leq t \leq 1$$

is said to be *Bézier curve* of $2^{-\mathrm{d}}$ power with basic $\triangle A_0A_1A_2$. It is also in use the denotation $\mathcal{C}_{\{A_0A_1A_2\}}$.

Generally, $\mathcal{C}_{\{A_0A_1\ldots A_n\}} : r(t) = \displaystyle\sum_{k=0}^{n} \binom{n}{k} t^k (1-t)^{n-k}.a_k,\ t \in [0,1]$ is Bézier curve of $n^{-\mathrm{th}}$ power with basic symplex $A_0A_1 \ldots A_n$, formed as the linear combination of $a_k = \overrightarrow{OA_k}$ with coefficients the basic Bernstein's polynomials $B_{k,n} = \binom{n}{k} t^k (1-t)^{n-k}$, $k = 0, 1, \ldots, n$. Point A_0 is called *initial*, A_n — *end-point*, all the rest A_k — *control points.* Any Bézier curve is invariant under the action of any affine transformation.

We use the following matrix forms for $\mathcal{C}_{\{A_0A_1A_2\}}$

$$\mathcal{C}_{\{A_0A_1A_2\}} : r(t) = (1,\, t,\, t^2) \cdot \begin{pmatrix} \overrightarrow{OA_0} \\ 2\overrightarrow{A_0A_1} \\ \overrightarrow{A_1A_0} + \overrightarrow{A_1A_2} \end{pmatrix} = (1,\, t,\, t^2) \cdot \begin{pmatrix} \overrightarrow{OA_0} \\ 2\overrightarrow{A_0A_1} \\ 2\overrightarrow{A_1M_1} \end{pmatrix},$$

where M_1 is the midpoint of A_0A_2, from which it's easy to calculate the derivatives $\dot{r}(t) = \frac{dr(t)}{dt}$ ets. and to proove the well known properties, used in the De'Casteljau algorithm:

Lemma 2.1. *Arbitrary Bézier curve* $\mathcal{C}_{\{A_0A_1A_2\}}$ *passes through* $A_0(t = 0)$ *and* $A_2(t = 1)$*, tangent to* $\overrightarrow{A_0A_1} \| \dot{r}(0)$*,* $\overrightarrow{A_1A_2} \| \dot{r}(1)$*, with curvatures at the* *initial point A_0 and the end-point A_2*

$$\varkappa(0) = \frac{S_{\triangle A_0A_1A_2}}{|\overrightarrow{A_0A_1}|^3}, \qquad \varkappa(1) = \frac{S_{\triangle A_0A_1A_2}}{|\overrightarrow{A_1A_2}|^3},$$

respectively, where $S_{\triangle A_0A_1A_2}$ is the area of $\triangle A_0A_1A_2$. Moreover $\mathcal{C}_{\{A_0A_1A_2\}}$ intersects the median A_1M_1 at the midpoint $M(t=\frac{1}{2})$ and is tangent at M to the midsegment: $\dot{r}(\frac{1}{2})\|A_0A_2$.

From barycentric point of view any Bézier curve $\mathcal{C}_{\{A_0A_1A_2\}}$ is the set of points P with b-coordinates with respect to basic $\triangle A_0A_1A_2$ — the basic Bernstein polinomials of second degree $P\Big((1-t)^2,\, 2t(1-t),\, t^2\Big)$, $t \in [0,1]$.

Lemma 2.2.

(i) *Let $P\Big((1-t)^2,\, 2t(1-t),\, t^2\Big)$, $t \in [0,1]$ be arbitrary point of the Bézier curve $\mathcal{C}_{\{A_0A_1A_2\}}$. The b-coordinates of the points: $B_0 = A_0P \cap A_1A_2$, $B_1 = A_1P \cap A_0A_2$, $B_2 = A_2P \cap A_1A_0$ are*

$$B_0\left(0;\, \frac{2(1-t)}{2-t};\, \frac{t}{2-t}\right), \qquad B_1\left(\frac{(1-t)^2}{2t^2-2t+1};\, 0;\, \frac{t^2}{2t^2-2t+1}\right),$$

$$B_2\left(\frac{1-t}{1+t};\, \frac{2t}{1+t};\, 0\right).$$

(ii) *Any Bézier curve $\mathcal{C}_{\{A_0A_1A_2\}}$ generates the functional matrix $A(t)$ from the L-stochastic-like Lie group $\mathcal{A}$ with*

$$\lambda(t) = \frac{t}{2(t-1)} \qquad \text{and} \qquad \mu(t) = \frac{t(t-2)}{(t-1)^2}, \qquad t \in [0,\, 1).$$

Proof. The statement (i) follows by applying Theorem 1.1. and (ii) — by calculating the affine ratios $\lambda(t) = (A_2A_1B_0)$ and $\mu(t) = (A_0B_0P)$. □

For $\triangle A_0A_1A_2$ we denote: $l_i = |\overrightarrow{A_jA_k}|$, $m_i = |\overrightarrow{A_iM_i}|$, M_i — the midpoints of A_jA_k for $i \neq j \neq k \in \{0,1,2\}$, $\varphi = \measuredangle A_0A_1A_2$ and $\psi = \measuredangle A_0A_1M_1$.

Theorem 2.1. *Let $\mathcal{C}_{\{A_0A_1A_2\}}$ be Bézier curve.*

(i) *For any $P \in \mathcal{C}_{\{A_0A_1A_2\}}$ the ratio of the areas*

$$s(t) = \frac{S_{\triangle B_0B_1B_2}}{S_{\triangle A_0A_1A_2}} = \frac{4t^2(1-t)^2}{(2-t)(1+t)(2t^2-2t+1)} \in \left[0;\, \frac{2}{9}\right],$$

$\max\limits_{[0,1]} s = s(\frac{1}{2}) = \frac{2}{9} \iff P \equiv M(\frac{1}{4};\, \frac{1}{2};\, \frac{1}{4})$ *— the midpoint of the median A_1M_1 for $\triangle A_0A_1A_2$;*

(ii) *The length of the curve is*

$$l_C = \frac{2m_1 - l_2\cos\psi}{2m_1}\sqrt{(2m_1 - l_2\cos\psi)^2 + l_2^2\sin^2\psi} + \frac{l_2^2\cos\psi}{2m_1}$$
$$+\frac{l_2^2\sin^2\psi}{2m_1}\ln\frac{\sqrt{(2m_1 - l_2\cos\psi)^2 + l_2^2\sin^2\psi} + 2m_1 - l_2\cos\psi}{l_2(1-\cos\psi)}.$$

Proof. The statement (i) follows from Theorem 1.1.-(ii) and standard obtaining of the $\max s(t)$ over $[0,1]$. For (ii) we apply the lenght-formula

$$l_C = \int_0^1 |\dot{r}(t)|dt \qquad \text{for} \qquad |\dot{r}(t)| = 2\sqrt{(1,2t).\Gamma(\overrightarrow{A_0A_1}\overrightarrow{A_1M_1}).\begin{pmatrix}1\\2t\end{pmatrix}}$$

where $\Gamma(\overrightarrow{A_0A_1}\overrightarrow{A_1M_1})$ is the Gramm-matrix for the pair $\{\overrightarrow{A_0A_1}, \overrightarrow{A_1M_1}\}$. □

Theorem 2.2. *The curvature $\varkappa(t)$ of any Bézier curve $\mathcal{C}_{\{A_0A_1A_2\}}$ is*

$$\varkappa(t) = S_{\triangle A_0A_1A_2}.[g(t)]^{-3/2},$$

where the function $g(t)$, defined over $[0,1]$, is of the form

$$g(t) = |(t-1)\overrightarrow{A_1A_0} + t\overrightarrow{A_1A_2}|^2 = (t-1\,;t)\begin{pmatrix} l_2^2 & l_0l_2\cos\varphi \\ l_0l_2\cos\varphi & l_0^2\end{pmatrix}\begin{pmatrix}t-1\\t\end{pmatrix}.$$

Proof. It follows by direct calculations, analogous to Theorem 2.1.-(ii), over the well known formula $\varkappa(t) = \dfrac{|\dot{r}\times\ddot{r}|}{|\dot{r}|^3}$. □

Theorem 2.2. implies immediately the well known

Corollary 2.1. *The Bézier curve $\mathcal{C}_{\{A_0A_1A_2\}}$ is a parabola with vertex at point K, such that $\overrightarrow{OK} = r(t_0)$, where*

$$t_0 = \frac{l_2^2 + l_0l_2\cos\varphi}{l_2^2 + 2l_0l_2\cos\varphi + l_0^2}$$

is the unique (global) maximum for $\varkappa(t)$: $\varkappa_{|K} = \varkappa(t_0) = \max\limits_{[0,1]}\varkappa(t)$.

Moreover, the result of Theorem 2.2. gives additional geometric information of the shape of Bézier curve $\mathcal{C}_{\{A_0A_1A_2\}}$, considered and discussed in [2] and generalized in [3] for cubic Bézier curve.

By cosidering the action of $\mathbb{RP}_2$-isotomic conjugation over each point of $\mathcal{C}_{\{A_0A_1A_2\}}\backslash\{A_0, A_2\}$, jointly with Lemma 2.2. (i), we get

Theorem 2.3. *The $\mathbb{RP}_2$-isotomic-conjugated image of $\mathcal{C}_{\{A_0A_1A_2\}}$ is*

$$\pi(\mathcal{C}_{\{A_0A_1A_2\}}) = \left\{p\,[2t^2;\, t(1-t);\, 2(1-t)^2] \,:\, t \in (0,1)\right\},$$

i.e. is the 1-parameter set of polar lines

$$p = \pi(P) : 2t^2 \cdot x + t(1-t) \cdot y + 2(1-t)^2 \cdot z = 0, \quad t \in (0,1),\ u, v \in \mathbb{R}.$$

The lines, corresponding to $t = \frac{1}{3}, \frac{1}{2}, \frac{2}{3}$, are the polar lines of the intersecting points of $\mathcal{C}_{\{A_0A_1A_2\}}$ with the medians through A_2, A_1, A_0 and are parallel to A_1A_0, A_0A_2, A_2A_1, respectively.

From projective point of view w.r. to the space coordinate system $\mathcal{K} = \{O, a_1, a_2, a_3\}$ the obtained 1-parameter family of lines p, give rise 1-parameter family of the planes $\sigma : \begin{cases} \text{z}\ O \\ \text{z}\ p \end{cases}$. The envelope of the last one is conic surface $\mathcal{S}$ with vertex O and generator $\mathcal{E}_{\pi(\mathcal{C})} = \mathcal{S} \cap \alpha : x+y+z = 1$ — the envelope of $\pi(\mathcal{C}_{\{A_0A_1A_2\}})$. More precisely, after standard calculations, we get the following

Theorem 2.4. *The envelope $\mathcal{S}$ of the set of planes $\sigma : \begin{cases} \text{z}\ O \\ \text{z}\ \pi(\mathcal{C}_{\{A_0A_1A_2\}}) \end{cases}$ with respect to $\mathcal{K} = \{O, a_1, a_2, a_3\}$ is the real hyperbolic conic surface*

$$\mathcal{S} : 16xz - y^2 = 0$$

with generator in the basic plane $\alpha : x + y + z = 1$, the envelope

$$\mathcal{E}_{\pi(\mathcal{C})} : \left(x = \frac{(1-t)^2}{6t^2 - 6t + 1}; y = \frac{4t(t-1)}{6t^2 - 6t + 1}; z = \frac{t^2}{6t^2 - 6t + 1} \right)$$

(in b-coordinates) and $t \in (0,\, 1) \backslash \{\frac{1}{2} \pm \frac{1}{2\sqrt{3}}\}$.

Proof. From Theorem 2.3. the b-equation $F(x,y,z,t) = 0$, where $F(x,y,z,t) = 2t^2.x + t(1-t).y + 2(1-t)^2.z$ for $t \in (0,1)$ describes the 1-parameter set of polar lines. Then the envelope is obtained by the solutions of

$$\mathcal{E}_{\pi(\mathcal{C})} : \begin{cases} F(x,y,z,t) = 0 \\ \dfrac{\partial F}{\partial t}(x,y,z,t) = 0 \\ x + y + z = 1 \end{cases} \iff \begin{pmatrix} 2t^2 & t(1-t) & 2(1-t)^2 \\ 4t & 1-2t & 4(t-1) \\ 1 & 1 & 1 \end{pmatrix} \cdot \begin{pmatrix} x \\ y \\ z \end{pmatrix} = \begin{pmatrix} 0 \\ 0 \\ 1 \end{pmatrix}.$$

By solving the SLE with respect to x, y, z we proof the theorem. □

Analogous to Lemma 2.2, we get

Lemma 2.3. *The envelope $\mathcal{E}_{\pi(C)}$ of $\mathbb{RP}_2$-isotomic image of any Bézier curve $\mathcal{C}_{\{A_0A_1A_2\}}$ generates the functional L-stochastic-like matrix $A_\varepsilon(t)$ with*

$$\lambda(t) = \frac{t}{4(1-t)}, \qquad \mu(t) = \frac{t(4-5t)}{(t-1)^2}, \qquad t \in (0,1) : t \neq \frac{4}{5}, \frac{1}{2} \pm \frac{1}{2\sqrt{3}}.$$

By applying (2) to Lemma 2.2 (ii) and Lemma 2.3, we calculate the commutators and obtaine

Lemma 2.4. *Let $A(t)$ and $A_\varepsilon(t)$ be the functional matrices in $\mathcal{A}$ of $\mathcal{C}_{\{A_0A_1A_2\}}$ and $\mathcal{E}_{\pi(C)}$ respectively. For $t_1 \neq t_2$ in $(0,1)$, the following relations for the commutators are valid*

$$[A(t_1), A(t_2)] = -2[A_\varepsilon(t_1), A_\varepsilon(t_2)] = \frac{t_1t_2(t_2-t_1)}{2(1-t_1)^2(1-t_2)^2}.D.$$

By the notion *b-trajectory* of a point we consider *b-trajectories* of the Bézier curve $\mathcal{C}_{\{A_0A_1A_2\}}$ as the family of lines $\mathcal{C}^{(n)}_{\{A_0A_1A_2\}}$ formed by the nth–b-image of any point of $\mathcal{C}_{\{A_0A_1A_2\}}$, $\forall\, n \in \mathbb{N}$.

Theorem 2.5. *The b-trajectories of the Bézier curve $\mathcal{C}_{\{A_0A_1A_2\}}$ are the elements of the sequence of curves*

$$\mathcal{C}^{(n)}_{\{A_0A_1A_2\}} : r_n(t) = x_n(t).a_0 + y_n(t).a_1 + z_n(t).a_2, \quad t \in [0,1],$$

with

$$(x_n(t); y_n(t); z_n(t)) = \left((t-1)^{2n};\ 2f(t)\begin{vmatrix} (t-1)^n & 2^n \\ 1 & (t-2)^n \end{vmatrix};\ 1 - f(t)\begin{vmatrix} (t-1)^{n+1} & 2^{n+1} \\ 1 & (t-2)^{n-1} \end{vmatrix} \right) \tag{4}$$

and where $f(t) = \dfrac{(1-t)^n}{(t-3)(2-t)^{n-1}}$.

(i) *For arbitrary $n \in \mathbb{N}$, $r_n(0) = a_0$, i.e. $A_0 \in \mathcal{C}^{(n)}_{\{A_0A_1A_2\}}$, $r_n(1) = a_2$ and $A_2 \in \mathcal{C}^{(n)}_{\{A_0A_1A_2\}}$.*

(ii) *The tangent vectors to each $\mathcal{C}^{(n)}_{A_0A_1A_2}$ at A_0 and A_2 are*

$$\dot{r}_n(0) = 2n \cdot (a_1 - a_0) = 2n \cdot \overrightarrow{A_0A_1}, \qquad \forall n \in \mathbb{N},$$

$$\dot{r}_n(1) = \dot{f}(1).2^{n+1}(a_2 - a_1) = \begin{cases} 0, & \text{for } n \geq 2 \\ 2(a_2 - a_1) = 2.\overrightarrow{A_1A_2}, & \text{for } n = 1. \end{cases}$$

(iii) *The envelope* $\mathcal{E}_n$ *of* $\pi(\mathcal{C}^{(n)}_{A_0A_1A_2})$, *in b-coordinates, is*

$$\mathcal{E}_n : \begin{cases} x = \Delta_1.\Delta^{-1} \\ y = \Delta_2.\Delta^{-1}, \\ z = \Delta_3.\Delta^{-1} \end{cases}$$

for all (t, n) *in order to have well defined*

$$\Delta_1 = \frac{-1}{y_n^2}\left(\frac{y_n}{z_n}\right)^{\bullet}, \quad \Delta_2 = \frac{-1}{z_n^2}\left(\frac{z_n}{x_n}\right)^{\bullet}, \quad \Delta_3 = \frac{-1}{x_n^2}\left(\frac{x_n}{y_n}\right)^{\bullet},$$

and $\Delta = \sum_{i=1}^{3} \Delta_i$.

Proof. By applying Lemma 2.2.(ii) jointly with (3) we get (4) and calculate (i). We express the derivatives $\dot{x}_n$, $\dot{y}_n$, $\dot{z}_n$ of (4) are as follows:

$$\dot{x}_n = 2n(t-1)^{2n-1},$$

$$\dot{y}_n = \frac{2f(t)}{(t-1)(t-2)(3-t)} \begin{vmatrix} (t-1)^n(t-2)^n & 2^n \\ t-1+n(t-3) & t-1+2n(t-2)(3-t) \end{vmatrix},$$

$$\dot{z}_n = \frac{2f(t)}{(t-1)(t-2)(3-t)} \begin{vmatrix} (t-1)^{n+1}(t-2)^{n-1} & -2^n \\ t-1+n(t-3) & (3-t)[1-n(t-2)]-1 \end{vmatrix}.$$

Thus we calculate (ii).

(iii) The $\mathbb{RP}_2-$ isotomic conjugated image of each $\mathcal{C}^{(n)}_{\{A_0A_1A_2\}}$ is the 1-parameter family of lines

$$\pi(\mathcal{C}^{(n)}_{\{A_0A_1A_2\}}) = p_n : \underbrace{\frac{1}{x_n(t)} \cdot x + \frac{1}{y_n(t)} \cdot y + \frac{1}{z_n(t)} \cdot z}_{=F_n(x,y,z,t)} = 0.$$

The envelope $\mathcal{E}_n$ of $\pi(\mathcal{C}^{(n)}_{\{A_0A_1A_2\}})$ is obtained by the solutions of the system

$$\mathcal{E}_n : \begin{cases} x + y + z = 1 \\ F_n(x, y, z, t) = 0 \\ \dfrac{\partial F_n}{\partial t}(x, y, z, t) = 0 \end{cases} \iff \mathcal{E}_n : \begin{cases} x + y + z = 1 \\ x_n^{-1} \cdot x + y_n^{-1} \cdot y + z_n^{-1} \cdot z = 0 \\ \dot{x}_n x_n^{-2} \cdot x + \dot{y}_n y_n^{-2} \cdot y + \dot{z}_n z_n^{-2} \cdot z = 0. \end{cases}$$

By applying the Cramer's rule for all t and n, each determinant to be well defined, we get the evelope $\mathcal{E}_n$ in b-coordinates. □

By (iii) of Theorem 2.5 in the space arises the sequence of conic surfaces $\mathcal{S}_n$ with vertex O and generators $\mathcal{E}_n$ — the envelopes of $\pi(\mathcal{C}^{(n)}_{\{A_0A_1A_2\}})$.

Acknowledgements

A support by Scientific researches fund of "St. Cyril and St. Methodius" University of V. Tarnovo under contract RD-491-08/2008 is gratefully acknowledged.

References

1. G. Farin, *Curves and Surfaces for Computer Aided Geometric Design*, 4th ed., Academic Press, London, 1997.
2. G. Georgiev, *Curve and Surface Design: Avignon 2006*, pp. 143, Nashboro Press, Brentwood, TN, 2007.
3. G. Georgiev, *Pure and Appl. Diff. Geom. PADGE 2007*, pp. 98, Shaker Verlag, Aachen, 2007.
4. K.R.S. Sastry, *Forum Geometricorum* **4**, 73 (2004).
5. V. Volenec, *Mathematical Communications* **8**, 55 (2003).
6. P. Yiu, *Int. J. Math. Educ. Sci. Technol.* **31**, 569 (2000).

HEISENBERG RELATIONS IN THE GENERAL CASE

BOZHIDAR Z. ILIEV

Institute for Nuclear Research and Nuclear Energy
Bulgarian Academy of Sciences
Boul. Tzarigradsko chaussée 72, 1784 Sofia, Bulgaria
E-mail: bozho@inrne.bas.bg

The Heisenberg relations are derived in a quite general setting when the field transformations are induced by three representations of a given group. They are considered also in the fibre bundle approach. The results are illustrated in a case of transformations induced by the Poincaré group.

Keywords: Heisenberg relations; Heisenberg equations; Quantum field theory.

1. Introduction

As Heisenberg relations or equations in quantum field theory are known a kind of commutation relations between the field operators and the generators (of a representation) of a group acting on system's Hilbert space of states. Their (global) origin is in equations like

$$\varphi_i'(r) = U \circ \varphi_i(r) \circ U^{-1}, \tag{1}$$

which connect the components φ_i and φ_i' of a quantum field φ with respect to two frames of reference. Here U is an operator acting on the state vectors of the quantum system considered and it is expected that the transformed field operators φ_i' can be expressed explicitly by means of φ_i via some equations. If the elements U (of the representation) of the group are labeled by $b = (b^1, \dots, b^s) \in \mathbb{K}^s$ for some $s \in \mathbb{N}$ (we are dealing, in fact, with a Lie group), i.e. we may write $U(b)$ for U, then the corresponding Heisenberg relations are obtained from Eq. (1) with $U(b)$ for U by differentiating it with respect to b^ω, $\omega = 1, \dots, s$, and then setting $b = b_0$, where $b_0 \in \mathbb{K}^s$ is such that $U(b_0)$ is the identity element.

The above shows that the Heisenberg relations are from pure geometric-group-theoretical origin and the only physics in them is the motivation leading to equations like Eq. (1). However, there are strong evidences that to the Heisenberg relations can be given dynamical/physical sense by identifying/replacing in them the generators (of the representation) of the group by the corresponding operators of conserved physical quantities if the system considered is invariant with respect to this group (see, e.g. the discussion in [1, §68]).

In sections 2-4, we consider Heisenberg relations in the non-bundle approach. At first (section 2), we derive the Heisenberg relation connected with the Poincaré group. Then (section 3) the Heisenberg relations arising from internal transformation, which are related with conserved charges, are investigated. At last, in section 4 are considered the Heisenberg relations in the most general case, when three representations of a group are involved. In section 5 are investigated the Heisenberg relation on the ground of fibre bundles. Section 6 closes the paper.

2. The Poincaré group

Suppose we study a quantum field with components φ_i relative to two reference frames connected by a general Poincaré transformation

$$u'(x) = \Lambda u(x) + a. \tag{2}$$

Here x is a point in the Minkowski spacetime M, u and u' are the coordinate homeomorphisms of some local charts in M, Λ is a Lorentz transformation (i.e. a matrix of a 4-rotation), and $a \in \mathbb{R}^4$ is fixed and represents the components of a 4-vector translation. The "global" version of the Heisenberg relations is expressed by the equation

$$U(\Lambda, a) \circ \varphi_i(x) \circ U^{-1}(\Lambda, a) = D_i^j(\Lambda, a)\varphi_j(\Lambda x + a), \tag{3}$$

where U (resp. D) is a representation of the Poincaré group on the space of state vectors (resp. on the space of field operators), $U(\Lambda, a)$ (resp. $\boldsymbol{D}(\Lambda, a) = [D_i^j(\Lambda, a)]$) is the mapping (resp. the matrix of the mapping) corresponding via U (resp. D) to Eq. (2). Note that here we have rigorously to write $\varphi_{u,i} := \varphi_i \circ u^{-1}$ for φ_i, i.e. we have omitted the index u. Besides, the point $x \in M$ is identified with $\boldsymbol{x} = u(x) \in \mathbb{R}^4$. Since for $\Lambda = \mathbf{1}$ and $a = \mathbf{0} \in \mathbb{R}^4$ is fulfilled $u'(x) = u(x)$, we have $U(\mathbf{1}, \mathbf{0}) = \mathsf{id}$, $\boldsymbol{D}(\mathbf{1}, \mathbf{0}) = \mathbf{1}$, where id is the corresponding identity mapping and $\mathbf{1}$ stands for the corresponding identity matrix. Let $\Lambda = [\Lambda^\mu{}_\nu]$, $\Lambda^{\mu\nu} := \eta^{\nu\lambda}\Lambda^\mu{}_\lambda$, with $\eta^{\mu\nu}$ being the components of the Lorentzian metric with signature $(-+++)$, and define

$$T_\mu := \frac{\partial U(\Lambda, a)}{\partial a^\mu}\Big|_{(\Lambda,a)=(\mathbf{1},\mathbf{0})}, \qquad S_{\mu\nu} := \frac{\partial U(\Lambda, a)}{\partial \Lambda^{\mu\nu}}\Big|_{(\Lambda,a)=(\mathbf{1},\mathbf{0})}, \tag{4a}$$

$$H^i_{j\mu} := \left.\frac{\partial D^i_j(\Lambda, a)}{\partial a^\mu}\right|_{(\Lambda,a)=(\mathbf{1},\mathbf{0})}, \qquad I^i_{j\mu\nu} := \left.\frac{\partial D^i_j(\Lambda, a)}{\partial \Lambda^{\mu\nu}}\right|_{(\Lambda,a)=(\mathbf{1},\mathbf{0})}. \tag{4b}$$

The particular form of the numbers $I^i_{j\mu\nu}$ depends on the field under consideration. In particular, we have $I^1_{1\mu\nu} = 0$ for spin-0 (scalar) field and $I^\sigma_{\rho\mu\nu} = \delta^\sigma_\mu \eta_{\nu\rho} - \delta^\sigma_\nu \eta_{\mu\rho}$ for spin-1 (vector) field.

Differentiating Eq. (3) relative to a^μ and setting after that $(\Lambda, a) = (\mathbf{1}, \mathbf{0})$, we find

$$[T_\mu, \varphi_i(x)]_- = \partial_\mu \varphi_i(x) + H^j_{i\mu} \varphi_j(x), \tag{5}$$

where $[A, B]_- := AB - BA$ is the commutator of some operators or matrices A and B. Since the field theories considered at the time being are invariant relative to spacetime translation of the coordinates, i.e. with respect to $\boldsymbol{x} \mapsto \boldsymbol{x} + a$, further we shall suppose that $H^i_{j\mu} = 0$. In this case equation Eq. (5) reduces to

$$[T_\mu, \varphi_i(x)]_- = \partial_\mu \varphi_i(x). \tag{6a}$$

Similarly, differentiation Eq. (3) with respect to $\Lambda^{\mu\nu}$ and putting after that $(\Lambda, a) = (\mathbf{1}, \mathbf{0})$, we obtain

$$[S_{\mu\nu}, \varphi_i(x)]_- = x_\mu \partial_\nu \varphi_i(x) - x_\nu \partial_\mu \varphi_i(x) + I^j_{i\mu\nu} \varphi_j(x) \tag{6b}$$

where $x_\mu := \eta_{\mu\nu} x^\nu$. The equations Eq. (6) are identical up to notation with [1, eqs.(11.70) and (11.73)]. Note that for complete correctness one should write $\varphi_{u,i}(\boldsymbol{x})$ instead of $\varphi_i(x)$ in Eq. (6), but we do not do this to keep our results near to the ones accepted in the physical literature [2–4].

As we have mentioned earlier, the particular Heisenberg relations Eq. (6) are from pure geometrical-group-theoretical origin. The following heuristic remark can give a dynamical sense to them. Recalling that the translation (resp. rotation) invariance of a (Lagrangian) field theory results in the conservation of system's momentum (resp. angular momentum) operator P_μ (resp. $M_{\mu\nu}$) and the correspondences

$$\mathrm{i}\hbar T_\mu \mapsto P_\mu \qquad \mathrm{i}\hbar S_{\mu\nu} \mapsto M_{\mu\nu}, \tag{7}$$

with $\hbar$ being the Planck's constant (divided by 2π), one may suppose the validity of the Heisenberg relations

$$[P_\mu, \varphi_i(x)]_- = \mathrm{i}\hbar \partial_\mu \varphi_i(x), \tag{8a}$$

$$[M_{\mu\nu}, \varphi_i(x)]_- = \mathrm{i}\hbar \{x_\mu \partial_\nu \varphi_i(x) - x_\nu \partial_\mu \varphi_i(x) + I^j_{i\mu\nu} \varphi_j(x)\}. \tag{8b}$$

However, one should be careful when applying the last two equations in the Lagrangian formalism as they are external to it and need a particular proof in this approach; e.g. they hold in the free field theory [3,5], but a general proof seems to be missing. In the axiomatic quantum field theory [2,6,7] these equations are identically valid as in it the generators of the translations (rotations) are identified up to a constant factor with the components of the (angular) momentum operator, $P_\mu = \mathrm{i}\hbar T_\mu$ ($M_{\mu\nu} = \mathrm{i}\hbar S_{\mu\nu}$).

3. Internal transformations

In our context, an internal transformation is a change of the reference frame $(u, \{e^i\})$, consisting of a local coordinate system u and a frame $\{e^i\}$ in some vector space V, such that the spacetime coordinates remain unchanged. We suppose that $e^i : x \in M \mapsto e^i(x) \in V$, where M is the Minkowski spacetime and the quantum field φ considered takes values in V, i.e. $\varphi: x \in M \mapsto \varphi(x) = \varphi_i(x)e^i(x) \in V$

Let G be a group whose elements g_b are labeled by $b \in \mathbb{K}^s$ for some $s \in \mathbb{N}$. [a] Consider two reference frames $(u, \{e^i\})$ and $(u', \{e'^i\})$, with $u' = u$ and $\{e^i\}$ and $\{e'^i\}$ being connected via a matrix $I^{-1}(b)$, where $I\colon G \mapsto \mathrm{GL}(\dim V, \mathbb{K})$ is a matrix representation of G and $I\colon G \ni g_b \mapsto I(b) \in \mathrm{GL}(\dim V, \mathbb{K})$. The components of the fields, known as field operators, transform into (cf. Eq. (1))

$$\varphi'_{u,i}(r) = U(b) \circ \varphi_{u,i}(r) \circ U^{-1}(b) \tag{9}$$

where U is a representation of G on the Hilbert space of state vectors and $U\colon G \ni g_b \mapsto U(b)$. Now the analogue of Eq. (3) reads

$$U(b) \circ \varphi_{u,i}(r) \circ U^{-1}(b) = I_i^j(b)\varphi_{u,j}(r) \tag{10}$$

due to $u' = u$ in the case under consideration.

Suppose $b_0 \in \mathbb{K}^s$ is such that g_{b_0} is the identity element of G and define

$$Q_\omega := \frac{\partial U(b)}{\partial b^\omega}\bigg|_{b=b_0}, \qquad I_{i\omega}^j := \frac{\partial I_i^j(b)}{\partial b^\omega}\bigg|_{b=b_0} \tag{11}$$

where $b = (b^1, \ldots, b^s)$ and $\omega = 1, \ldots, s$. Then, differentiation Eq. (10) with respect to b^ω and putting in the result $b = b_0$, we get the following Heisenberg relation

$$[Q_\omega, \varphi_{u,i}(r)]_{\_} = I_{i\omega}^j \varphi_{u,j}(r) \tag{12}$$

---

[a] In fact, we are dealing with an s-dimensional Lie group and $b \in \mathbb{K}^s$ are the (local) coordinates of g_b in some chart on G containing g_b in its domain.

or, if we identify $x \in M$ with $r = u(x)$ and omit the subscript u,

$$[Q_\omega, \varphi_i(x)]_- = I^j_{i\omega} \varphi_j(x). \tag{13}$$

To make the situation more familiar, consider the case of one-dimensional group G, $s = 1$, when $\omega = 1$ due to which we shall identify b^1 with $b = (b^1)$. Besides, let us suppose that

$$I(b) = \mathbf{1} \exp(f(b) - f(b_0)) \tag{14}$$

for some C^1 function f. Then Eq. (13) reduces to

$$[Q_1, \varphi_i(x)]_- = f'(b_0)\varphi_i(x), \tag{15}$$

where $f'(b) := df(b)/db$. In particular, if we are dealing with phase transformations, i.e.

$$U(b) = e^{bQ_1/(ie)}, \quad I(b) = \mathbf{1}e^{-qb/(ie)}, \qquad b \in \mathbb{R} \tag{16}$$

for some constants q and e (having a meaning of charge and unit charge, respectively) and operator Q_1 on system's Hilbert space of states (having a meaning of a charge operator), then Eq. (10) and Eq. (15) take the familiar form [2, eqs. (2.81) and (2-80)]

$$\varphi'_i(x) = e^{bQ_1/(ie)} \circ \varphi_i(x) \circ e^{-bQ_1/(ie)} = e^{-qb/(ie)} \varphi(x), \tag{17}$$

$$[Q_1, \varphi_i(x)]_- = -q\varphi_i(x). \tag{18}$$

The considerations in the framework of Lagrangian formalism invariant under phase transformations [2–4] implies conservation of the charge operator Q and suggests the correspondence (cf. Eq. (7)) $Q_1 \mapsto Q$ which in turn suggests the Heisenberg relation $[Q, \varphi_i(x)]_- = -q\varphi_i(x)$. We should note that this equation is external to the Lagrangian formalism and requires a proof in it [5].

4. The general case

The corner stone of the (global) Heisenberg relations is the equation

$$U \circ \varphi_{u,i}(r) \circ U^{-1} = \frac{\partial(u' \circ u^{-1})(r)}{\partial r} (A^{-1}(u^{-1}(r)))^j_i \varphi_{u,j}((u' \circ u^{-1})(r)) \tag{19}$$

representing the components $\varphi'_{u',i}$ of a quantum field φ in a reference frame $(u, \{e'^i = A^i_j e^j\})$ via its components $\varphi_{u,i}$ in a frame $(u, \{e^i\})$ in two different way. Here $A = [A^j_i]$ is a non-degenerate matrix-valued function, $r \in \mathbb{R}^4$ and $\varphi_{u,i} := \varphi_i \circ u^{-1}$. Now, following the ideas at the beginning of section 1, we shall demonstrate how from the last relation can be derived Heisenberg relations in the general case.

Let G be an s-dimensional, $s \in \mathbb{N}$, Lie group. Without going into details, we admit that its elements are labeled by $b = (b^1, \dots, b^s) \in \mathbb{K}^s$ and g_{b_0} is the identity element of G for some fixed $b_0 \in \mathbb{K}^s$. Suppose that there are given three representations H, I and U of G and consider frames of reference with the following properties:

(1) $H\colon G \ni g_b \mapsto H_b\colon \mathbb{R}^{\dim M} \to \mathbb{R}^{\dim M}$ and any change $(U, u) \mapsto (U', u')$ of the charts of M is such that $u' \circ u^{-1} = H_b$ for some $b \in \mathbb{K}^s$.
(2) $I\colon G \ni g_b \mapsto I(b) \in \mathrm{GL}(\dim V, \mathbb{K})$ and any change $\{e^i\} \mapsto \{e'^i = A^i_j e^j\}$ of the frames in V is such that $A^{-1}(x) = I(b)$ for all $x \in M$ and some $b \in \mathbb{K}^s$.
(3) $U\colon G \ni g_b \mapsto U(b)$, where $U(b)$ is an operator on the space of state vectors, and the changes $(u, \{e^i\}) \mapsto (u', \{e'^i\})$ of the reference frames entail Eq. (1) with $U(b)$ for U.

Under the above hypotheses equation Eq. (19) transforms into

$$U(b) \circ \varphi_{u,i}(r) \circ U^{-1}(b) = \det\Big[\frac{\partial (H_b(r))^i}{\partial r^j}\Big] I^j_i(b) \varphi_{u,j}(H_b(r)) \qquad (20)$$

which can be called global Heisenberg relation in the particular situation. The next step is to differentiate this equation with respect to b^ω, $\omega = 1, \dots, s$, and then to put $b = b_0$ in the result. In this way we obtain the following (local) Heisenberg relation

$$[U_\omega, \varphi_{u,i}(r)]_- = \Delta_\omega(r)\varphi_{u,i}(r) + I^j_{i\omega}\varphi_{u,j}(r) + (h_{\omega(r)})^k \frac{\partial \varphi_{u,i}(r)}{\partial r^k}, \qquad (21)$$

where

$$U_\omega := \frac{\partial U(b)}{\partial b^\omega}\Big|_{b=b_0}, \qquad \Delta_\omega(r) := \frac{\partial \det\Big[\frac{\partial (H_b(r))^j}{\partial r^j}\Big]}{\partial b^\omega}\Bigg|_{b=b_0} \in \mathbb{R}^{\dim M}, \quad (22a)$$

$$I^j_{i\omega} := \frac{\partial I^j_i(b)}{\partial b^\omega}\Big|_{b=b_0} \in \mathbb{K}, \quad h_\omega := \frac{\partial H_b}{\partial b^\omega}\Big|_{b=b_0} : \mathbb{R}^{\dim M} \to \mathbb{R}^{\dim M}. \qquad (22b)$$

In particular, if H_b is linear and non-homogeneous, i.e. $H_b(r) = H(b)\cdot r + a(b)$ for some $H(b) \in \mathrm{GL}(\dim M, \mathbb{R})$ and $a(b) \in \mathbb{K}^{\dim M}$ with $H(b_0) = \mathbf{1}$ and $a(b_0) = \mathbf{0}$, then (Tr means trace of a matrix or operator)

$$\Delta_\omega(r) = \frac{\partial \det(H(b))}{\partial b_\omega}\Big|_{b=b_0} = \frac{\partial \mathrm{Tr}(H(b))}{\partial b_\omega}\Big|_{b=b_0}$$

and

$$h_\omega(\cdot) = \frac{\partial H(b)}{\partial b_\omega}\Big|_{b=b_0} \cdot (\cdot) + \frac{\partial a(b)}{\partial b_\omega}\Big|_{b=b_0}$$

as $\partial \det B/\partial b^j_i\big|_{B=\mathbf{1}} = \delta^i_j$ for any square matrix $B = [b^j_i]$. In this setting the Heisenberg relations corresponding to Poincaré transformations (see subsection 2) are described by $b \mapsto (\Lambda^{\mu\nu}, a^\lambda)$, $H(b) \mapsto \Lambda$, $a(b) \mapsto a$ and $I(b) \mapsto I(\Lambda)$, so that $U_\omega \mapsto (S_{\mu\nu,T_\lambda})$, $\Delta_\omega(r) \equiv 0$, $I^j_{i\omega} \mapsto (I^j_{i\mu\nu}, 0)$ and

$$(h_\omega(r))^k \frac{\partial}{\partial r^k} \mapsto r_\mu \frac{\partial}{\partial r^\nu} - r_\nu \frac{\partial}{\partial r^\mu}.$$

The case of internal transformations, considered in the previous subsection, corresponds to $H_b = \mathsf{id}_{\mathbb{R}^{\dim M}}$ and, consequently, in it $\Delta_\omega(r) \equiv 0$ and $h_\omega = 0$.

5. Fibre bundle approach

Suppose a physical field is described as a section $\varphi\colon M \to E$ of a vector bundle (E, π, M). Here M is a real differentiable (4-)manifold (of class at least C^1), serving as a spacetime model, E is the bundle space and $\pi\colon M \to E$ is the projection; the fibres $\pi^{-1}(x)$, $x \in M$, are isomorphic vector spaces.

Let (U, u) be a chart of M and $\{e^i\}$ be a (vector) frame in the bundle with domain containing U, i.e. $e^i\colon x \mapsto e^i(x) \in \pi^{-1}(x)$ with x in the domain of $\{e^i\}$ and $\{e^i(x)\}$ being a basis in $\pi^{-1}(x)$. Below we assume $x \in U \subseteq M$. Thus, we have

$$\varphi\colon M \ni x \mapsto \varphi(x) = \varphi_i(x)e^i(x) = \varphi_{u,i}(\boldsymbol{x})e^i(u^{-1}(\boldsymbol{x})), \tag{23}$$

where

$$\boldsymbol{x} := u(x), \qquad \varphi_{u,i} := \varphi_i \circ u^{-1} \tag{24}$$

and $\varphi_i(x)$ are the components of the vector $\varphi(x) \in \pi^{-1}(x)$ relative to the basis $\{e^i(x)\}$ in $\pi^{-1}(x)$.

The origin of the Heisenberg relations on the background of fibre bundle setting is in the equivalent equations

$$U \circ \varphi_i(x) \circ U^{-1} = (A^{-1})^j_i(x)\varphi_j(x), \tag{25}$$

$$U \circ \varphi_{u,i}(\boldsymbol{x}) \circ U^{-1} = (A^{-1})^j_i(x)\varphi_{u,j}(\boldsymbol{x}). \tag{25$'$}$$

Similarly to subsection 4, consider a Lie group G, its representations I and U and reference frames with the following properties:

(1) $I\colon G \ni g_b \mapsto I(b) \in \mathrm{GL}(\dim V, \mathbb{K})$ and the changes $\{e^i\} \mapsto \{e'^i = A^i_j e^j\}$ of the frames in V are such that $A^{-1}(x) = I(b)$ for all $x \in M$ and some $b \in \mathbb{K}^s$.

(2) $U: g \ni g_b \mapsto U(b)$, where $U(b)$ is an operator on the space of state vectors, and the changes $(u, \{e^i\}) \mapsto (u', \{e'^i\})$ of the reference frames entail Eq. (1) with $U(b)$ for U.

Remark 5.1. One can consider also simultaneous coordinate changes $u \mapsto u' = H_b \circ u$ induced by a representation $H: G \ni g_b \mapsto H_b: \mathbb{R}^{\dim M} \to \mathbb{R}^{\dim M}$, as in subsection 4. However such a supposition does not influence our results as the basic equations Eq. (26) and Eq. (26′) below are independent from it; in fact, equation Eq. (26) is coordinate-independent, while Eq. (26′) is its version valid in any local chart (U, u) as $\varphi_u := \varphi \circ u^{-1}$ and $\boldsymbol{x} := u(x)$.

Thus equations Eq. (25) and Eq. (25′) transform into (cf. Eq. (20))

$$U(b) \circ \varphi_i(x) \circ U^{-1}(b) = I_i^j(b)\varphi_j(x), \tag{26}$$

$$U(b) \circ \varphi_{u,i}(\boldsymbol{x}) \circ U^{-1}(b) = I_i^j(b)\varphi_{u,j}(\boldsymbol{x}). \tag{26′}$$

Differentiating Eq. (26) with respect to b^ω and then putting $b = b_0$, we derive the following Heisenberg relation

$$[U_\omega, \varphi_i(x)]_- = I_{i\omega}^j \varphi_j(x) \tag{27}$$

or its equivalent version (cf. Eq. (21))

$$[U_\omega, \varphi_{u,i}(\boldsymbol{x})]_- = I_{i\omega}^j \varphi_{u,j}(\boldsymbol{x}), \tag{27′}$$

where

$$U_\omega := \frac{\partial U(b)}{\partial b^\omega}\bigg|_{b=b_0}, \tag{28a}$$

$$I_{i\omega}^j := \frac{\partial I_i^j(b)}{\partial b^\omega}\bigg|_{b=b_0}. \tag{28b}$$

We can rewire the Heisenberg relations obtained as

$$[U_\omega, \varphi]_- = I_{i\omega}^j \varphi_j e^i. \tag{29}$$

One can prove that the r.h.s. of this equation is independent of the particular frame $\{e^i\}$ in which it is represented.

The case of Poncaré transformations is described by the replacements $b \mapsto (\Lambda^{\mu\nu}, a^\lambda)$, $U_\omega \mapsto (S_{\mu\nu}, T_\lambda)$ and $I_{i\omega}^j \mapsto (I_{i\mu\nu}^j, 0)$ and, consequently, the equations Eq. (26) and Eq. (26′) now read

$$U(\Lambda, a) \circ \varphi_i(x) \circ U^{-1}(\Lambda, a) = I_i^j(\Lambda, a)\varphi_j(x), \tag{30}$$

$$U(\Lambda, a) \circ \varphi_{u,i}(\boldsymbol{x}) \circ U^{-1}(\Lambda, a) = I_i^j(\Lambda, a)\varphi_{u,j}(\boldsymbol{x}). \tag{30′}$$

Hence, for instance, the Heisenberg relations Eq. (27) now takes the form (cf. Eq. (6))

$$[T_\mu, \varphi_i(x)]_- = 0, \tag{31a}$$

$$[S_{\mu\nu}, \varphi_i(x)]_- = I^j_{i\mu\nu}\varphi_j(x). \tag{31b}$$

Respectively, the correspondences in Eq. (7) transform these equations into

$$[P_\mu, \varphi_i(x)]_- = 0, \tag{32a}$$

$$[M_{\mu\nu}, \varphi_i(x)]_- = I^j_{i\mu\nu}\varphi_j(x) \tag{32b}$$

which now replace Eq. (8).

Since equation Eq. (8a) (and partially equation Eq. (8b)) is (are) the corner stone for the particle interpretation of quantum field theory [3–5], the equation Eq. (32a) (and partially equation Eq. (32b)) is (are) physically unacceptable if one wants to retain the particle interpretation in the fibre bundle approach to the theory. For this reason, it seems that the correspondences Eq. (7) should *not* be accepted in the fibre bundle approach to quantum field theory, in which Eq. (6) transform into Eq. (31). However, for retaining the particle interpretation one can impose Eq. (8) as subsidiary restrictions on the theory in the fibre bundle approach. It is almost evident that this is possible if the frames used are connected by linear homogeneous transformations with spacetime constant matrices, $A(x) = \text{const}$ or $\partial_\mu A(x) = 0$. Consequently, if one wants to retain the particle interpretation of the theory, one should suppose the validity of Eq. (8) in some frame and, then, it will hold in the whole class of frames obtained from one other by transformations with spacetime independent matrices.

Since the general setting investigated above is independent of any (local) coordinates, it describes also the fibre bundle version of the case of internal transformations considered in section 3. This explains why equations like Eq. (12) and Eq. (27′) are identical but the meaning of the quantities $\varphi_{u,i}$ and $I^j_{i\omega}$ in them is different. [b] In particular, in the case of phase transformations

$$U(b) = e^{bQ_1/(ie)}, \quad I(b) = \mathbf{1}e^{-bq/(ie)}, \qquad b \in \mathbb{R} \tag{33}$$

the Heisenberg relations Eq. (27) reduce to

$$[Q_1, \varphi_i(x)]_- = -q\varphi_i(x), \tag{34}$$

[b] Note, now $I(b)$ is the matrix defining transformations of frames in the bundle space, while in Eq. (16) it serves a similar role for frames in the vector space V.

which is identical with Eq. (18), but now φ_i are the components of the section φ in $\{e^i\}$. The invariant form of the last relations is

$$[Q_1, \varphi]_- = -q\varphi \tag{35}$$

which is also a consequence from Eq. (29) and Eq. (17).

6. Conclusion

In this paper we have shown how the Heisenberg equations arise in the general case and in particular situations. They are from pure geometrical origin and one should be careful when applying them to the Lagrangian formalism in which they are subsidiary conditions, like the Lorenz gauge in the electrodynamics. In the general case they need not to be consistent with the Lagrangian formalism and their validity should carefully be checked. For instance, if one starts with field operators in the Lagrangian formalism of free fields and adds to it the Heisenberg relations Eq. (8a) concerning the momentum operator, then the arising scheme is not consistent as in it start to appear distributions, like the Dirac delta function. This conclusion leads to the consideration of the quantum fields as operator-valued distribution in the Lagrangian formalism even for free fields. In the last case, the Heisenberg relations concerning the momentum operator are consistent with the Lagrangian formalism. Besides, they play an important role in the particle interpretation of the so-arising theory.

Acknowledgments

This work was partially supported by the National Science Fund of Bulgaria under Grant No. F 1515/2005.

References

1. J. D. Bjorken and S.D. Drell, *Relativistic quantum fields*, McGraw-Hill Book Company, New York, 1965; Russian translation: Nauka, Moscow, 1978.
2. P. Roman, *Introduction to quantum field theory*, John Wiley & Sons, Inc., New York-London-Sydney-Toronto, 1969.
3. J.D. Bjorken and S.D. Drell, *Relativistic quantum mechanics*, McGraw-Hill Book Company, New York, 1964, 1965; Russian translation: Nauka, Moscow, 1978.
4. N.N. Bogolyubov and D.V. Shirkov, *Introduction to the theory of quantized fields*, third edn., Nauka, Moscow, 1976, In Russian; English translation: Wiley, New York, 1980.

5. B.Z. Iliev, *Langrangian Quantum Field theory in Momentum Picture. Free fields*, Nova Science Publishers, New York, 2008, 306 pages, Hardcover, ISBN-13: 978-1-60456-170-8, ISBN-10: 1-60456-170-X.
6. N.N. Bogolubov, A.A. Logunov and I.T. Todorov, *Introduction to axiomatic quantum field theory*, W.A. Benjamin, Inc., London, 1975; Translation from Russian: Nauka, Moscow, 1969.
7. N.N. Bogolubov, A.A. Logunov, A.I. Oksak and I.T. Todorov, *General principles of quantum field theory*, Nauka, Moscow, 1987, In Russian; English translation: Kluwer Academic Publishers, Dordrecht, 1989.

POISSON STRUCTURES OF EQUATIONS ASSOCIATED WITH GROUPS OF DIFFEOMORPHISMS

ROSSEN I. IVANOV

School of Mathematical Sciences, Dublin Institute of Technology, Kevin Street, Dublin 8, Ireland
E-mail: rivanov@dit.ie

A class of equations describing the geodesic flow for a right-invariant metric on the group of diffeomorphisms of $\mathbb{R}^n$ is reviewed from the viewpoint of their Lie-Poisson structures. A subclass of these equations is analogous to the Euler equations in hydrodynamics (for $n = 3$), preserving the volume element of the domain of fluid flow. An example in $n = 1$ dimension is the Camassa-Holm equation, which is a geodesic flow equation on the group of diffeomorphisms, preserving the H^1 metric.

Keywords: Lie group; Virasoro group; Group of diffeomorphisms; Lie-Poisson bracket; Vector fields.

1. Camassa-Holm equation

The Camassa-Holm (CH) equation can be considered as a member of the family of EPDiff equations, that is, Euler-Poincaré equations, associated with the diffeomorphism group in n-dimensions [18]. Let us consider first the CH equation in the form

$$q_t + 2u_x q + uq_x = 0, \qquad q = u - u_{xx} + \omega, \tag{1}$$

with ω an arbitrary parameter. The traveling wave solutions of (1) are smooth solitons [5] if $\omega > 0$ and peaked solitons (peakons) if $\omega = 0$ [4,13, 14,24,28], assuming that $u(x;t)$ is rapidly decaying to zero at $x \to \pm\infty$.

CH is a bi-Hamiltonian equation, i.e. it admits two compatible Hamiltonian structures [4,15] $J_1 = -(q\partial + \partial q)$, $J_2 = -(\partial - \partial^3)$:

$$q_t = J_2 \frac{\delta H_2[q]}{\delta q} = J_1 \frac{\delta H_1[q]}{\delta q}, \tag{2}$$

$$H_1 = \frac{1}{2} \int qu \, \mathrm{d}x, \tag{3}$$

$$H_2 = \frac{1}{2}\int (u^3 + uu_x^2 + 2\omega u^2)\,\mathrm{d}x. \tag{4}$$

If $\omega \neq 0$ the invariance group of the Hamiltonian is the Virasoro group, Vir $=$ Diff$(\mathbb{S}^1) \times \mathbb{R}$ and the central extension of the corresponding Virasoro algebra is proportional to ω [10–12,19,25,30]. Thus CH has various conformal properties [21]. It is also completely integrable, possesses bi-Hamiltonian form and infinite sequence of conservation laws [4,8,9,22,32].

The soliton solution has the form

$$q(x,t) = \int_0^{\infty} \delta(x - X(\xi,t))P(\xi,t)\,d\xi, \tag{5}$$

where $X(\xi,t)$ and $P(\xi,t)$ are quantities well defined in terms of the scattering data [7,8,10] ($q(x,0) > 0$ is assumed, otherwise wave breaking occurs [6]). From (5) one can easily compute $u = (1-\partial^2)^{-1}(q-\omega)$,

$$u(x,t) = \frac{1}{2}\int_0^{\infty} e^{-|x-X(\xi,t)|}P(\xi,t)\,d\xi - \omega. \tag{6}$$

Substitution of (5) and (6) into the equation (1) and using the fact that

$$f(x)\delta'(x-x_0) = f(x_0)\delta'(x-x_0) - f'(x_0)\delta(x-x_0)$$

we derive a system of integral equations for X and P:

$$X_t(\xi,t) = \int G(X(\xi,t) - X(\underline{\xi},t))P(\underline{\xi},t)\,\mathrm{d}\underline{\xi} - \omega, \tag{7}$$

$$P_t(\xi,t) = -\int G'(X(\xi,t) - X(\underline{\xi},t))P(\xi,t)P(\underline{\xi},t)\,\mathrm{d}\underline{\xi}, \tag{8}$$

where $G(x) \equiv \frac{1}{2}e^{-|x|}$. From (5) and (6) the Hamiltonian H_1 can be expressed as

$$H_1(X,P) = \frac{1}{2}\int G(X(\xi_1,t) - X(\xi_2,t))P(\xi_1,t)P(\xi_2,t)\,\mathrm{d}\xi_1\mathrm{d}\xi_2 - \omega\int P(\xi,t)\,\mathrm{d}\xi$$

and the equations (7) and (8) as

$$X_t(\xi,t) = \frac{\delta H}{\delta P(\xi,t)}, \qquad P_t(\xi,t) = -\frac{\delta H}{\delta X(\xi,t)}, \tag{9}$$

i.e. these equations are Hamiltonian, with respect to the canonical Poisson bracket

$$\{A,B\}_c = \int \left(\frac{\delta A}{\delta X(\xi,t)}\frac{\delta B}{\delta P(\xi,t)} - \frac{\delta B}{\delta X(\xi,t)}\frac{\delta A}{\delta P(\xi,t)} \right) \mathrm{d}\xi \tag{10}$$

and the canonical variables are $X(\xi,t)$, $P(\xi,t)$:

$$\{X(\xi_1,t),P(\xi_2,t)\}_c = \delta(\xi_1-\xi_2), \tag{11}$$

$$\{P(\xi_1,t),P(\xi_2,t)\}_c = \{X(\xi_1,t),X(\xi_2,t)\}_c = 0. \tag{12}$$

Now we can show that (5) is a Clebsh parametrization that produces the Poisson brackets, given by the Hamiltonian operator J_1. To do this we will use the canonical Poisson brackets (11), (12) to compute $\{q(x_1),q(x_2)\}_c$.

Indeed with $X_i \equiv X(\xi_i,t)$, and $P_i \equiv P(\xi_i,t)$, $i=1,2$ we get:

$$\begin{aligned}
&\{q(x_1,t),q(x_2,t)\}_c \\
&= \left\{\int_0^\infty \delta(x_1-X_1)P_1\,d\xi_1\,,\ \int_0^\infty \delta(x_2-X_2)P_2\,d\xi_2\right\}_c \\
&= -\int_0^\infty\int_0^\infty \{X_1,P_2\}_c\delta'(x_1-X_1)P_1\delta(x_2-X_2)\,d\xi_1 d\xi_2 \\
&\quad -\int_0^\infty\int_0^\infty \{P_1,X_2\}_c\delta(x_1-X_1)\delta'(x_2-X_2)P_2\,d\xi_1 d\xi_2 \\
&= -\delta'(x_1-x_2)\int_0^\infty P_2\delta(x_2-X_2)\,d\xi_2 + \delta'(x_2-x_1)\int_0^\infty P_1\delta(x_1-X_1)\,d\xi_1 \\
&= -q(x_2,t)\delta'(x_1-x_2) + q(x_1,t)\delta'(x_2-x_1) \\
&= -\left(q(x_1,t)\frac{\partial}{\partial x_1} + \frac{\partial}{\partial x_1}q(x_1,t)\right)\delta(x_1-x_2) \\
&= J_1(x_1)\delta(x_1-x_2).
\end{aligned}$$

Now it is straightforward to check, using (2), that (1) can be written in a Hamiltonian form as

$$q_t = \{q, H_1\}_c,$$

with the Poisson bracket, generated by J_1:

$$\begin{aligned}
\{A,B\}_c &= \int \frac{\delta A}{\delta q(x)} J_1(x) \frac{\delta B}{\delta q(x)}\,\mathrm{d}x \\
&= -\int q(x)\left(\frac{\delta A}{\delta q(x)}\frac{\partial}{\partial x}\frac{\delta B}{\delta q(x)} - \frac{\delta B}{\delta q(x)}\frac{\partial}{\partial x}\frac{\delta A}{\delta q(x)}\right)\mathrm{d}x.
\end{aligned} \tag{13}$$

A singular version of (5) is used [18] for the construction of peakon, filament and sheet singular solutions for higher dimensional EPDiff equations.

The parallel with the geometric interpretation of the integrable SO(3) top can be made explicit by a discretization of CH equation based on Fourier modes expansion [23]. Since the Virasoro algebra is an infinite-dimensional algebra, the obtained equation represents an 'integrable top' with infinitely many momentum components.

If we compare (6) and (7) we have

$$X_t(\xi,t)=u(X(\xi,t),t), \tag{14}$$

i.e. $X(\xi,t)$ is explicitly the diffeomorphism related to the geodesic curve [10,27,30], i.e. $X(x,t)$ is an one-parameter curve of diffeomorphisms of $\mathbb{R}$ (or, with periodic boundary conditions, of the circle $\mathbb{S}^1$), depending on a parameter t and associated with a right-invariant metric given by the Hamiltonian H_1.

For the peakon solutions ($\omega=0$) the dependence on the scattering data is also known. For completeness and comparison we mention the analogous results for this case. The N-peakon solution has the form [2,4]

$$u(x,t)=\frac{1}{2}\sum_{i=1}^{N}p_i(t)\exp(-|x-x_i(t)|), \tag{15}$$

provided p_i and x_i evolve according to the following system of ordinary differential equations:

$$\dot{x}_i=\frac{\partial H}{\partial p_i}, \qquad \dot{p}_i=-\frac{\partial H}{\partial x_i}, \tag{16}$$

where the Hamiltonian is $H=1/4\sum_{i,j=1}^{N}p_ip_j\exp(-|x_i-x_j|)$. Now one can see immediately the analogy between $X(\xi,t)$ and $x_i(t)$; $P(\xi,t)$ and $p_i(t)$ due to the fact that the N-soliton solution with the limit $\omega\to 0$ converges to the N-peakon solution [3].

2. n-dimensional EPDiff equations

Let us consider motion in $\mathbb{R}^n$ with a velocity field $\mathbf{u}(\mathbf{x},t)$: $\mathbb{R}^n\times\mathbb{R}\to\mathbb{R}^n$ and define a momentum variable $\mathbf{m}=Q\mathbf{u}$ for some (inertia) operator Q (for CH generalizations Q is the Helmholtz operator $Q=1-\partial_i\partial_i=1-\Delta$, where $\partial_i=\partial/\partial x^i$). The kinetic energy defines a Lagrangian

$$L[\mathbf{u}]=\frac{1}{2}\int\mathbf{m}\cdot\mathbf{u}\,\mathrm{d}^n\mathbf{x}. \tag{17}$$

Since the velocity $\mathbf{u}=u^i\partial_i$ is a vector field, $\mathbf{m}=m_i\mathrm{d}x^i\otimes\mathrm{d}^n\mathbf{x}$ is a $n+1$-form density, we have a natural bilinear form

$$\langle\mathbf{m},\mathbf{u}\rangle=\int\mathbf{m}\cdot\mathbf{u}\,\mathrm{d}^n\mathbf{x}. \tag{18}$$

The Euler-Poincaré equation for the geodesic motion is [18,19]

$$\frac{d}{dt}\frac{\delta L}{\delta\mathbf{u}}+\mathrm{ad}^*_{\mathbf{u}}\frac{\delta L}{\delta\mathbf{u}}=0, \qquad \mathbf{u}=G*\mathbf{m}, \tag{19}$$

where G is the Green function for the operator Q. The corresponding Hamiltonian is

$$H[\mathbf{m}] = \langle \mathbf{m}, \mathbf{u} \rangle - L[\mathbf{u}] = \frac{1}{2} \int \mathbf{m} \cdot G * \mathbf{m} \, \mathrm{d}^n\mathbf{x}, \tag{20}$$

and the equation in Hamiltonian form ($\mathbf{u} = \delta H / \delta \mathbf{m}$) is

$$\frac{\partial \mathbf{m}}{\partial t} = -\mathrm{ad}^*_{\delta H/\delta \mathbf{m}} \mathbf{m}. \tag{21}$$

The left Lie algebra of vector fields is $[\mathbf{u}, \mathbf{v}] = -(u^k(\partial_k v^p) - v^k(\partial_k u^p))\partial_p$. For an arbitrary vector field $\mathbf{v}$ one can write [19]

$$\begin{aligned}
&\langle \mathrm{ad}^*_{\mathbf{u}} \mathbf{m}, \mathbf{v} \rangle = \langle \mathbf{m}, \mathrm{ad}_{\mathbf{u}} \mathbf{v} \rangle = \langle \mathbf{m}, [\mathbf{u}, \mathbf{v}] \rangle \\
&= -\left\langle m_l \, \mathrm{d}x^l \otimes \mathrm{d}^n\mathbf{x}, \left(u^k \left(\partial_k v^p\right) - v^k \left(\partial_k u^p\right)\right) \partial_p \right\rangle \\
&= -\int m_p \left(u^k \left(\partial_k v^p\right) - v^k \left(\partial_k u^p\right)\right) \mathrm{d}^n\mathbf{x} \\
&= -\int \left(\partial_k \left(m_p u^k v^p\right) - \left(\partial_k m_p\right) u^k v^p - m_p v^p \left(\partial_k u^k\right) - m_p v^k \left(\partial_k u^p\right)\right) \mathrm{d}^n\mathbf{x} \\
&= \int v^p \left(u^k \left(\partial_k m_p\right) + m_p \left(\partial_k u^k\right) + m_k \left(\partial_p u^k\right)\right) \mathrm{d}^n\mathbf{x} \\
&= \left\langle \left(\left(\mathbf{u} \cdot \nabla\right) m_p + \mathbf{m} \cdot \partial_p \mathbf{u} + m_p \mathrm{div}\, \mathbf{u}\right) \mathrm{d}x^p \otimes \mathrm{d}^n\mathbf{x}, \mathbf{v} \right\rangle,
\end{aligned}$$

and therefore (21) has the form

$$\frac{\partial m_p}{\partial t} + (\mathbf{u} \cdot \nabla) m_p + \mathbf{m} \cdot \partial_p \mathbf{u} + m_p \mathrm{div}\, \mathbf{u} = 0. \tag{22}$$

Let us now define an one-parametric group of diffeomorphisms of $\mathbb{R}^n$, with elements that satisfy

$$\frac{\partial \mathbf{X}(\mathbf{x}, t)}{\partial t} = \mathbf{u}(\mathbf{X}(\mathbf{x}, t), t), \qquad \mathbf{X}(\mathbf{x}, 0) = \mathbf{x}. \tag{23}$$

Due to the invariance of the Hamiltonian under the action of the group there is a momentum conservation law:

$$m_i(\mathbf{X}(\mathbf{x}, t), t) \partial_j X^i(\mathbf{x}, t) \det \left(\frac{\partial \mathbf{X}}{\partial \mathbf{x}} \right) = m_j(x, 0), \tag{24}$$

where $(\partial \mathbf{X} / \partial \mathbf{x})_{ij} = \partial X^i / \partial x^j$ is the Jacobian matrix.

The Lie-Poisson bracket is

$$\begin{aligned}
\{A, B\}(\mathbf{m}) &= \left\langle \mathbf{m}, \left[\frac{\delta A}{\delta \mathbf{m}}, \frac{\delta B}{\delta \mathbf{m}} \right] \right\rangle \\
&= -\int m_i \left(\frac{\delta A}{\delta m_k} \partial_k \frac{\delta B}{\delta m_i} - \frac{\delta B}{\delta m_k} \partial_k \frac{\delta A}{\delta m_i} \right) \mathrm{d}^n\mathbf{x}.
\end{aligned} \tag{25}$$

When $n = 1$ clearly (25) gives (13) and the algebra, associated with the bracket is the algebra of vector fields on the circle. This algebra admits a generalization with a central extension, which is the famous Virasoro algebra [10–12,19,25,30]. In two dimensions, $n = 2$, the algebra, associated with the bracket is the algebra of vector fields on a torus [1,16,33]. This algebra also admits central extensions [16,20].

3. Reduction to the subgroup of volume-preserving diffeomorphisms

In the case of volume-preserving diffeomorphisms we consider vector fields, further restricted by the condition $\operatorname{div}\mathbf{u} = 0$. Let us restrict ourselves to the three-dimensional case ($n = 3$) and let us assume that $\mathbf{m} = (1 - \Delta)\mathbf{u}$, so that $\operatorname{div}\mathbf{m} = 0$ as well. According to the Helmholtz decomposition theorem for vector fields, $\mathbf{m}$ can be determined only by the quantity $\boldsymbol{\Omega} = \nabla \times \mathbf{m}$. Therefore we can write the Lie-Poisson brackets (25) in terms of $\boldsymbol{\Omega}$. Indeed, one can compute that

$$\frac{\delta A}{\delta \mathbf{m}} = \nabla \times \frac{\delta A}{\delta \boldsymbol{\Omega}}. \tag{26}$$

Thus

$$\nabla \cdot \frac{\delta A}{\delta \mathbf{m}} = \nabla \cdot \left(\nabla \times \frac{\delta A}{\delta \boldsymbol{\Omega}} \right) = 0, \tag{27}$$

i.e. the vector fields $\frac{\delta A}{\delta \mathbf{m}}$ are divergence-free. Therefore

$$m_i \frac{\delta A}{\delta m_k} \partial_k \frac{\delta B}{\delta m_i} = \partial_k \left(m_i \frac{\delta A}{\delta m_k} \frac{\delta B}{\delta m_i} \right) - (\partial_k m_i) \frac{\delta A}{\delta m_k} \frac{\delta B}{\delta m_i}$$

and from (25) we obtain

$$\begin{aligned} \{A, B\} &= \int (\partial_k m_i) \left(\frac{\delta A}{\delta m_k} \frac{\delta B}{\delta m_i} - \frac{\delta B}{\delta m_k} \frac{\delta A}{\delta m_i} \right) \mathrm{d}^3\mathbf{x} \\ &= \int \boldsymbol{\Omega} \cdot \left(\frac{\delta A}{\delta \mathbf{m}} \times \frac{\delta B}{\delta \mathbf{m}} \right) \mathrm{d}^3\mathbf{x} \\ &= \int \boldsymbol{\Omega} \cdot \left(\left(\nabla \times \frac{\delta A}{\delta \boldsymbol{\Omega}} \right) \times \left(\nabla \times \frac{\delta B}{\delta \boldsymbol{\Omega}} \right) \right) \mathrm{d}^3\mathbf{x}. \end{aligned} \tag{28}$$

This is the well known Poisson bracket used in fluid mechanics [1,17,26,29, 31,34]. The curl of the equation (22) gives the following equation for $\boldsymbol{\Omega}$ [19]:

$$\boldsymbol{\Omega}_t + (\mathbf{u} \cdot \nabla)\boldsymbol{\Omega} - (\boldsymbol{\Omega} \cdot \nabla)\mathbf{u} = 0, \qquad \boldsymbol{\Omega} = (1 - \Delta)(\nabla \times \mathbf{u}). \tag{29}$$

Note that $\mathbf{u}$ can be expressed through $\mathbf{\Omega}$:

$$\mathbf{u}[\mathbf{\Omega}] = (1-\Delta)^{-1}\left(\nabla \times \int \frac{\mathbf{\Omega}(\mathbf{x}')}{4\pi|\mathbf{x}-\mathbf{x}'|}\,\mathrm{d}^3\mathbf{x}'\right),$$

thus

$$H[\mathbf{\Omega}] = \frac{1}{2}\int \mathbf{u}[\mathbf{\Omega}]\cdot(1-\Delta)\mathbf{u}[\mathbf{\Omega}]\,\mathrm{d}^3\mathbf{x}.$$

The vector $\mathbf{\Omega}$ is always perpendicular to $\mathbf{u}$. The further reduction to an equation in $n = 2$ dimensions is straightforward. Introducing a (scalar) stream function $\psi(x^1, x^2)$ we have

$$\mathbf{u}(x^1, x^2) = (-\partial_2\psi, \partial_1\psi, 0) = \mathbf{e}_3 \times \nabla\psi, \tag{30}$$

$$\mathbf{\Omega}(x^1, x^2) = (1-\Delta)(\nabla \times \mathbf{u}) = (1-\Delta)\Delta\psi\mathbf{e}_3, \tag{31}$$

where $\mathbf{e}_3$ is the unit vector in the direction of x^3. Since $(\mathbf{\Omega}\cdot\nabla)\mathbf{u} = \Omega\partial_3\mathbf{u} = 0$, (29) leads to the equation

$$\mathbf{\Omega}_t + (\mathbf{u}\cdot\nabla)\mathbf{\Omega} = 0,$$

which produces a scalar equation for the stream function ψ due to (30) and (31), or alternatively for $\Omega \equiv \mathbf{\Omega}\cdot\mathbf{e}_3$. The Poisson bracket that one can find from (28) is

$$\begin{aligned}\{A, B\} &= \int \Omega\left(\partial_1\left(\frac{\delta A}{\delta\Omega}\right)\partial_2\left(\frac{\delta B}{\delta\Omega}\right) - \partial_2\left(\frac{\delta A}{\delta\Omega}\right)\partial_1\left(\frac{\delta B}{\delta\Omega}\right)\right)\mathrm{d}^2\mathbf{x}\\ &= \int \mathbf{\Omega}\cdot\left(\nabla\left(\frac{\delta A}{\delta\Omega}\right)\times\nabla\left(\frac{\delta B}{\delta\Omega}\right)\right)\mathrm{d}^2\mathbf{x}\end{aligned} \tag{32}$$

and the Hamiltonian

$$H = \frac{1}{2}\int \nabla\psi\cdot(1-\Delta)\nabla\psi\,\mathrm{d}^2\mathbf{x}.$$

Acknowledgments

The author is thankful to Prof. A. Constantin, Prof. V. Gerdjikov and Dr G. Grahovski for stimulating discussions. Partial support from INTAS grant No 05-1000008-7883 is acknowledged.

References

1. V. Arnold and B. Khesin, *Topological Methods in Hydrodynamics*, Springer Verlag, New York, 1998.
2. R. Beals, D. Sattinger and J. Szmigielski, *Inv. Problems* **15**, L1 (1999).

3. A. Boutet de Monvel and D. Shepelsky, *C. R. Math. Acad. Sci. Paris* **343**, 627 (2006).
4. R. Camassa and D. Holm, *Phys. Rev. Lett.* **71**, 1661 (1993).
5. A. Constantin, *Proc. R. Soc. Lond.* **A457**, 953 (2001).
6. A. Constantin and J. Escher, *Acta Math.* **181**, 229 (1998).
7. A. Constantin, V. Gerdjikov and R. Ivanov, *Inv. Problems* **22**, 2197 (2006); nlin. SI/0603019.
8. A. Constantin, V.S. Gerdjikov and R.I. Ivanov, *Inv. Problems* **23**, 1565 (2007); nlin. SI/0707. 2048.
9. A. Constantin and R. Ivanov, *Lett. Math. Phys.* **76**, 93 (2006).
10. A. Constantin and R. Ivanov, in *Topics in Contemporary Differential Geometry, Complex Analysis and Mathematical Physics*, eds: S. Dimiev and K. Sekigawa, World Scientific, London, 2007, 33–41; arXiv:0706. 3810. [nlin. SI].
11. A. Constantin, B. Kolev, *Comment. Math. Helv.* **78**, 787 (2003).
12. A. Constantin, B. Kolev, *J. Nonlinear Sci.* **16**, 109 (2006).
13. A. Constantin, W. Strauss, *Comm. Pure Appl. Math.* **53**, 603 (2000).
14. A. Constantin, W. Strauss, *J. Nonlinear Sci.* **12**, 415 (2002).
15. A. Fokas and B. Fuchssteiner, *Physica D* **4**, 47 (1981).
16. H. Guo, J. Shen, S. Wang and K. Xu, *Chinese Phys. Lett.* **6**, 53 (1989).
17. D. Holm, *Physica D* **133**, 215 (1999).
18. D. Holm, J. Marsden, in *The breadth of symplectic and Poisson geometry, Progr. Math.* **232**, Birkhäuser, Boston, MA, 2005.
19. D. Holm, J. Marsden and T. Ratiu, *Adv. Math.* **137**, 1 (1998).
20. J. Hoppe, *Phys. Lett. B* **215**, 706 (1988).
21. R. Ivanov, *Phys. Lett. A* **345**, 235 (2005); nlin. SI/0507005.
22. R. Ivanov, *Zeitschrift für Naturforschung* **61a**, 133 (2006); nlin. SI/0601066.
23. R. Ivanov, *Journal of Nonlinear Mathematical Physics* **15**, supplement 2, 1 (2008).
24. R. Johnson, *J. Fluid. Mech.* **457**, 63 (2002).
25. B. Khesin, G. Misiołek, *Adv. Math.* **176**, 116 (2003).
26. E.A. Kuznetsov and A.V. Mikhailov, *Phys. Lett. A* **77**, 37 (1980).
27. B. Kolev, *J. Nonlinear Math. Phys.* **11**, 480 (2004).
28. J. Lenells, *J. Diff. Eq.* **217**, 393 (2005).
29. D. Lewis, J. Marsden, R. Montgomery and T. Ratiu, *Physica D* **18**, 391 (1986).
30. G. Misiołek, *J. Geom. Phys.* **24**, 203 (1998).
31. P.J. Morrison, *Rev. Mod. Phys.* **70**, 467 (1998).
32. E. Reyes, *Lett. Math. Phys.* **59**, 117 (2002).
33. V. Zeitlin, *Phys. D* **49**, 353 (1991).
34. V. Zeitlin, *Phys. Lett. A* **164**, 177 (1992).

HYPERBOLIC GAUSS MAPS AND PARALLEL SURFACES IN HYPERBOLIC THREE-SPACE

MASATOSHI KOKUBU

Department of Mathematics, School of Engineering, Tokyo Denki University, Chiyoda-Ku, Tokyo 101-8457, Japan
E-mail: kokubu@cck.dendai.ac.jp

Keywords: Hyperbolic Gauss map; Parallel surface; Linear Weingarten surface.

1. Introduction

In the differential geometry of surfaces in hyperbolic three-space H^3, surfaces of constant mean curvature one (CMC-1 surfaces) and surfaces of constant Gaussian curvature zero (flat surfaces) are well studied. (cf. Refs. 1,2,4,6–13, etc.) These surfaces have representation formulas which turn the complex function theory into the efficacious tool. They play the role such as the Weierstrass-Enneper formula in the minimal surface theory of Euclidean space. Gálvez, Martínez and Milán (Ref. 3) also derived a representation formula for a certain class of Weingarten surfaces in H^3, which contains CMC-1 surfaces and flat surfaces. These Weingarten surfaces are the ones satisfying $\alpha(H-1)=\beta K$ for some constants α and β, where K denotes the Gaussian curvature and H the mean curvature. In Ref. 5, the author gave a refinement for their representation formula.

In these formulas, the hyperbolic Gauss maps (see Section 2 for the definition) play an important role. Moreover, one of the remarkable thing is that this class of Weingarten surfaces is closed under taking the parallel surfaces. The author thinks it rather important to investigate hyperbolic Gauss maps and parallel surfaces themselves, before representation formulas. For this reason, the purpose of this note is to collect some elementary properties of hyperbolic Gauss maps and parallel surfaces.

The author was supported by Grant-in-Aid for Scientific Research (C) No. 18540096 from the Japan Society for the Promotion of Science.

2. Basic items

Let L^4 be the Minkowski 4-space with the Lorentzian metric $\langle\ ,\ \rangle_L$, i.e., $\langle x,y\rangle_L = -x^0y^0 + \sum x^iy^i$ for $x=(x^\alpha), y=(y^\alpha)\in L^4$. As is well known,

$$H^3 := \{x\in L^4\,;\ \langle x,x\rangle_L = -1,\ x^0>0\}$$

is a simply-connected, complete 3-dimensional Riemannian manifold of constant negative curvature -1, which is called *hyperbolic 3-space.*

Let M^2 be an oriented connected surface and $f\colon M^2\to H^3$ an immersion. We denote by n the unit normal field along f. Let e_1, e_2 be a local orthonormal frame defined on $U\subset M^2$.

Structure equations. We consider f, e_1, e_2, n as L^4-valued functions and local frame field $(e_0=f, e_1, e_2, e_3=n)\colon U(\subset M^2)\to SO(1,3)$. The 1-form (ω_α^β) defined by $de_\alpha = \sum_{\beta=0}^3 e_\beta\otimes\omega_\alpha^\beta$ is valued in $\mathfrak{o}(1,3)$, the Lie algebra of $SO(1,3)$. In other words, 1-forms ω_α^β satisfy

$$\omega_0^0=0,\quad \omega_\alpha^\alpha=0,\quad -\omega_\alpha^0+\omega_0^\alpha=0,\quad \omega_\alpha^\beta+\omega_\beta^\alpha=0\quad (1\le\alpha,\beta\le 3).$$

Moreover, $\omega^3(=\omega_0^3)=0$ and the following equations hold:

$$de_0 = e_i\otimes\omega^i, \tag{1}$$
$$de_i = e_0\otimes\omega^i + e_j\otimes\omega_i^j + e_3\otimes\omega_i^3, \tag{2}$$
$$de_3 = e_j\otimes\omega_3^j, \tag{3}$$
$$d\omega^i = -\omega_j^i\wedge\omega^j, \tag{4}$$
$$0(=d\omega^3) = -\omega_j^3\wedge\omega^j, \tag{5}$$
$$d\omega_2^1 = -\omega_3^1\wedge\omega_2^3 - \omega^1\wedge\omega^2, \tag{6}$$
$$d\omega_j^3 = -\omega_k^3\wedge\omega_j^k. \tag{7}$$

Here, the indices i,j,k run over the range $1\le i,j,k\le 2$ and the Einstein's summation convention is used, i.e., the notation $\sum$ is omitted. Moreover, ω_0^i is denoted by ω^i for simplicity.

Fundamental forms, curvatures. The first and second fundamental forms I, II are, by definition,

$$\mathrm{I} = (\omega^1)^2 + (\omega^2)^2,\quad \mathrm{II} = \omega_1^3\omega^1 + \omega_2^3\omega^2.$$

Determining the locally defined functions h_{ij} by $\omega_i^3 = h_{ij}\omega^j$, we have $h_{ij}=h_{ji}$ and can write $\mathrm{II} = h_{ij}\omega^i\omega^j$. On the other hand, the *Gaussian curvature* K is intrinsically defined by the equation $d\omega_2^1 = K\omega^1\wedge\omega^2$. It follows from (6) that $K=-1+h_{11}h_{22}-(h_{12})^2$. It is usually called the *Gauss equation.* The eigenvalues κ_1, κ_2 and their eigenvectors of the symmetric matrix (h_{ij}) are called *principal curvatures* and *principal directions.* The Gaussian curvature K equals $-1+\kappa_1\kappa_2$. The average $H := (\kappa_1+\kappa_2)/2$ of the principal curvatures is called *mean curvature.* H equals $(h_{11}+h_{22})/2$.

Recall that the *third fundamental form* is $\mathrm{III} = (\omega_1^3)^2 + (\omega_2^3)^2$. We also set $\mathrm{IV} := \det(f, df, n, dn) = \omega^2\omega_1^3 - \omega^1\omega_2^3$, and call it the *fourth fundamental form.* A tangent vector v is a principal vector if and only if $\mathrm{IV}(v,v) = 0$. A curve $\gamma\colon I \to M^2$ on the surface is called the *curvature line* or the *line of curvature* if the tangent vectors are always principal. In other words, a curve γ satisfying $\gamma^*\mathrm{IV} = 0$ is the curvature line.

Hyperbolic Gauss maps. Let us consider the lightcone LC in L^4;

$$LC := \{x \in L^4 \setminus \{0\};\ \langle x, x\rangle_L = 0\}.$$

Moreover, we consider the projectification $\mathbb{P}(LC)$ of the lightcone LC. Then $\mathbb{P}(LC)$ is diffeomorphic to the 2-sphere S^2. It is naturally endowed with a conformal structure induced by $\langle\,,\,\rangle_L$. This conformal 2-sphere S^2 is called the *ideal boundary* of H^3, and is also denoted by ∂H^3. Since $f \pm n$ are lightlike vectors, we can define two maps $G^\pm = [f \pm n]\colon M^2 \to S^2 = \partial H^3$. These maps $G^\pm$ are called the *hyperbolic Gauss maps*.

Center surfaces. If one of the principal curvatures κ_i of the surface $f\colon M^2 \to H^3$ satisfies $|\kappa_i| > 1$ at $p \in M^2$, then the *radius of principal curvature* corresponding to κ_i can be defined. The *radius of principal curvature* is, by definition, the real number r_i determined by $\coth r_i = \kappa_i$. Using this r_i at every point $p \in M^2$, we can define another surface

$$C_i = \cosh r_i\, e_0 + \sinh r_i\, e_3\colon M^2 \to H^3 \subset L^4,$$

called the *center surface* or the *caustic* or the *focal surface* of f. C_i draws the locus of the centers of the principal curvature corresponding to κ_i. Note that, in general, the center surface C_i is not defined whole on M^2. The center surface C_i is defined only on the domain $U_i = \{\, p \in M^2\,;\ |\kappa_i(p)| > 1\,\}$.

Finally we introduce formulas for the surface $f = e_0$ and the unit normal field $n = e_3$ recovering from two center surfaces if they exist:

Lemma 2.1.

$$f = \frac{1}{\sinh(r_1 - r_2)}\left(C_2 \sinh r_1 - C_1 \sinh r_2\right), \tag{8}$$

$$n = \frac{1}{\sinh(r_1 - r_2)}\left(C_1 \cosh r_2 - C_2 \cosh r_1\right). \tag{9}$$

Proof. Since $C_1 = \cosh r_1\, e_0 + \sinh r_1\, e_3$ and $C_2 = \cosh r_2\, e_0 + \sinh r_2\, e_3$ hold, we have

$$\begin{aligned} C_1 \sinh r_2 - C_2 \sinh r_1 &= (\cosh r_1 \sinh r_2 - \sinh r_1 \cosh r_2) f \\ &= -\sinh(r_1 - r_2) f, \\ C_1 \cosh r_2 - C_2 \cosh r_1 &= (\sinh r_1 \cosh r_2 - \cosh r_1 \sinh r_2) n \\ &= \sinh(r_1 - r_2) n. \end{aligned}$$

□

3. Parallel surface

For a surface $f\colon M^2 \to H^3$ and a real number t, the *parallel surface* at the distance t is defined to be a surface $f_t = \cosh t \cdot f + \sinh t \cdot n\colon M^2 \to H^3$. The unit normal field n_t of f_t is given by $n_t = \sinh t \cdot f + \cosh t \cdot n$.

Singularities of parallel surfaces. In general, the parallel surface f_t may fail to be an immersion. In fact, since

$$\begin{aligned} df_t &= \cosh t\; de_0 + \sinh t\; de_3 = \cosh t\; e_i \omega^i + \sinh t\; e_i \omega_3^i \\ &= e_i (\cosh t\; \omega^i + \sinh t\; \omega_3^i), \end{aligned}$$

the first fundamental form $\mathrm{I}_t = \langle df_t, df_t \rangle_L$ is given by

$$\mathrm{I}_t = (\cosh t\; \omega^1 + \sinh t\; \omega_3^1)^2 + (\cosh t\; \omega^2 + \sinh t\; \omega_3^2)^2.$$

If we set $\theta^i = \cosh t\; \omega^i + \sinh t\; \omega_3^i$, the singularities of f_t appear at the point where $\theta^1 \wedge \theta^2 = 0$. At regular points, i.e., the point where $\theta^1 \wedge \theta^2 \neq 0$, θ^1 and θ^2 form an orthonormal coframe of f_t.

Proposition 3.1. *A point $p \in M^2$ is a singularity of f_t if and only if t coincides with the radius of principal curvature of f at p.*

Proof. $p \in M^2$ is a singularity of f_t if and only if $(\cosh t\, \omega^1 + \sinh t\, \omega_3^1) \wedge (\cosh t\, \omega^2 + \sinh t\, \omega_3^2) = 0$ at p. We will examine this condition taking a frame e_1, e_2 that are principal vectors at p. Because $\omega_1^3 = \kappa_1 \omega^1$ and $\omega_2^3 = \kappa_2 \omega^2$, we find that the condition p a singularity is equivalent to $\kappa_i = \coth t$ at p at least for one of $i = 1, 2$. □

Corollary 3.1. *Assume that an immersion $f\colon M^2 \to H^3$ has principal curvatures κ_1, κ_2 such that $|\kappa_i| \leq 1$. Then singularities never appear in its parallel surfaces.*

Corollary 3.2. *The singular locus $S_i = \{\, f_t(p) \mid p \in M^2, \kappa_i(p) = \coth t\}$ coincides with the image of the center surface C_i $(i = 1, 2)$.*

Invariants of parallel surfaces. The fundamental forms of f_t are written in terms of the fundamental forms of the original surface f as follows:

$$\mathrm{I}_t = \cosh^2 t\,\mathrm{I} - 2\cosh t \sinh t\,\mathrm{II} + \sinh^2 t\,\mathrm{III} \tag{10}$$

$$\mathrm{II}_t = -\cosh t \sinh t\,\mathrm{I} + (\cosh^2 t + \sinh^2 t)\,\mathrm{II} - \cosh t \sinh t\,\mathrm{III} \tag{11}$$

$$\mathrm{III}_t = \sinh^2 t\,\mathrm{I} - 2\cosh t \sinh t\,\mathrm{II} + \cosh^2 t\,\mathrm{III} \tag{12}$$

$$\mathrm{IV}_t = \mathrm{IV} \tag{13}$$

(13) asserts that the fourth fundamental form IV is invariant under the parallel surface transform. There also exist invariants other than IV.

Proposition 3.2. *The following (i)–(vii) are the common geometric invariants of parallel surfaces:*

(i) *hyperbolic Gauss maps* $G^{\pm}$
(ii) KdA*, where* dA *is the are element*
(iii) *the ratio* $[\kappa_1 - \kappa_2 : 1 - \kappa_1\kappa_2]$
(iv) *umbilical points*
(v) *lines of curvature*
(vi) *the difference* $r_1 - r_2$ *of radii of principal curvatures*
(vii) *center surfaces*

Proof.

(i) It is obvious from the following equation:

$$\begin{aligned}[f_t \pm n_t] &= [(\cosh t\, e_0 + \sinh t\, e_3) \pm (\sinh t\, e_0 + \cosh t\, e_3)] \\ &= [(\cosh t \pm \sinh t)(e_0 \pm e_3)] = [e_0 \pm e_3].\end{aligned}$$

(ii) Recall that $\mathrm{I}_t = (\theta^1)^2 + (\theta^2)^2$ and $\theta^i = \cosh t\, \omega^i + \sinh t\, \omega^i_3$. It follows from (4) and (7) that $d\theta^i = -\omega^i_j \wedge \theta^j$. It implies that ω^1_2 is also the connection form of f_t. Therefore $(d\omega^1_2 =) K_t \theta^1 \wedge \theta^2 = K\omega^1 \wedge \omega^2$.

(iii) Let p be an arbitrary point. Take a frame e_1, e_2 so that it is in the principal direction at p. Then $\mathrm{I}(p) = (\omega^1)^2 + (\omega^2)^2$, $\mathrm{II}(p) = \kappa_1(\omega^1)^2 + \kappa_2(\omega^2)^2$, $\mathrm{III}(p) = \kappa_1^2(\omega^1)^2 + \kappa_2^2(\omega^2)^2$. Hence, by (10), (11), we have

$$\mathrm{I}_t(p) = \sum_i \{\cosh^2 t - 2\kappa_i \cosh t \sinh t + \kappa_i^2 \sinh^2 t\}(\omega^i)^2$$

$$\mathrm{II}_t(p) = \sum_i \{-\cosh t \sinh t + \kappa_i(\cosh^2 t + \sinh^2 t) - \kappa_i^2 \cosh t \sinh t\}(\omega^i)^2$$

Therefore the principal curvature $\kappa_i^{(t)}$ of f_t is given by

$$\kappa_i^{(t)} = \frac{-\cosh t \sinh t + \kappa_i(\cosh^2 t + \sinh^2 t) - \kappa_i^2 \cosh t \sinh t}{\cosh^2 t - 2\kappa_i \cosh t \sinh t + \kappa_i^2 \sinh^2 t}$$

$$= \frac{-(\cosh t - \kappa_i \sinh t)(\sinh t - \kappa_i \cosh t)}{(\cosh t - \kappa_i \sinh t)^2}$$

$$= \frac{\kappa_i \cosh t - \sinh t}{-\kappa_i \sinh t + \cosh t} \tag{14}$$

at p. Since p is arbitrary, the equation (14) holds on M^2.

$\kappa_i^{(t)}$ is also written as $\kappa_i^{(t)} = (\kappa_i - \tanh t)/(-\kappa_i \tanh t + 1)$. Solving this for $\tanh t$, we have $\tanh t = (\kappa_i - \kappa_i^{(t)})(1 - \kappa_i \kappa_i^{(t)})$. Hence we have

$$\frac{\kappa_1 - \kappa_1^{(t)}}{1 - \kappa_1 \kappa_1^{(t)}} = \frac{\kappa_2 - \kappa_2^{(t)}}{1 - \kappa_2 \kappa_2^{(t)}} \quad \text{i.e.,} \quad \frac{\kappa_1 - \kappa_2}{1 - \kappa_1 \kappa_2} = \frac{\kappa_1^{(t)} - \kappa_2^{(t)}}{1 - \kappa_1^{(t)} \kappa_2^{(t)}}.$$

(iv) It immediately follows from (iii).

(v) It immediately follows from (13).

(vi) By (14), the radius $r_j^{(t)}$ of principal curvature of f_t satisfies

$$\begin{aligned}
\coth r_j^{(t)} &= \frac{\kappa_j \cosh t - \sinh t}{-\kappa_j \sinh t + \cosh t} = \frac{\coth r_j \cosh t - \sinh t}{-\coth r_j \sinh t + \cosh t} \\
&= \frac{\cosh r_j \cosh t - \sinh r_j \sinh t}{-\cosh r_j \sinh t + \sinh r_j \cosh t} = \frac{\cosh(r_j - t)}{\sinh(r_j - t)} \\
&= \coth(r_j - t).
\end{aligned}$$

Thus we have $r_j^{(t)} = r_j - t$. It implies that $r_1^{(t)} - r_2^{(t)} = r_1 - r_2$.

(vii) By $r_j^{(t)} = r_j - t$, the center surface $C_i^{(t)} = \cosh r_i^{(t)} \cdot f_t + \sinh r_i^{(t)} \cdot n_t$ of f_t is given by

$$\begin{aligned}
C_i^{(t)} &= \cosh(r_i - t)\{\cosh t \cdot f + \sinh t \cdot n\} \\
&\quad + \sinh(r_i - t)\{\sinh t \cdot f + \cosh t \cdot n\} \\
&= \cosh((r_i - t) + t) \cdot f + \sinh((r_i - t) + t) \cdot n \\
&= \cosh r_i \cdot f + \sinh r_i \cdot n = C_i.
\end{aligned}$$

□

It is natural to ask the converse problem of Proposition 3.2.

Theorem 3.1. *Suppose that there are two surfaces* $f, \tilde{f}\colon M^2 \to H^3$. *If* $C_1 = \tilde{C}_1$ *(or* $C_2 = \tilde{C}_2$*) and* $G^\pm = \tilde{G}^\pm$, *then* f *and* $\tilde{f}$ *belong to the common parallel family.*

Proof. Since the hyperbolic Gauss maps coincide, $e_0 + e_3 = \lambda(\tilde{e}_0 + \tilde{e}_3)$ holds for some λ. It follows that the condition $C_1 = \tilde{C}_1$, i.e., $\cosh r_1 \, e_0 + \sinh r_1 \, e_3 = \cosh \tilde{r}_1 \, \tilde{e}_0 + \sinh \tilde{r}_1 \, \tilde{e}_3$ is equivalent to

$$\begin{aligned}
&\cosh r_1 \, e_0 + \sinh r_1 \{\lambda(\tilde{e}_0 + \tilde{e}_3) - e_0\} = \cosh \tilde{r}_1 \, \tilde{e}_0 + \sinh \tilde{r}_1 \, \tilde{e}_3, \\
&(\cosh r_1 - \sinh r_1) \, e_0 = (\cosh \tilde{r}_1 - \lambda \sinh r_1) \, \tilde{e}_0 + (\sinh \tilde{r}_1 - \lambda \sinh r_1) \, \tilde{e}_3.
\end{aligned}$$

Hence, we can set $e_0 = a\tilde{e}_0 + b\tilde{e}_3$. Taking the norm of both side of $e_0 = a\tilde{e}_0 + b\tilde{e}_3$, we find that $a^2 - b^2 = 1$.

On the one hand, by $\langle de_0, e_0\rangle_L = \langle de_0, e_3\rangle_L = 0$, we have $\langle de_0, e_0 \pm e_3\rangle_L = 0$. Thus, $\langle de_0, \tilde{e}_0 \pm \tilde{e}_3\rangle_L = 0$. Hence, $\langle de_0, \tilde{e}_0\rangle_L = \langle de_0, \tilde{e}_3\rangle_L = 0$.

On the other hand, differentiating $e_0 = a\tilde{e}_0 + b\tilde{e}_3$, we have $de_0 = da\tilde{e}_0 + ad\tilde{e}_0 + db\tilde{e}_3 + bd\tilde{e}_3$ and then

$$-da + b\langle d\tilde{e}_3, \tilde{e}_0\rangle_L = 0, \quad a\langle d\tilde{e}_0, \tilde{e}_3\rangle_L + db = 0.$$

Using $\langle d\tilde{e}_3, \tilde{e}_0\rangle_L = 0$ and $\langle d\tilde{e}_0, \tilde{e}_3\rangle_L = 0$, we have $da = db = 0$. Thus a and b are constants. It means that e_0 is a parallel surface of $\tilde{e}_0$. □

Theorem 3.2. *Suppose that there are two surfaces $f, \tilde{f}\colon M^2 \to H^3$. If they have the common center surfaces (i.e., $C_1 = \tilde{C}_1$, $C_2 = \tilde{C}_2$) and the same difference of radii of principal curvature (i.e., $r_1 - r_2 = \tilde{r}_1 - \tilde{r}_2$), then f and $\tilde{f}$ belong the same parallel family.*

Proof. It follows from (8) and (9) that

$$\tilde{f} = \frac{1}{\sinh(r_1 - r_2)}\left(C_2 \sinh \tilde{r}_1 - C_1 \sinh \tilde{r}_2\right),$$
$$\tilde{n} = \frac{1}{\sinh(r_1 - r_2)}\left(C_1 \cosh \tilde{r}_2 - C_2 \cosh \tilde{r}_1\right).$$

It implies that

$$\begin{aligned}[\tilde{f} \pm \tilde{n}] &= [C_1(\pm \cosh \tilde{r}_2 - \sinh \tilde{r}_2) + C_2(\sinh \tilde{r}_1 \mp \cosh \tilde{r}_1)] \\ &= [C_1 + C_2(-e^{\tilde{r}_2 - \tilde{r}_1})] \text{ or } [C_1 - C_2 e^{-\tilde{r}_2 + \tilde{r}_1}] \\ &= [C_1 + C_2(-e^{r_2 - r_1})] \text{ or } [C_1 - C_2 e^{-r_2 + r_1}] = [f \pm n],\end{aligned}$$

that is, the hyperbolic Gauss maps coincide. Therefore Theorem 3.1 completes the proof. □

4. Linear Weingarten surface

Recall that the surface is called a *Weingarten surface* if the Gaussian curvature K and the mean curvature H are dependent, i.e., $dK \wedge dH = 0$. One can also say that a surface is Weingarten if and only if $d\kappa_1 \wedge d\kappa_2 = 0$.

The formula (14) immediately implies the following proposition:

Proposition 4.1. *Any parallel surface of Weingarten surface is also Weingarten.*

A surface satisfying $aK + bH + c = 0$ for some $[a : b : c] \in \mathbb{R}P^2$ is called a *linear Weingarten surface.*

Proposition 4.2. *Any parallel surface of a linear Weingarten surface is also linear Weingarten.*

Proof. It follows from (14) that the Gaussian curvature K_t and the mean curvature H_t of the parallel surface f_t is given by

$$K_t = \frac{K}{K\sinh^2 t - 2H\cosh t \sinh t + \cosh^2 t + \sinh^2 t}, \tag{15}$$

$$H_t = \frac{H(\cosh^2 t + \sinh^2 t) - (2 + K)\cosh t \sinh t}{K\sinh^2 t - 2H\cosh t \sinh t + \cosh^2 t + \sinh^2 t}, \tag{16}$$

where K, H are Gaussian, mean curvatures of f. By (15), (16), we have

$$K = (K_t)_{-t} = \frac{K_t}{K_t \sinh^2 t + 2H_t \cosh t \sinh t + \cosh^2 t + \sinh^2 t},$$

$$H = (H_t)_{-t} = \frac{H_t(\cosh^2 t + \sinh^2 t) + (2 + K_t)\cosh t \sinh t}{K_t \sinh^2 t + 2H_t \cosh t \sinh t + \cosh^2 t + \sinh^2 t}.$$

It follows from the condition $aK + bH + c = 0$ that

$$\begin{aligned} 0 = &\{a + b\cosh t \sinh t + c\sinh^2 t\}K_t \\ &+ \{b(\cosh^2 t + \sinh^2 t) + 2c\cosh t \sinh t\}H_t \\ &+ 2b\cosh t \sinh t + c(\cosh^2 t + \sinh^2 t). \end{aligned} \tag{17}$$

It means that f_t is linear Weingarten. □

In the right-hand side of the equality (17) above, we shall denote each coefficient by a_t, b_t, c_t, respectively:

$$\begin{aligned} a_t &= \{a + b\cosh t \sinh t + c\sinh^2 t\}, \\ b_t &= \{b(\cosh^2 t + \sinh^2 t) + 2c\cosh t \sinh t\}, \\ c_t &= 2b\cosh t \sinh t + c(\cosh^2 t + \sinh^2 t). \end{aligned}$$

The ratio $[a_t : b_t : c_t]$ can be written as

$$\begin{aligned} [a_t : b_t : c_t] &= [a : b : c] \begin{bmatrix} 1 & 0 & 0 \\ \cosh t \sinh t & \cosh^2 t + \sinh^2 t & 2\cosh t \sinh t \\ \sinh^2 t & 2\cosh t \sinh t & \cosh^2 t + \sinh^2 t \end{bmatrix} \\ &= [1 : \cosh 2t : \sinh 2t] \begin{bmatrix} 2a - c & 0 & 0 \\ c & 2b & 2c \\ b & 2c & 2b \end{bmatrix}. \end{aligned}$$

The rank of the matrix $\begin{bmatrix} 2a-c & 0 & 0 \\ c & 2b & 2c \\ b & 2c & 2b \end{bmatrix}$ is given by

$$\mathrm{rk} = \begin{cases} 3 & \text{when } c \neq 2a, b^2 \neq c^2 \\ 2 & \text{when } (c \neq 2a, b^2 = c^2 \neq 0) \text{ or } (c = 2a, b^2 \neq c^2) \\ 1 & \text{when } (c \neq 2a, b = c = 0) \text{ or } (c = 2a, b^2 = c^2) \end{cases}$$

$$= \begin{cases} 1 & \text{when } [a:b:c] = [1:0:0] \text{ or } [1:\pm 2:2] \\ 2 & \text{when } [a:b:c] = [a:\pm b:b](2a \neq b \neq 0) \text{ or} \\ & \quad [a:b:2a](4a^2 \neq b^2) \\ 3 & \text{otherwise} \end{cases}$$

$$= \begin{cases} 1 & \text{when } K = 0 \text{ or } K \pm 2H + 2 = 0 \\ 2 & \text{when } aK + b(1 \pm H) = 0 (2a \neq b \neq 0) \text{ or} \\ & \quad a(K+2) = bH (4a^2 \neq b^2) \\ 3 & \text{otherwise.} \end{cases}$$

Here, rank 3 means the curve $r\colon t \mapsto [a_t : b_t : c_t]$ is full, rank 2 means the curve r lies in the one-dimensional subspace, and rank 1 means that r degenerates to a single point.

The observation above leads us to the following proposition:

Proposition 4.3.

(i) *Any parallel surface of a flat surface is also flat.*

(ii) *Any parallel surface of a surface satisfying $2H = \pm(K+2)$ also satisfies $2H = \pm(K+2)$.*

(iii) *Let f be a surface satisfying $bH = a(K+2)$ for some ratio $[a:b](\neq [\pm 1 : 2])$. Then the parallel surfaces of f also satisfy the condition $b'H = a'(K+2)$.*

(iv) *Let f be a surface satisfying $aK + b(1 \pm H) = 0$ for some ratio $[a:b]$ $(\neq [1:0],[1:2])$. Then the parallel surfaces of f also satisfy the condition $a'K + b'(1 \pm H) = 0$.*

References

1. R. Bryant, *Surfaces of mean curvature one in hyperbolic space*, in *Théorie des variétés minimales et applications*, Astérisque, **154–155** (1988), 321–347.
2. J.A. Gálvez, A. Martínez and F. Milán, *Flat surfaces in hyperbolic 3-space*, Math. Ann. **316** (2000), 419–435.

3. J.A. Gálvez, A. Martínez and F. Milán, *Complete linear Weingarten surfaces of Bryant type. A Plateau problem at infinity*, Trans. Amer. Math. Soc. **356** (2004), no. 9, 3405–3428.
4. Y. Kawakami, *Ramification estimates for the hyperbolic Gauss map*, preprint.
5. M. Kokubu, *Surfaces and fronts with harmonic-mean curvature one in hyperbolic three-space*, preprint.
6. M. Kokubu, W. Rossman, K. Saji, M. Umehara and K. Yamada, *Singularities of flat fronts in hyperbolic space*, Pacific J. Math. **221** 303–351.
7. M. Kokubu, W. Rossman, M. Umehara and K. Yamada, *Flat fronts in hyperbolic 3-space and their caustics*, J. Math. Soc. Japan **59** (2007), no. 1, 265–299.
8. M. Kokubu, W. Rossman, M. Umehara and K. Yamada, *Asymptotic behavior of flat surfaces in hyperbolic 3-space*, preprint.
9. M. Kokubu, M. Umehara and K. Yamada, *An elementary proof of Small's formula for null curves in* $\mathrm{PSL}(2,\mathbf{C})$ *and an analogue for Legendrian curves in* $\mathrm{PSL}(2,\mathbf{C})$, Osaka J. Math. **40** (2003), no. 3, 697–715.
10. M. Kokubu, M. Umehara and K. Yamada, *Flat fronts in hyperbolic 3-space*, Pacific J. Math. **216** (2004), no. 1, 149–175.
11. P. Roitman, *Flat surfaces in hyperbolic 3-space as normal surfaces to a congruence of geodesics*, Tohoku Math. J. **59** (2007), no. 1, 21–37.
12. W. Rossman, M. Umehara and K. Yamada, *A new flux for mean curvature 1 surfaces in hyperbolic 3-space, and applications*, Proc. Amer. Math. Soc. **127** (1999), 2147–2154.
13. M. Umehara and K. Yamada, *Complete surfaces of constant mean curvature 1 in the hyperbolic 3-space*, Ann. of Math. (2) **137** (1993), no. 3, 611–638.

ON THE LAX PAIR FOR TWO AND THREE WAVE INTERACTION SYSTEM

N. A. KOSTOV

Institute for Nuclear Research and Nuclear Energy,
Bulgarian Academy of Sciences,
72 Tsarigradsko chaussee, 1784 Sofia, Bulgaria
E-mail: nakostov@inrne.bas.bg

We study two wave interacting system by the inverse scattering method. We develop the algebraic approach in terms of classical r-matrix and give an interpretation of the Poisson brackets as linear r-matrix algebra. The solutions are expressed in terms of Weierstrass functions and Jacobian functions.

1. Two-wave interaction system and monomer

Several studies have appeared recently on discrete systems of coupled quadratic nonlinear oscillators, each with two frequencies, a fundamental one W_n and a second harmonic V_n close to resonance [1,2]. A prototype of these systems is of the form

$$i\partial_\xi W_n + (W_{n+1} + W_{n-1}) + W_n^* V_n = 0, \tag{1}$$

$$i\partial_\xi V_n + \eta(V_{n+1} + V_{n-1}) - \alpha V_n + \frac{1}{2} W_n^2 = 0, \tag{2}$$

where ξ is the evolution coordinate and η is the ratio between nearest neighbor coupling strength of the second harmonic V_n and the fundamental W_n. The quantity α corresponds to the normalized wave number mismatch. This system is used in optics to describe arrays of quadratic nonlinear waveguides [2] and in solid state physics to describe nonlinear interface waves between two media close to Fermi resonance. For a single waveguide ($n = 1$) Eqs. (1), (2) reduce to second harmonic generation, which is one of earliest and most well studied effects of nonlinear optics [3,4]. The case $n = 1$ we shall refer as *monomer* [1] or explicitly we have

$$i\frac{dW_1}{d\xi} + W_1^* V_1 = 0, \tag{3}$$

$$i\frac{dV_1}{d\xi} - \alpha V_1 + \frac{1}{2}W_1^2 = 0 \tag{4}$$

and introduce the following Lax representation

$$\frac{dL}{d\xi} = [L, M], \tag{5}$$

where L, M are 2×2 matrices satisfying the following linear system:

$$\frac{d\psi}{d\xi} = M(\xi,\lambda)\psi(\xi,\lambda) \quad L(\xi,\lambda)\psi(\xi,\lambda) = 0. \tag{6}$$

In explicit terms we have that

$$L(\xi,\lambda) = \begin{pmatrix} i\lambda - i|W_1|^2/(4\lambda) - i\alpha/2 & V_1 - W_1^2/(4\lambda) \\ V_1^* - W_1^{*2}/(4\lambda) & -i\lambda + i|W_1|^2/(4\lambda) + i\alpha/2 \end{pmatrix}, \tag{7}$$

$$M(\xi,\lambda) = \begin{pmatrix} i\lambda & V_1 \\ V_1^* & -i\lambda \end{pmatrix}. \tag{8}$$

To integrate the system (3), (4) we introduce new variable

$$\mu = (i\frac{dV_1}{d\xi} - \alpha V_1)/2V_1 = -\frac{W_1^2}{4V_1}, \tag{9}$$

in terms of which our equations can be written as

$$\frac{d\mu}{d\xi} = 2i\sqrt{R(\mu)}, \tag{10}$$

where

$$\begin{aligned} R(\lambda) &= \lambda^4 - \alpha_1\lambda^3 + \alpha_2\lambda^2 - \alpha_3\lambda + \alpha_4 \\ &= \left(\lambda^2 + \frac{1}{2}\alpha\lambda - \frac{1}{4}|W_1|^2\right)^2 - |V_1|^2(\lambda-\mu)(\lambda-\mu^*). \end{aligned}$$

The μ variable and V_1, W_1 obey the equations

$$-\alpha = \alpha_1, \qquad \frac{\alpha^2}{4} - |V_1|^2 - \frac{1}{2}|W_1|^2 = \alpha_2,$$

$$\frac{\alpha}{4}|W_1|^2 - |V_1|^2(\mu + \mu^*) = \alpha_3, \tag{11}$$

$$\frac{1}{16}|W_1|^4 - \mu\mu^*|V_1|^2 = \alpha_4, \tag{12}$$

which are related to the integrals of motion of the monomer system with $\alpha_4 = 0$. The equation of motion is then

$$\left(\frac{d\mu}{d\xi}\right)^2 + 4\left(\mu^4 + \alpha\mu^3\right) + (\alpha^2 - 2N)\mu^2 - H_1\mu = 0, \tag{13}$$

where the system (3), (4) conserves the dimensionless power N and the Hamiltonian H

$$N = |W_1|^2 + 2|V_1|^2, \qquad H_1 = \alpha N - 2H,$$
$$H = \alpha |V_1|^2 - \frac{1}{2} W_1^2 V_1^* - \frac{1}{2} V_1 W_1^{*2}.$$

Solving Eqs. (12) for μ variable we obtain

$$\mu = -\alpha/4 + \frac{1}{4\nu}\left(H + i\sqrt{P(\nu)}\right), \tag{14}$$

where

$$P(\nu) = 4\nu^3 - (4N + \alpha^2)\nu^2 + (2\alpha H + N^2)\nu - H^2. \tag{15}$$

We seek the solution V_1 in the following form

$$V_1 = \sqrt{\nu(\xi)} \exp\left(-i\alpha\xi/2 + iC \int_0^\xi \frac{d\xi'}{\nu(\xi')}\right) = \sqrt{\nu} \exp(i\psi(\xi)), \tag{16}$$

where $\nu = |V_1|^2 = \wp(\xi + \omega') + C_1$, $\wp$ is the Weierstrass function, and ω' is half period.Using Eq. (14) and the following equation

$$\frac{d\nu}{d\xi} = -2i\nu(\mu - \mu^*), \tag{17}$$

derived from (9) and the monomer equations we obtain

$$\left(\frac{d\nu}{d\xi}\right)^2 = 4\nu^3 - (\alpha^2 + 4N)\nu^2 + (2\alpha H + N^2)\nu - H^2, \tag{18}$$

whose solution can be expressed in terms of the Weierstrass elliptic functions as

$$\nu = \wp(\xi + \omega') + (\alpha^2 + 4N)/12. \tag{19}$$

Substituting this expression in Eq. (16) we obtain

$$V_1 = \sqrt{\wp(\xi + \omega') + (\alpha^2 + 4N)/12} \exp(i\psi(\xi)), \tag{20}$$

where the phase $\psi(\xi)$ is given by

$$\psi(\xi) = -\alpha\xi/2 - \frac{H}{2\wp'(\kappa)}\left(\ln \frac{\sigma(\xi + \omega' - \kappa)}{\sigma(\xi + \omega' + \kappa)} + 2\zeta(\kappa)\xi\right) + \psi_0, \tag{21}$$

and ψ_0 is initial constant phase. Here we are using the well known relation from elliptic functions theory [5]

$$\int \frac{dz}{\wp(z) - \wp(\kappa)} = \frac{1}{\wp'(\kappa)}\left(2z\zeta(\kappa) + \ln \frac{\sigma(z - \kappa)}{\sigma(z + \kappa)}\right),$$

with the parameters κ, C_1, C satisfying

$$\wp'(\kappa) = iH, \quad C_1 = -\wp(\kappa) = \frac{1}{12}(\alpha^2 + 4N), \quad C = -\frac{1}{2}H. \tag{22}$$

Using the relation between $\wp$ and sn functions we have

$$\nu(\xi) = \nu_1 + (\nu_2 - \nu_1)\mathrm{sn}^2\sqrt{\nu_3 - \nu_1}(\xi + \xi_0, k). \tag{23}$$

Here $\mathrm{sn}(W, k)$ is the Jacobi elliptic function determined by the three real roots $0 \leq \nu_1 \leq \nu_2 \leq \nu_3 \leq N/2$ and with modulus $k^2 = (\nu_2 - \nu_1)/(\nu_3 - \nu_1)$, ξ_0 is determined by initial condition. These results are in agreement with results obtained in [1,6,7]. The integrability of this system can also be seen in terms of the existence of a classical r-matrix algebra. To be specific, introduce standard Poisson bracket, $\{\cdot;\cdot\}$

$$\{f;g\} = -i\left(\frac{\partial f}{\partial V_1}\frac{\partial g}{\partial V_1^*} - \frac{\partial f}{\partial V_1^*}\frac{\partial g}{\partial V_1}\right) - i\left(\frac{\partial f}{\partial W_1}\frac{\partial g}{\partial W_1^*} - \frac{\partial f}{\partial W_1^*}\frac{\partial g}{\partial W_1}\right),$$

and denote the entries of L-matrix as follows,

$$L = \begin{pmatrix} A(\lambda) & B(\lambda) \\ C(\lambda) & D(\lambda) \end{pmatrix}. \tag{24}$$

Then the L operator yields the linear r-matrix algebra

$$\{A(\lambda); A(\mu)\} = \{B(\lambda); B(\mu)\} = \{C(\lambda); C(\mu)\} = \{D(\lambda); D(\mu)\} = 0,$$

$$\{A(\lambda); B(\mu)\} = \{B(\lambda); D(\mu)\} = -\frac{1}{2(\lambda - \mu)}(B(\lambda) - B(\mu)),$$

$$\{A(\lambda); C(\mu)\} = \{C(\lambda); D(\mu)\} = \frac{1}{2(\lambda - \mu)}(C(\lambda) - C(\mu)),$$

$$\{B(\lambda); C(\mu)\} = -\frac{1}{(\lambda - \mu)}(A(\lambda) - A(\mu)),$$

$$\{A(\lambda); D(\mu)\} = 0,$$

which leads to the r-matrix representation. Indeed, by introducing the 4×4-matrices $L_1(\lambda) = L(\lambda) \otimes 1_2$ and $L_2(\lambda) = 1_2 \otimes L(\lambda)$ one recognize that

$$\{L_1(\lambda) \underset{,}{\otimes} L_2(\mu)\} = [r(\lambda - \mu), L_1(\lambda) + L_2(\mu)] \tag{25}$$

where the 4×4-matrix$\{L_1(\lambda) \underset{,}{\otimes} L_2(\mu)\}$ represents itself the direct product of the matrices $L(\lambda)$ and $L(\mu)$ with the products of two matrix elements replaced by the Poisson bracket and the r-matrix has the form

$$r(\lambda - \mu) = \frac{1}{2}\frac{P}{\lambda - \mu}, \qquad P = \begin{pmatrix} 1 & 0 & 0 & 0 \\ 0 & 0 & 1 & 0 \\ 0 & 1 & 0 & 0 \\ 0 & 0 & 0 & 1 \end{pmatrix}.$$

This algebra can be extended at the quantum level (quantum linear R-matrix algebra [8]) this leading to the quantization of the system in the same way as done in [9] for the DST dimer. The bihamiltonian structures for different Hamiltonian systems are given in [11,12].

2. Three-wave interaction system

Let us consider three wave system [10],

$$i\frac{dA_1}{d\xi} = \epsilon A_3 A_2^*, \qquad i\frac{dA_2}{d\xi} = \epsilon A_1^* A_3, \qquad i\frac{dA_3}{d\xi} = \epsilon A_1 A_2. \tag{26}$$

The corresponding elements of Lax matrices are

$$L(\xi,\lambda) = \begin{pmatrix} -i\lambda/2 + i\mathcal{A}/(2\lambda) & -i\epsilon A_1 - iA_3 A_2^*/\lambda \\ -i\epsilon A_1^* - iA_2 A_3^*/\lambda & i\lambda/2 - i\mathcal{A}/(2\lambda) \end{pmatrix}, \tag{27}$$

$$M(\xi,\lambda) = \begin{pmatrix} -i\lambda/2 & -i\epsilon A_1 \\ -i\epsilon A_1^* & -i\lambda/2 \end{pmatrix}, \quad \mathcal{A} = |A_2|^2 - |A_3|^2. \tag{28}$$

To integrate the system (26) we introduce new variable with $\epsilon = 1$

$$\mu = i\frac{1}{A_1}\frac{dA_1}{d\xi} = \frac{A_3 A_2^*}{A_1}, \tag{29}$$

in terms of which our equations can be written as

$$\frac{d\mu}{d\xi} = 2i\sqrt{R(\mu)}, \tag{30}$$

where

$$R(\lambda) = \left(\frac{1}{4}\lambda^2 - \frac{1}{2}\mathcal{A}\right)^2 - |A_1|^2(\lambda-\mu)(\lambda-\mu^*) \tag{31}$$

$$= \frac{1}{4}\lambda^4 - \alpha_1\lambda^3 + \alpha_2\lambda^2 - \alpha_3\lambda + \alpha_4. \tag{32}$$

The μ variable and $A_j, j = 1,\dots 3$ obey the equations

$$\alpha_1 = 0, \qquad |A_1|^2 - \frac{1}{2}\mathcal{A} = \alpha_2, \tag{33}$$

$$-|A_1|^2(\mu+\mu^*) = \alpha_3, \qquad \frac{1}{4}\mathcal{A} - \mu\mu^*|A_1|^2 = \alpha_4. \tag{34}$$

The equation of motion is then

$$\left(\frac{d\mu}{d\xi}\right)^2 = -4\left(\mu^4 + N\mu^2 - H\mu\right) \tag{35}$$

where the system (26) conserves the dimensionless variable N and the Hamiltonian H

$$N = |A_1|^2 - \frac{1}{2}\mathcal{A}, \qquad H = A_1 A_2 A_3^* + A_3 A_1^* A_2^*. \tag{36}$$

Solving Eqs. (34) for μ variable we obtain

$$\mu = \frac{1}{4\nu}\left(H + i\sqrt{P(\nu)}\right), \tag{37}$$

where

$$P(\nu) = 4\nu^3 - 4N\nu^2 + N^2\nu - H^2. \tag{38}$$

We seek the solution A_1 in the following form

$$A_1 = \sqrt{\nu(\xi)}\exp\left(iC\int_0^\xi \frac{d\xi'}{\nu(\xi')}\right) = \sqrt{\nu(\xi)}\exp(i\psi(\xi)), \tag{39}$$

where $\nu = |A_1|^2 = \wp(\xi + \omega') + C_1$, $\wp$ is the Weierstrass function, and ω' is half period.Using Eq. (37) and the following equation

$$\frac{d\nu}{d\xi} = -2i\nu(\mu - \mu^*), \tag{40}$$

derived from (29) and three wave equations we obtain

$$\left(\frac{d\nu}{d\xi}\right)^2 = 4\nu^3 - 4N\nu^2 + N^2\nu - H^2, \tag{41}$$

whose solution can be expressed in terms of the Weierstrass elliptic function $\wp$ as

$$\nu = \wp(\xi + \omega') + \frac{N}{3}. \tag{42}$$

Substituting this expression in Eq. (39) we obtain

$$A_1 = \sqrt{\wp(\xi + \omega') + \frac{N}{3}}\exp(i\psi(\xi)), \tag{43}$$

where the phase $\psi(\xi)$ is given by

$$\psi(\xi) = \frac{H}{2\wp'(\kappa)}\left(\ln\frac{\sigma(\xi + \omega' - \kappa)}{\sigma(\xi + \omega' + \kappa)} + 2\zeta(\kappa)\xi\right) + \psi_0, \tag{44}$$

where σ, ζ are Weierstrass functions and ψ_0 is initial constant phase.

Acknowledgements

It is my pleasure to thank Prof. M. Salerno and Prof. V.Z. Enolskii for useful and stimulating discussions.

References

1. O. Bang, P.L. Christiansen, C.B. Clausen, *Stationary solutions and self-trapping in discrete quadratic nonlinear systems*, Phys. Rev E **56**, 7257–7266, 1997.
2. T. Peschel, U. Peschel, F. Lederer and B.A. Malomed, *Solitary waves in Bragg gratings with a quadratic nonlinearity*, Phys. Rev. E **57**, 1127–1133, 1998.
3. J.A. Armstrong, N. Bloembergen, J. Ducuing, P.S. Pershan, *Interactions between light waves in a nonlinear dielectric*, Phys. Rev. **127**, 1918–1939, 1962.
4. S.A. Akhmanov , R.V. Khokhlov, *Problems of Nonlinear optics*, Gordon and Breach, New York, 1972; Russian edition, 1965.
5. M. Abramowitz, I.A. Stegun, *Handbook of Mathematical functions*, National Bureau, New Work, 1964.
6. A.M. Kamchatnov, *On improving the effectiveness of periodic solutions of the NLS and DNLS equations*, J. Phys. A: Math. Gen. **23**, 2945–2960, 1990.
7. K.R. Khusnutdinova, H. Steudel, *Second harmonic generation:Hamiltonian structures and particular solutions*, J. Math. Phys. **39**, 3754–3764, 1998.
8. P.P. Kulish and E.K. Sklyanin, in: Lecture Notes in Physics vol. 151, Springer, Berlin, 61, 1982.
9. V.Z. Enolskii, M. Salerno, N.A. Kostov, A.C. Scott, *Alternative quantization of the discrete self-trapping dimer system*, Physica Scripta **42**, 229–235, 1991.
10. N.A. Kostov, A.V. Tsiganov, *New Lax pair for restricted multiple three wave interaction system, quasiperiodic solutions and bi-hamiltonian structure*, arXiv:0705.4363, 2007.
11. A.V. Tsiganov, *A family of the Poisson brackets compatible with the Sklyanin bracket*, J. Phys.A: Math. Theor. **40**, 4803–4816, 2007.
12. A.V. Tsiganov, *On the two different bi-Hamiltonian structures for the Toda lattice*, J. Phys. A: Math. Theor. **40**, 6395–6406, 2007.

MATHEMATICAL OUTLOOK OF FRACTALS AND CHAOS RELATED TO SIMPLE ORTHORHOMBIC ISING-ONSAGER-ZHANG LATTICES

Dedicated to Professor Stancho Dimiev on the occasion of his seventy fifth birthday

JULIAN ŁAWRYNOWICZ*

Institute of Physics, University of Łodź, Pomorska 149/153, PL-9236 Łodź, Poland and Institute of Mathematics, Polish Academy of Sciences, Łodź Branch, Banacha 22, PL-90-236 Łodź, Poland
E-mail: jlawryno@uni.lodz.pl,

STEFANO MARCHIAFAVA

Departimento di Matematica "Guido Castelnuovo", Universita di Roma I "La Sapienza", Piazzale Aldo Moro, 2, I-00-185 Roma, Italia
E-mail: marchiaf@mat.uniroma1.it

MAŁGORZATA NOWAK-KĘPCZYK

High School of Business, Kolejowa 22, PL-26-600 Radom, Poland
E-mail: gosianmk@poczta.onet.pl

The paper is inspired by an elegant generalization by Z.-D. Zhang (2007) of the exact solution by L. Onsager (1944) to the problem of description of the E. Ising lattices. We start with mentioning the relationship with Bethe-type fractals. We characterize then the differences when taking into account the first, second, and third nearest neighbours of an atom in an alloy. In turn we pass to discussing the Ising-Onsager-Zhang lattices in the context of possible applications of the Jordan-von Neumann-Wigner (1934) approach (Theorem 1).

2000 *Mathematics Subject Classification*: 28A78, 37F45, 17C90, 11R52, 15A66, 30D40, 82D25.

*Research of the author partially supported by the Ministry of Science and Higher Education grant P03A 001 26 (Sections 1-2 of the paper) and partially by the grant of the University of Łódź no. 505/692 (Sections 3–5).

We decide to use some fractals of the algebraic Clifford structure in relation with the quaternionic sequence of Jordan algebras, and this appears to be quite promising (Theorem 2). Finally we outline further perspectives of investigation in this direction.

Keywords: Fractals; Chaos; Ising lattice; Clifford structure; Quaternion; Jordan algebra; Bilinear form; Quadratic form.

1. Introduction and relationship with Bethe-type fractals

The idea of fractal modeling of crystals comes back to Bethe [3] who observed its convenience when coming to first, second and third nearest neighbours etc. Taking into account that an atom is a neighbour of two or more other atoms, even in the case of one layer with a lattice formed by squares, one naturally comes to the notion of *cluster* [4,12]. It is then natural to cut the plane of lattice (dashed lines – – on Fig. 1) correspondingly to the cluster involved and construct a *Riemann surface* or a *Bethe lattice* — a fractal set of the *branch type* [7,14] (Fig. 2). The construction is parallel to that related to the holomorphic function $f(z) = \exp z^2$ in $\mathbb{C}$ (Fig. 3); cf. also [1].

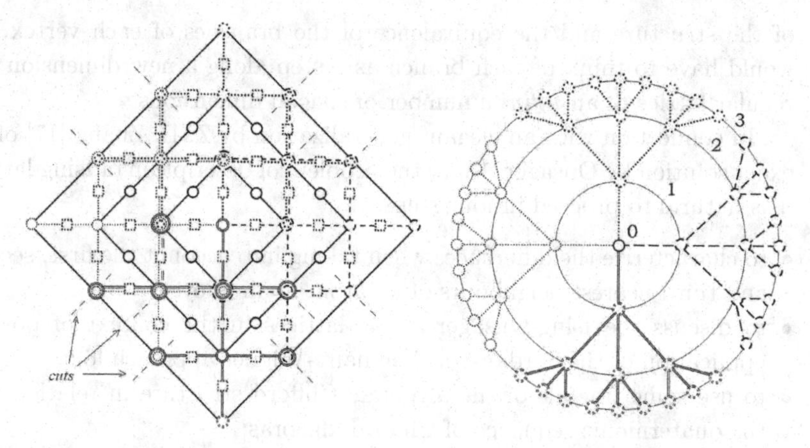

Fig. 1. Cuts – – in the plane of lattice. Fig. 2. A Bethe lattice.

The Bethe method allows us to discuss the order at large and small distances, distinguishing two degrees of order, influence of the number of dimensions, qualitative discussion of the transition point, and approximation for vanishing long distance order; it gives an exact solution for the Ising model [5] on a Bethe lattice. Unfortunately, the lattice cannot be realized as a physical system in the sense that in order to preserve the homogeneity

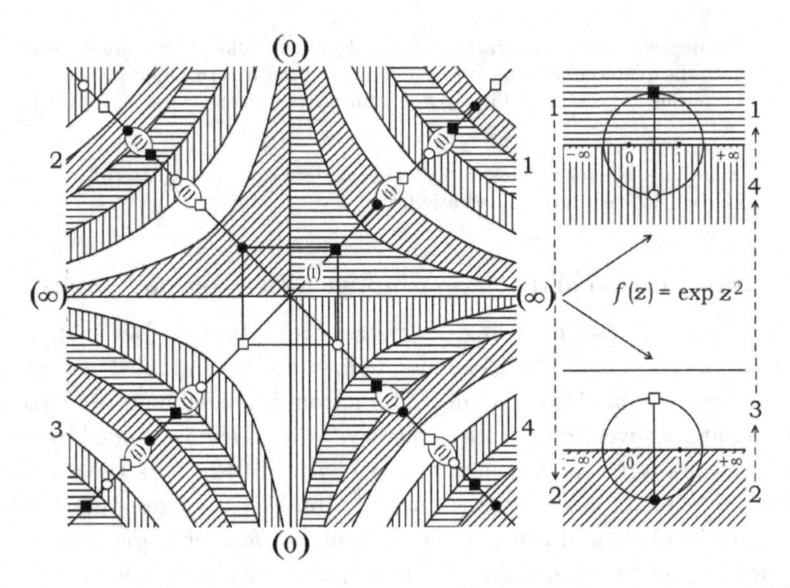

Fig. 3. Construction of the Riemann surface corresponding to the function $f(z) = \exp z^2$ in $\mathbb{C}$.

of the structure and the equivalence of the branches of each vertex, one would have to think of each branch as a step along a new dimension in a regular lattice of an *infinite* number of spacial dimensions.

In connection with an elegant generalization by Z.-D. Zhang [17] of the exact solution by Onsager [13] to the problem of description of Ising lattices it is natural to proceed in four steps:

- to characterize the differences when taking into account the first, second, and third nearest neighbours of an atom in an alloy;
- to discuss the Ising-Onsager-Zhang lattices in the context of possible application of the Jordan-von Neumann-Wigner approach [6];
- to use some fractals of the algebraic Clifford structure in relation with the quaternionic sequence of Clifford algebras;
- to analyze further perspectives of investigation in this direction.

Wu, McCoy, Fisher, and Chayes [16] have downloaded their comment on Zhang's work with criticism of correctness of the Zhang's extension. Some other criticism has been downloaded by Perk [15]; cf. also Zhang's replies. We believe we are referring to the well-motivated results of Zhang only.

2. First, second, and third nearest neighbours of an atom in an alloy

Mathematically, an alloy is an ordered pair of foliation of a sample space by layers and a lattice determined by various atoms of type A (e.g. $A =$ Au - gold), B (e.g. $B =$ Cu - copper), ..., situated in the crystallographic lattice sites which, in the stochiometric and ordered case, are denoted by α, β, ..., respectively [3]. Some of the sites may correspond to vacancies and, because of fluctuations, it is possible, that for instance an atom A appears in a site denoted by β, etc.

Choosing a binary AB_3 alloy of fcc lattice and (111) surface orientation we can determine the first, second, and third nearest neighbours of an α- or β-site (Fig. 4). To this end we have to consider the spheres of radii $R_1 = \frac{1}{2}\sqrt{2}a$, $R_2 = a$, $R_3 = \frac{1}{2}\sqrt{6}a$, where a is the lattice constant. Atoms of the i-th ($i+1$-st, $i+2$-nd, etc.) layer are denoted by i, $i+1$, $i+2$, etc. In analogy we can consider the case of AB_3 fcc (100) alloy (Fig. 5). In the case of a ternary ABC_2 alloy of bcc lattice and (111) surface orientation we have to consider the spheres of radii $R_1 = \frac{\sqrt{3}}{4}a$, $R_2 = \frac{1}{2}a$, and $R_3 = \frac{1}{2}\sqrt{2}a$ and to distinguish three cases at least, corresponding to the nearest neighbours of an α-, or β-, or γ-site (Fig. 6). A configuration for AB_3 fcc with vacancies is shown in Fig. 7. It can be regarded a particular case of a ternary alloy or even of a quaternary alloy because a vacancy may be situated in an α-site or in a β-site.

3. Ising-Onsager-Zhang lattices vs. Jordan-von Neumann-Wigner approach

We observe that the very elegant Zhang's basis is determined by the formulae (8a), (8b), (8c) and (12) of [17], which can be equivalently written as

$$
\begin{aligned}
s^1_{r,s} &= \mathbf{1}\otimes\mathbf{1}\otimes\cdots\otimes\mathbf{1}\otimes\sigma_1\otimes\mathbf{1}\otimes\cdots\otimes\mathbf{1};\\
s^2_{r,s} &= \mathbf{1}\otimes\mathbf{1}\otimes\cdots\otimes\mathbf{1}\otimes\frac{1}{i}\sigma_2\otimes\mathbf{1}\otimes\cdots\otimes\mathbf{1} \quad \text{with} \quad \mathbf{1} = \begin{pmatrix}1&0\\0&1\end{pmatrix};\\
s^3_{r,s} &= \mathbf{1}\otimes\underbrace{\mathbf{1}\otimes\cdots\otimes\mathbf{1}}_{r-2,\ s-2}\otimes\sigma_3\otimes\underbrace{\mathbf{1}\otimes\cdots\otimes\mathbf{1}}_{n-r,\ \ell-s};
\end{aligned}
\tag{1}
$$

$$Z = (2\sinh 2K_1)^{\frac{1}{2}m\cdot n\cdot l}\cdot\operatorname{trace}(V_2V_3V_1)^m = (2\sinh 2K_1)^{\frac{1}{2}m\cdot n\cdot \ell}\sum_{j=1}^{2^{(n\cdot\ell)}}\lambda_j^m.$$

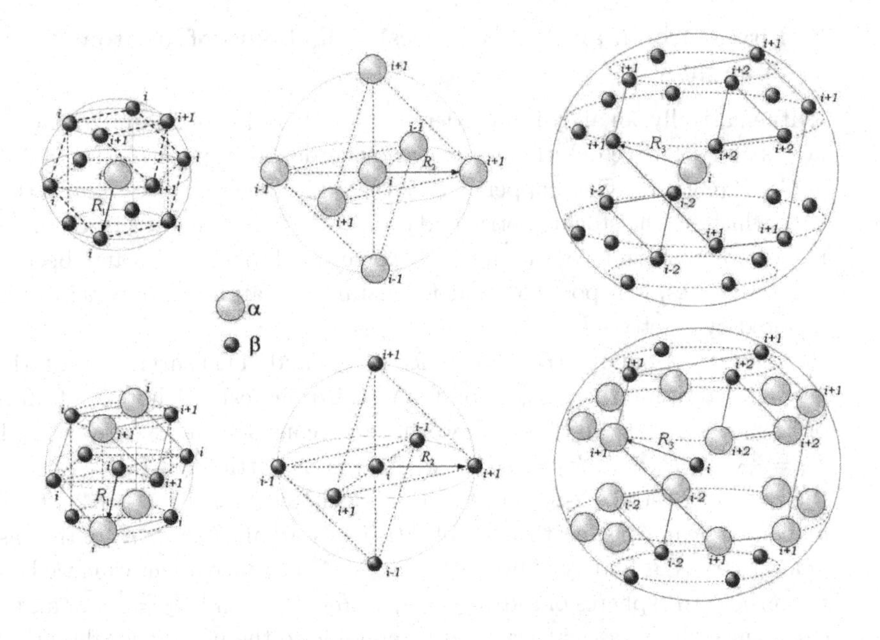

Fig. 4. First, second, and third neighbours of an α- or β-site in an AB_3 fcc (111) alloy in the stochiometric case (A in α, B in β).

Here, if $E_1(\nu_j,\nu_{j+1})$ denotes the energy due to interaction between two contiguous rows ($j=1,2,\dots,m$) in all the monoatomic layers represented by planes ($i=1,2,\dots,\ell$), $E_2(\mu_j)$ denotes the energy due to interaction between two j-th rows in two contiguous monoatomic layers, and $E_3(\mu_j)$ stands for the energy due to interactions within the j-th row in all the monoatomic layers (we assume that each row contains n sites with particles or vacancies: $k=1,2,\dots,n$), then the matrices $[s^\alpha_{r,s}]$, $r=1,2,\dots,n$; $s=1,2,\dots,\ell$; $\alpha=1,2,3$, satisfy some relations

$$V_1(E_1(\nu_j,\nu_{j+1}))=\Phi^1_j(K_1,s^1_{r,s})),\quad V_\alpha(E_\alpha(\nu_j,\nu_{j+1}))=\Phi^\alpha_j(K_\alpha,s^\alpha_{r,s})),$$

$\alpha=2,3$; whereas λ_j are the eigenvalues of $\mathbf{V}=\mathbf{V}_2\cdot\mathbf{V}_3\cdot\mathbf{V}_1$. It is very important that $s^\alpha_{r,s}$ in (1), $\alpha=1,2,3$, represent $2^{n\cdot\ell}$-dimensional *quaternion matrices* and σ_α are the familiar Pauli matrices.

Therefore the relations (1) suggest a relationship with the Jordan-von Neumann-Wigner approach [6]. The idea of Jordan was to consider the family of algebras $\mathcal{A}$ with addition $+$ and multiplication $\circ$ such that

$$\lambda A\in\mathcal{A}\quad\text{for}\quad A\in\mathcal{A},\ \lambda\in\mathbb{R};\tag{2}$$

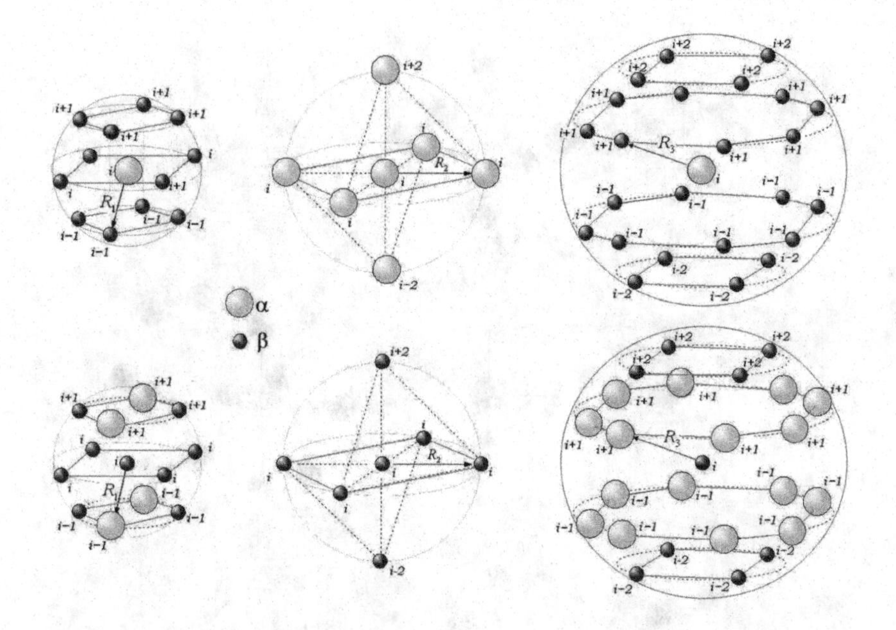

Fig. 5. First, second, and third neighbours of an α- or β-site in an AB_3 fcc (100) alloy in the stochiometric case (A in α, B in β).

$$((A \circ A) \circ B) \circ A = (A \circ A) \circ (B \circ A) \quad \text{for} \quad A,\, B \in \mathcal{A}; \tag{3}$$

$$\begin{aligned} \text{if} \quad (A \circ A) + (B \circ B) + (C \circ C) + \cdots = 0 \quad \text{for} \quad A, B, C \in \mathcal{A}, \\ \text{then} \quad A = B = C = \cdots = 0. \end{aligned} \tag{4}$$

He was looking for matrix algebras $\mathcal{A} \ni A, B$ such that

$$A \circ B = \frac{1}{2}(AB + BA), \tag{5}$$

where AB represents the usual matrix multiplication.

In the quoted paper Jordan, von Neumann and Wigner had proved what follows:

Theorem 1. *The only irreducible algebras satisfying the conditions* (2)–(4) *are the following:*

- *the algebra of real numbers with* $A + B$, λA, $A \circ B$ *defined in the usual way;*
- $\mathcal{C}_n$, $n = 3, 4, \ldots$; $\mathcal{C}$ *being the algebra with the linear basis* $1, s_1, \ldots, s_{n-1}$, *where* $A + B$ *and* λA *are defined in the usual way, but* $A \circ B$ *is defined*

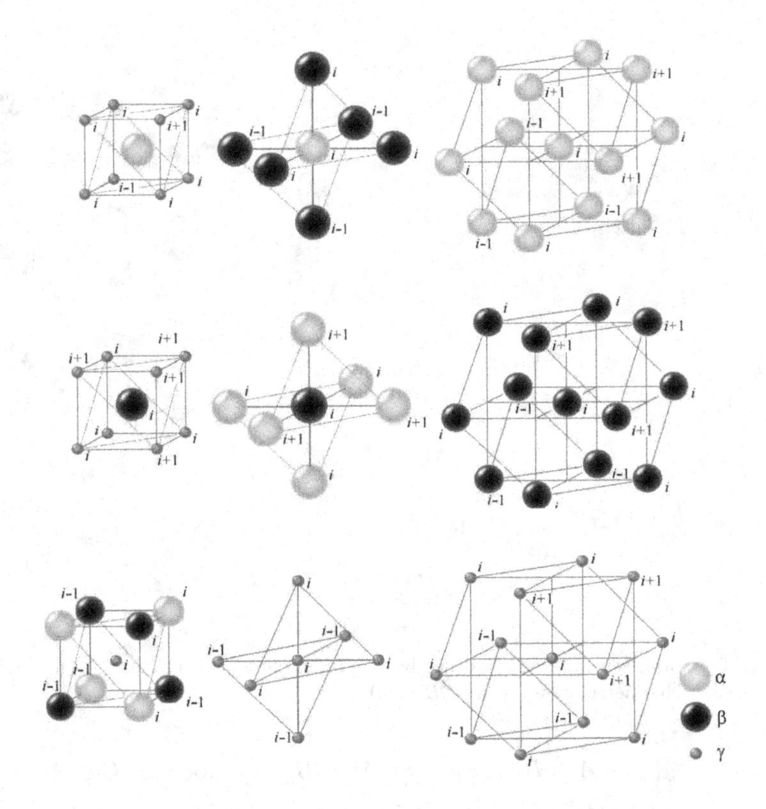

Fig. 6. First, second, and third neighbours of an α-, or β-, or γ-site in an ABC_2 bcc (111) alloy in the stochiometric case (A in α, B in β, and C in γ).

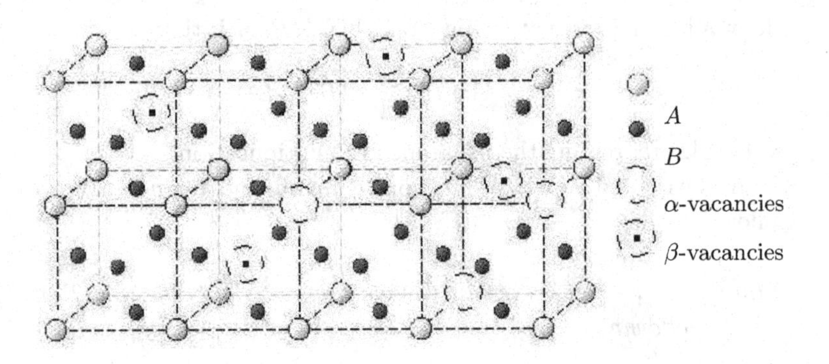

Fig. 7. An example of a binary AB_3 fcc alloy with α- and β-vacancies.

by

$$1 \circ 1 = 1,\ 1 \circ s_j = s_j \text{ and } s_j \circ s_k = \delta_{j,k} \circ 1,\ j,k = 1,2,\ldots,n-1$$

and $\delta_{j,k}$ denoting the Kronecker delta;

- $\mathcal{H}_q^p$, $p = 1,2,4,8$ *and* $q = 3$, *or* $p = 1,2,4$ *and* $q = 4,5,\ldots$, $\mathcal{H}_q^p$ *being the algebra of Hermitian matrices of order q whose elements are:*
  — *real numbers for $p = 1$,*
  — *complex numbers for $p = 2$,*
  — ***quaternions*** *for $p = 4$,*
  — *octonions for $p = 8$;*

  $A + B$ and λA are defined in the usual way, but for $\circ$ we have (5).

From our point of view, because of (1) with $s_{r,s}^{\alpha}$, $\alpha = 1,2,3$, representing $2^{n \cdot \ell}$-dimensional quaternion matrices, the $\mathbb{H}$-sequence $(\mathcal{H}_q^4)$ is of a particular importance.

4. The use of fractals of the algebraic structure

In [8-11] we are dealing with and applying the interaction and approximation process of constructing subsequent structure fractals related to the foliated manifold endowed with a lattice joining vertices being distinguished points of n kinds (or of one kind with n gradation functions). We can extend our geometry to $n = \frac{1}{2}, 1, 1\frac{1}{2}, \ldots$, and consider the iteration process

$$p(n) \mapsto p(n+2) \quad \text{with} \quad p(n) = 4n + 1,$$

involving generators of a *Clifford algebra*. The objects constructed, *graded fractal bundles* Σ_n, appear to have an interesting property of periodicity related to the fractal sets situated on the diagonals of the matrices represented by squares — generators of Clifford algebras, normalized to unit squares.

Explicitly, given generators A_1^1, A_2^1, $\ldots, A_{2p-1}^1$ of a Clifford algebra $C\ell_{2p-1}(\mathbb{C})$, $p = 2,3,\ldots$, consider the sequence

$$A_\alpha^{q+1} = \sigma_3 \otimes A_\alpha^q \equiv \begin{pmatrix} A_\alpha^q & 0 \\ 0 & -A_\alpha^q \end{pmatrix}, \quad \alpha = 1,2,\ldots,2p+2q-3;$$

$$A_{2p+2q-2}^{q+1} = \sigma_1 \otimes I_{p,q} \equiv \begin{pmatrix} 0 & I_{p,q} \\ I_{p,q} & 0 \end{pmatrix},$$

$$A_{2p+2q-1}^{q+1} = -\sigma_2 \otimes I_{p,q} \equiv \begin{pmatrix} 0 & iI_{p,q} \\ -iI_{p,q} & 0 \end{pmatrix},$$

of generators of Clifford algebras $C\ell_{2p+2q-1}(\mathbb{C})$, $q = 1, 2, \ldots$, and the sequence of corresponding systems of closed squares Q_q^α of diameter 1, centered at the origin of $\mathbb{C}$, where $I_{p,q} = I_{2^{p+q-2}}$, the unit matrix of order 2^{p+q-2}. It is convenient to start with q always from 1, i.e., to shift q for $\alpha \geq 2p$ correspondingly. For understanding the idea of this choice in details we refer to Fig. 3 in [8] and the related description. In particular we may take $A_\alpha^1 = \sigma_\alpha$, $\alpha = 1, 2, 3$; $p = 2$, so that

$$A_\alpha^{q+1} = \sigma_3 \otimes A_\alpha^q, \quad \alpha = 1, 2, \ldots, 2q+1; \tag{6}$$

$$A_{2q-2}^{q+1} = \sigma_1 \otimes I_{2,q}, \quad A_{2q-1}^{q+1} = -\sigma_2 \otimes I_{2,q}, \quad q = 1, 2, \ldots. \tag{7}$$

Leaving details for a subsequent paper, we announce

Theorem 2. *The fractal bundles obtained correspond in a canonical way to* $(\mathcal{H}_q^4)$, *the "quaternionic" sequence of Jordan algebras, with* $q = 2^{2n-2}(2n+1)$.

Physically, we get at least five models:

- *Melting model* with $\Sigma_2 \sim \mathcal{H}_{4\cdot 5}^4 = \mathcal{H}_{20}^4$ and 5 degrees of freedom: 3 co-ordinates, time, entropy.
- *Binary alloys model* with $\Sigma_4 \sim \mathcal{H}_{64\cdot 9}^4 = \mathcal{H}_{576}^4$ and 9 degrees of freedom: 6 co-ordinates, time, entropy, long range order parameter.
- *Ising-Onsager model* of order q and number of degrees of freedom to be calculated from (1)–(3), with an arbitrary $\ell \in \mathbb{N}$.
- *Ternary alloy* with $\Sigma_4 \sim \mathcal{H}_{1024\cdot 13}^4 = \mathcal{H}_{13\,312}^4$ and 13 degrees of freedom: 9 co-ordinates, time, entropy, 2 long range order parameters.
- *Zhang model* of order q and number of degrees of freedom to be calculated from the (1)–(3), with an arbitrary $\ell \in \mathbb{N}$.

5. Further perspectives

We can see the following perspectives of the present research:

- relationship with Kikuchi-type fractals [3];
- relationship with duality for fractal sets;
- relationship with lattice models on fractal sets;
- fractal renormalization and the renormalized Dirac operator;
- relationship with Schauder and Haar bases on fractal sets;
- meromorphic Schauder basis and hyperfunctions on fractal boundaries;
- the role of noncommutative isomorphisms between the function spaces;

- application of globally defined meromorphic functions to studying chaotic dynamical systems;
- the role of periodic orbits and dense orbits;
- an analysis of the conclusion that fluctuations observed in the nature can be realized as the image of fluctuation mappings of smooth functions without fluctuations.

The major part of these remarks is due to Professor Osamu Suzuki (Tokyo).

References

1. A.L. Barabasi and M.E. Stanley, *Fractal Concepts in Surface Growth*, Cambridge Univ. Press, Cambridge, 1975.
2. H.A. Bethe, *Statistical Theory of Superlattices*, Proc. Roy. Soc. London A **150** (1935), 552–75.
3. F.L. Castillo Alvarado, J. Ławrynowicz, and M. Nowak-Kępczyk, *Fractal modelling of alloy thin films. Temperatures characteristic for local phase transitionsions*, in: Applied Complex and Quaternionic Approximation. Ed. by R. Kovacheva, J. Ławrynowicz, and S. Marchiafava, World Scientific, Singapore, 2009, pp. 207–235, to appear.
4. E.F. Collingwood and J. Lohwater, *The Theory of Cluster Sets*, Cambridge Tracts in Mathematics and Mathematical Physics 56, Cambridge Univ. Press, Cambridge, 1966.
5. E. Ising, *Beitrag zur Theorie des Ferromagnetismus*, Zschr. f. Phys. **31** (1925), 253–258.
6. P. Jordan, J. von Neumann and E. Wigner, *On an algebraic generalization of the quantum mechanical formalism*, Ann. of Math. **35** (1934), 29–64.
7. J. Kigami, *Analysis on fractals*, Cambridge Tracts in Mathematics 143, Cambridge University Press, Cambridge, 2001.
8. J. Ławrynowicz, S. Marchiafava and M. Nowak-Kępczyk, *Periodicity theorem for structure fractals in quaternionic formulation*, Internat. J. of Geom. Meth. in Modern Phys. **3** (2006), 1167–1197.
9. J. Ławrynowicz, S. Marchiafava and M. Nowak-Kępczyk, *Cluster sets and periodicity in some structure fractals*, in: Topics in Contemporary Differential Geometry, Complex Analysis and Mathematical Physics. Ed. by S. Dimiev and K. Sekigawa, World Scientific New Jersey-London-Singapore, 2007, pp. 179–195.
10. J. Ławrynowicz, M. Nowak-Kępczyk and Osamu Suzuki, *A duality theorem for inoculated graded fractal bundles vs. Cuntz algebras and their central extensions*, Internat. J. of Pure and Appl. Math. **27** (2008), to appear.
11. J. Ławrynowicz and O. Suzuki, *Periodicity theorems for graded fractal bundles related to Clifford structures*, Internat. J. of Theor. Phys. **40** (2001), 387–397.
12. K. Noshiro, *Cluster Sets*, Springer, Berlin-Göttingen-Heidelberg 1960.
13. L. Onsager, *Crystal statistics I. A two-dimensional model with an order-disorder transition*, Phys. Rev. **65** (1944), 117–149.

14. G. Peano, *Formulario matematico*, vol. 5, Bocca, Torino 1908, esp. pp. 239–240; reprinted in: G. Peano, *Selected Works*. Ed. by H. C. Kennedy, Toronto Univ. Press, Toronto, 1983.
15. J.H.H. Perk, *Comment on "Conjectures on exact solution of three-dimensional (3D) simple orthorhombic Ising lattices"*, arXiv:0811.1802v2; to appear in Phil. Mag.
16. F-Y. Wu, B. M. McCoy, M. E. Fisher and L. Chayes, *Comment on a recent conjectured solution of the three-dimensional Ising model*, Phil. Mag. 88 (2008), 3093–3095 and 3103.
17. Z.-D. Zhang, *Conjectures on exact solution of three-dimensional (3D) simple orthorhombic Ising lattices*, and *Responses to comments by F-Y. Wu, B.M. McCoy, M.E. Fisher, and L. Chayes*, also *Response to "Comment on 'Conjectures on exact solution of three-dimensional (3D) simple orthorhombic Ising lattices'" by Perk*, Phil. Mag. **87** (2007), 5309–5419, and **88** (2008), 3097, also arXiv:0812.0194.

A CHARACTERIZATION OF CLIFFORD MINIMAL HYPERSURFACES OF A SPHERE IN TERMS OF THEIR GEODESICS

SADAHIRO MAEDA

Department of Mathematics, Saga University,
1 Honzyo, Saga 840-8502 JAPAN
E-mail: smaeda@ms.saga-u.ac.jp

We characterize Clifford minimal hypersurfaces $S^r(nc/r) \times S^{n-r}(nc/(n-r))$ with $1 \leqq r \leqq n-1$ in a sphere $S^{n+1}(c)$ of constant sectional curvature c by observing their geodesics from this ambient sphere.

Keywords: Spheres; Clifford minimal hypersurfaces; Geodesics; Small circles; Great circles; Principal curvatures.

1. Introduction

In an $(n+1)$-dimensional sphere $S^{n+1}(c)$ $(n \geqq 2)$ of constant sectional curvature c, a *Clifford hypersurface* $M^n_{r,n-r}(c_1, c_2)$ is defined as a Riemannian product of two spheres $S^r(c_1)$ and $S^{n-r}(c_2)$ with $1 \leqq r \leqq n-1$ and $1/c_1 + 1/c_2 = 1/c$. It is known that $M^n_{r,n-r}(c_1, c_2)$ has two constant principal curvatures $\lambda_1 = c_1/\sqrt{c_1 + c_2}$ and $\lambda_2 = -c_2/\sqrt{c_1 + c_2}$ whose multiplicities are r and $n-r$, respectively. Moreover, they are the only examples of non totally umbilic hypersurfaces with parallel shape operator in $S^{n+1}(c)$.

We here recall the fact that a hypersurface M^n of $S^{n+1}(c)$ is totally umbilic if and only if all geodesics on M^n are mapped to circles, namely great circles and small circles, in the ambient space $S^{n+1}(c)$. In this context, we investigate geodesics on $M^n_{r,n-r}(c_1, c_2)$ which are mapped to circles in $S^{n+1}(c)$ (see Lemma 2.1). Lemma 2.1 shows that *some* geodesics on each $M^n_{r,n-r}(c_1, c_2)$ are mapped to circles in $S^{n+1}(c)$. Considering the converse of Lemma 2.1, we characterize all Clifford hypersurfaces $M^n_{r,n-r}(c_1, c_2)$ $(1 \leqq r \leqq n-1,\ 1/c_1 + 1/c_2 = 1/c)$ and all minimal Clifford hypersurfaces

The author is partially supported by Grant-in-Aid for Scientific Research (C) (No. 19540084), Japan Society for the Promotion of Science.

$M^n_{r,n-r}(nc/r, nc/(n-r))$ $(1 \leqq r \leqq n-1)$ (see Proposition 3.1 and Theorem 3.1).

2. Geodesics on Clifford hypersurfaces

Let $TM^n_{r,n-r}(c_1, c_2) = V_{\lambda_1} \oplus V_{\lambda_2}$ be the decomposition of the tangent bundle $TM^n_{r,n-r}(c_1, c_2)$ into principal distributions $V_{\lambda_1} = \{X \in TM^n_{r,n-r}(c_1, c_2) \mid AX = \lambda_1 X\}$ and $V_{\lambda_2} = \{X \in TM^n_{r,n-r}(c_1, c_2) \mid AX = \lambda_2 X\}$. Our aim here is to classify geodesics on the hypersurface $M^n_{r,n-r}(c_1, c_2)$ which are mapped to circles in the ambient sphere $S^{n+1}(c)$.

Lemma 2.1. *For each Clifford hypersurface $M^n_{r,n-r}(c_1, c_2)$ $(1 \leqq r \leqq n-1$, $1/c_1 + 1/c_2 = 1/c)$ of $S^{n+1}(c)$, the following hold:*

(1) *A geodesic γ on $M^n_{r,n-r}(c_1, c_2)$ is mapped to a small circle in $S^{n+1}(c)$ if and only if the initial vector $\dot{\gamma}(0)$ satisfies either $\dot{\gamma}(0) \in V_{\lambda_1}$ or $\dot{\gamma}(0) \in V_{\lambda_2}$;*

(2) *A geodesic γ on $M^n_{r,n-r}(c_1, c_2)$ is mapped to a great circle in $S^{n+1}(c)$ if and only if the initial vector $\dot{\gamma}(0)$ is expressed as $\dot{\gamma}(0) = \sqrt{c_2/(c_1+c_2)}\, u + \sqrt{c_1/(c_1+c_2)}\, v$ for some unit vectors $u \in V_{\lambda_1}$ and $v \in V_{\lambda_2}$.*

Proof. We denote by $\widetilde{\nabla}$ and ∇ the Riemannian connections of $S^{n+1}(c)$ and $M^n_{r,n-r}(c_1, c_2)$, respectively. $TM^n_{r,n-r}(c_1, c_2)$ is decomposed as: $TM^n_{r,n-r}(c_1, c_2) = V_{\lambda_1} \oplus V_{\lambda_2}$.

We find that $\nabla_X Y \in V_{\lambda_i}$ holds for any $X, Y \in V_{\lambda_i}$ $(i = 1, 2)$, namely the principal distribution V_{λ_i} is integrable and its each leaf L_{λ_i} is a totally geodesic submanifold of our hypersurface $M^n_{r,n-r}(c_1, c_2)$. Indeed,

$$A(\nabla_X Y) = \nabla_X(AY) - (\nabla_X A)Y = \nabla_X(\lambda_i Y) = \lambda_i(\nabla_X Y),$$

where we have used the fact that the shape operator A of $M^n_{r,n-r}(c_1, c_2)$ in $S^{n+1}(c)$ is parallel. Also, our leaf L_{λ_i} $(i = 1, 2)$ is a real space form of constant sectional curvature d_i with $d_i = c + \lambda_i^2$, which is a totally umbilic but non totally geodesic submanifold in the ambient sphere $S^{n+1}(c)$. Then we see that every geodesic on L_{λ_i} $(i = 1, 2)$ is also a geodesic on $M^n_{r,n-r}(c_1, c_2)$ and a small circle (of positive curvature $|\lambda_i|$) on $S^{n+1}(c)$. These, together with the existence and uniqueness theorem for geodesics, yield the "if part" of Statement (1). That is, we can see that every geodesic $\gamma = \gamma(s)$ on $M^n_{r,n-r}(c_1, c_2)$ satisfying $\dot{\gamma}(0) \in V_{\lambda_1}$ (resp. $\dot{\gamma}(0) \in V_{\lambda_2}$) is mapped to a small circle of positive curvature $|\lambda_1|$ (resp. $|\lambda_2|$) on $S^{n+1}(c)$.

Conversely, we take a geodesic $\gamma = \gamma(s)$ on $M^n_{r,n-r}(c_1, c_2)$ which is mapped to a small circle of curvature (, say) k in $S^{n+1}(c)$. Then this curve satisfies

$$\widetilde{\nabla}_{\dot\gamma}\widetilde{\nabla}_{\dot\gamma}\dot\gamma = -k^2\dot\gamma. \tag{1}$$

On the other hand, using the Gauss formula $\widetilde{\nabla}_X Y = \nabla_X Y + \langle AX, Y\rangle\mathcal{N}$ and the Weingarten formula $\widetilde{\nabla}_X\mathcal{N} = -AX$ for the hypersurface $M^n_{r,n-r}(c_1, c_2)$, we see that

$$\widetilde{\nabla}_{\dot\gamma}\widetilde{\nabla}_{\dot\gamma}\dot\gamma = -\langle A\dot\gamma, \dot\gamma\rangle A\dot\gamma + \langle(\nabla_{\dot\gamma}A)\dot\gamma, \dot\gamma\rangle\mathcal{N}. \tag{2}$$

Here $\mathcal{N}$ is a unit normal vector field on $M^n_{r,n-r}(c_1, c_2)$. Comparing the tangential components of Equations (1) and (2) for $M^n_{r,n-r}(c_1, c_2)$, we have $\langle A\dot\gamma, \dot\gamma\rangle A\dot\gamma = k^2\dot\gamma$. This, combiend with $k \neq 0$, shows that $A\dot\gamma(s) = k\dot\gamma(s)$ for every s or $A\dot\gamma(s) = -k\dot\gamma(s)$ for every s. Hence, in particular the initial vector $\dot\gamma(0)$ is a principal curvature vector. Thus we can check the "only if" part of Statement (1).

Next, we shall check Statement (2). For a geodesic γ on $M^n_{r,n-r}(c_1, c_2)$ we have $\widetilde{\nabla}_{\dot\gamma}\dot\gamma = \langle A\dot\gamma, \dot\gamma\rangle\mathcal{N}$. We remark that $\langle A\dot\gamma(s), \dot\gamma(s)\rangle$ is a constant function along the geodesic γ, since $\nabla A = 0$. This tells us that a geodesic γ on the submanifold $M^n_{r,n-r}(c_1, c_2)$ is a great circle on $S^{n+1}(c)$ if and only if $\langle A\dot\gamma(0), \dot\gamma(0)\rangle = 0$. Here we set $\dot\gamma(0) = (\cos t)u + (\sin t)v$ $(0 < t < \pi/2)$, where u, v are unit vectors with $u \in V_{\lambda_1}$ and $v \in V_{\lambda_2}$. Then the equation $\langle A\dot\gamma(0), \dot\gamma(0)\rangle = 0$ gives

$$\cos^2 t \cdot \frac{c_1}{\sqrt{c_1 + c_2}} - \sin^2 t \cdot \frac{c_2}{\sqrt{c_1 + c_2}} = 0,$$

so that we obtain $\cos t = \sqrt{\frac{c_2}{c_1+c_2}}$, $\sin t = \sqrt{\frac{c_1}{c_1+c_2}}$. Hence we get Statement (2). $\square$

3. Statements of resuls

Motivated by Lemma 2.1, we characterize all Clifford hypersurfaces.

Proposition 3.1. *A connected hypersurface M^n in $S^{n+1}(c)$ is locally congruent to a Clifford hypersurface $M^n_{r,n-r}(c_1, c_2)$ $(1 \leqq r \leqq n-1$, $1/c_1 + 1/c_2 = 1/c)$ if and only if there exist a function $d : M \to \mathbb{N}$, a constant α $(0 < \alpha < 1)$ and an orthonormal basis $\{v_1, \dots, v_n\}$ of T_xM at each point $x \in M$ satisfying the following two conditions:*

(*i*) *All geodesics γ_i on M^n with initial vector v_i $(1 \leqq i \leqq n)$ are small circles in $S^{n+1}(c)$;*

(*ii*) *All geodesics* γ_{ij} *on* M^n *with initial vector* $\alpha v_i + \sqrt{1-\alpha^2}\, v_j$ $(1 \leqq i \leqq d_x < j \leqq n)$ *are great circles in* $S^{n+1}(c)$.

In this case d *is a constant function with* $d \equiv r$ *and*

$$M^n = M^n_{r,n-r}(c/\alpha^2, c/(1-\alpha^2)).$$

Proof. Lemma 2.1 shows that all Clifford hypersurfaces $M^n_{r,n-r}(c_1, c_2)$ satisfy Conditions (*i*) and (*ii*) in our Proposition by taking an othonormal basis $\{v_1, \dots, v_n\}$ of $T_x M^n_{r,n-r}(c_1, c_2)$ in such a way that each v_i is a principal curvature vector of $M^n_{r,n-r}(c_1, c_2)$ in $S^{n+1}(c)$.

Conversely, we consider a connected hypersurface M^n satisfying Conditions (*i*) and (*ii*). We explain the discussion in [1,4] in detail. We first concentrate our attention on Condition (*i*). We study on an open dense subset

$$\mathcal{U} = \left\{ x \in M^n \;\middle|\; \begin{array}{l} \text{the multiplicity of each principal curvature of } M^n \text{ in} \\ S^{n+1}(c) \text{ is constant on some neighborhood } \mathcal{V}_x (\subset \mathcal{U}) \text{ of } x \end{array} \right\}$$

of M^n. For an orthonormal basis $\{v_1, \dots, v_n\}$ of $T_x M^n$ at $x \in \mathcal{U}$ which satisfies the conditions, we take geodesics γ_i $(1 \leqq i \leqq n)$ on M^n with initial vector v_i. Since the geodesic γ_i is a circle of positive curvature (, say) k_i, the same discussion as that in the proof of Lemma 2.1 yields that $Av_i = k_i v_i$ or $Av_i = -k_i v_i$ for $1 \leqq i \leqq n$. This means that the tangent space $T_x M^n$ is decomposed as:

$$\begin{aligned} T_x M = \{v \in T_x M \mid Av = -k_{i_1} v\} &\oplus \{v \in T_x M \mid Av = k_{i_1} v\} \\ \oplus \cdots \oplus \{v \in T_x M \mid Av = -k_{i_g} v\} &\oplus \{v \in T_x M \mid Av = k_{i_g} v\}, \end{aligned}$$

where $0 < k_{i_1} < k_{i_2} < \cdots < k_{i_g}$ and g is the number of distinct positive k_i $(i = 1, \dots, n)$. We decompose $T_x M$ in such a way at each point $x \in \mathcal{U}$. Note that each k_{i_j} is a smooth function on $\mathcal{V}_x$ for each $x \in \mathcal{U}$. We shall show the constancy of each k_{i_j}. It suffices to check the case of $Av_{i_j} = k_{i_j} v_{i_j}$. As k_{i_j} is a constant function along the curve γ_{i_j} in $S^{n+1}(c)$, we have $v_{i_j} k_{i_j} = 0$. For any v_ℓ $(1 \leqq \ell \neq i_j \leqq n)$, since A is symmetric, we have

$$\langle (\nabla_{v_{i_j}} A) v_\ell, v_{i_j} \rangle = \langle v_\ell, (\nabla_{v_{i_j}} A) v_{i_j} \rangle. \tag{3}$$

We extend an orthonormal basis $\{v_1, \dots, v_n\}$ of $T_x M^n$ to a local field of orthonomal frames (, say) $\{V_1, \dots, V_n\}$ on some open neighborhood $\mathcal{W}_x (\subset \mathcal{V}_x)$. In order to compute Equation (3) easily, we extend the vector v_{i_j} to the vector field V_{i_j} satisfying that $AV_{i_j} = k_{i_j} V_{i_j}$ on $\mathcal{W}_x$ and $\nabla_{V_{i_j}} V_{i_j} = 0$ at the point $x \in \mathcal{U}$. Indeed, such a principal curvature smooth unit vector field V_{i_j} can be obtained in the following manner.

First of all we define a smooth unit vector field W_{i_j} on some "sufficiently small" neighborhood $\mathcal{W}_x(\subset \mathcal{V}_x)$ by using parallel displacement for the vector v_{i_j} along each geodesic with origin x. We note that in general W_{i_j} is not principal on $\mathcal{W}_x$, but $AW_{i_j} = k_{i_j} W_{i_j}$ on the geodesic $\gamma = \gamma(s)$ with $\gamma(0) = x$ and $\dot\gamma(0) = v_{i_j}$. We here define the vector field U_{i_j} on $\mathcal{W}_x$ as: $U_{i_j} = \prod_{\alpha\neq k_{i_j}}(A-\alpha I)W_{i_j}$, where α runs over the set of all distinct principal curvatures of M^n except for the principal curvature k_{i_j}. We remark that $U_{i_j} \neq 0$ on the neighborhood $\mathcal{W}_x$, because $(U_{i_j})_x \neq 0$. Moreover, the vector field U_{i_j} satisfies $AU_{i_j} = k_{i_j} U_{i_j}$ on $\mathcal{W}_{i_j}$. We define V_{i_j} by normalizing U_{i_j} in some sense. That is, when $\prod_{\alpha\neq k_{i_j}}(k_{i_j}-\alpha)(x) > 0$ (resp. $\prod_{\alpha\neq k_{i_j}}(k_{i_j}-\alpha)(x) < 0$), we define $V_{i_j} = U_{i_j}/\|U_{i_j}\|$ (resp. $V_{i_j} = -U_{i_j}/\|U_{i_j}\|$). Then we know that $AV_{i_j} = k_{i_j} V_{i_j}$ on $\mathcal{W}_x$ and $(V_{i_j})_x = v_{i_j}$. Furthermore, our constrction shows that the integral curve of V_{i_j} through the point x is a geodesic on M^n, so that in particular $\nabla_{V_{i_j}} V_{i_j} = 0$ at the point x.

Thanks to the Codazzi equation $\langle(\nabla_X A)Y, Z\rangle = \langle(\nabla_Y A)X, Z\rangle$, at the point x we have

$$\begin{aligned}
\text{(the left-hand side of (3))} &= \langle(\nabla_{v_\ell} A)v_{i_j}, v_{i_j}\rangle = \langle(\nabla_{V_\ell} A)V_{i_j}, V_{i_j}\rangle \\
&= \langle\nabla_{V_\ell}(k_{i_j}V_{i_j}) - A\nabla_{V_\ell}V_{i_j}, V_{i_j}\rangle \\
&= \langle(V_\ell k_{i_j})V_{i_j} + (k_{i_j}I - A)\nabla_{V_\ell}V_{i_j}, V_{i_j}\rangle = v_\ell k_{i_j}
\end{aligned}$$

and

$$\begin{aligned}
\text{(the right-hand side of (3))} &= \langle V_\ell, (\nabla_{V_{i_j}} A)V_{i_j}\rangle \\
&= \langle V_\ell, \nabla_{V_{i_j}}(k_{i_j}V_{i_j}) - A\nabla_{V_{i_j}}V_{i_j}\rangle \\
&= \langle v_\ell, (v_{i_j}k_{i_j})v_{i_j}\rangle = 0.
\end{aligned}$$

Thus we can see that the differential dk_{i_j} of k_{i_j} vanishes at the point x, which shows that every k_{i_j} (> 0) is constant on $\mathcal{W}_x$, since we can take the point x as an arbitrarily fixed point of $\mathcal{W}_x$. So the principal curvature function k_{i_j} is locally constant on the open dense subset $\mathcal{U}$ of M^n. This, together with the continuity of k_{i_j} and the connectivity of M^n, implies that k_{i_j} is constant on the hypersurface M^n. Hence all principal curvatures of M^n are (nonzero) constant if M^n satisfies Condition (i).

Next, we consider Condition (ii). The above argument tells us that each v_i $(1 \leqq i \leqq n)$ is principal. We set $Av_i = \lambda_i v_i$. Then the condition (ii) yields $\alpha\lambda_i + \sqrt{1-\alpha^2}\,\lambda_j = 0$ for $1 \leqq i \leqq d_x < j \leqq n$. Hence M has just two distinct constant principal curvatures. Therefore we conclude that the hypersurface M^n is locally congruent to a Clifford hypersurface $M^n_{r,n-r}(c_1, c_2)$ $(1 \leqq r \leqq n-1,\ 1/c_1 + 1/c_2 = 1/c)$ with $d_x = r$. □

As an immediate consequence of Proposition 3.1 we establish the following:

Theorem 3.1. *A connected hypersurface M^n in $S^{n+1}(c)$ is locally congruent to a minimal Clifford hypersurface $M^n_{r,n-r}(nc/r, nc/(n-r))$ $(1 \leqq r \leqq n-1)$ if and only if there exist a function $d : M \to \mathbb{N}$ and an orthonormal basis $\{v_1, \ldots, v_n\}$ of T_xM at each point $x \in M$ satisfying the following two conditions*:

(*i*) *All geodesics γ_i on M^n with initial vector v_i $(1 \leqq i \leqq n)$ are small circles in $S^{n+1}(c)$;*

(*ii*) *All geodesics γ_{ij} on M^n with initial vector $\sqrt{r/n}\, v_i + \sqrt{(n-r)/n}\, v_j$ $(1 \leqq i \leqq d_x < j \leqq n)$ are great circles in $S^{n+1}(c)$.*

In this case d is a constant function with $d \equiv r$ and

$$M^n = M^n_{r,n-r}(nc/r, nc/(n-r)).$$

Proof. By virtue of Proposition 3.1 we find that the hypersurface M^n satisfying Conditions (*i*), (*ii*) of Theorem 3.1 is locally congruent to some Clifford hypersurface $M^n_{r,n-r}(c_1, c_2)$. Next, we shall investigate the case that $M^n_{r,n-r}(c_1, c_2)$ is minimal in the ambient space $S^{n+1}(c)$. Equation $\operatorname{Tr} A = 0$ holds if and only if $rc_1 - (n-r)c_2 = 0$. This, combiend with $1/c_1 + 1/c_2 = 1/c$, implies that $c_1 = nc/r$, $c_2 = nc/(n-r)$. As the constant α in Proposition 3.1 is given by $\sqrt{c_2/(c_1+c_2)}$ (see the proof of Lemma 2.1), we know that $\alpha = \sqrt{r/n}$. Thus we obtain the desirable conclusion. □

Remark 3.1.

(1) In Proposition 3.1 and Theorem 3.1, we need Condition (*ii*) at some point $x \in M^n$.

(2) If we add a condition that M^n is complete to assumptions of Proposition 3.1 and Theorem 3.1, then these results characterize globally $M^n_{r,n-r}(c_1, c_2)$.

(3) In assumptions of Proposition 3.1 and Theorem 3.1, we do not need to take the vectors $\{v_1, \ldots, v_n\}$ as a local field of orthonormal frames of M^n. However, for each Clifford hypersurface $M^n_{r,n-r}(c_1, c_2)$, we can take a global smooth field of orthonormal frames $\{v_1, \ldots, v_n\}$ satisfying these assumptions.

(4) A connected hypersurface M^n of $S^{n+1}(c)$ is isoparametric without null principal curvatures, namely all principal curvatures of M^n are nonzero constans, if and only if M^n satisfies Condition (*i*) of Proposition 3.1.

In fact, the discussion in the proof of Proposition 3.1 gives the "if" part and the "only if" part can be proved by the following well-known lemma ([3]):

Lemma 3.1. *For each isoparametric hypersurface M_g^n of $S^{n+1}(c)$ with g distinct principal curvatures, every principal distribution V_{λ_i} $(i = 1, \ldots, g)$ is integrable and its any leaf L_{λ_i} is totally umbilic in the ambient sphere $S^{n+1}(c)$, moreover such a leaf is totally geodesic in the hypersurface M_g^n.*

Proof. It suffices to verify that $\nabla_X Y \in V_{\lambda_i}$ for $\forall X, Y \in V_{\lambda_i}$. By the same discussion as in the proof of Lemma 2.1 we see that

$$A(\nabla_X Y) = \nabla_X(AY) - (\nabla_X A)Y = \lambda_i(\nabla_X Y) - (\nabla_X A)Y.$$

On the other hand, for every $Z \in TM_g^n$ the Codazzi equation shows

$$\begin{aligned}\langle(\nabla_X A)Y, Z\rangle &= \langle(\nabla_Z A)Y, X\rangle = \langle\nabla_Z(AY) - A\nabla_Z Y, X\rangle \\ &= \langle(\lambda_i I - A)\nabla_Z Y, X\rangle = \langle\nabla_Z Y, (\lambda_i I - A)X\rangle = 0.\end{aligned}$$

Hence we get Statement of Lemma 3.1. □

At the end of this paper we refer to some papers on Clifford minimal hypersurfaces ([2,5,6,7,8]).

References

1. T. Adachi and S. Maeda, *Isoparametric hypersurfaces with less than four principal curvatures in a sphere*, Colloquium Math. **105** (2006), 143–148.
2. S.S. Chern, M.P. doCarmo and S. Kobayashi, *Minimal submanifolds of a sphere with second fundamental form of constant length*, Functional analysis and related fields, Springer (1970), 59–75.
3. T.E. Cecil and P.J. Ryan, *Tight and Taut immersions of manifolds*, Res. Notes Math. **107** Pitman, Boston, MA, (1985).
4. M. Kimura and S. Maeda, *Geometric meaning of isoparametric hypersurfaces in a real space form*, Canadian Math. Bull. **43** (2000), 74–78.
5. H.B. Lawson, *Local rigidity theorems for minimal hypersurfaces*, Ann. Math. **89** (1969), 187–197.
6. K. Nomizu and B. Smyth, *A formula of Simons' type and hypersurfaces with constant mean curvature*, J. Diff.Geom. **3** (1969), 367-377.
7. T. Otsuki, *Minimal hypersurfaces in a Riemannian manifold of constant curvature*, Amer. J. Math. **92** (1970), 145–173.
8. T. Otsuki, *On periods of solutions of a certain nonlinear differential equation and the Riemannian manifold O_n^2*, Proc. Symp. Pure. Math. **27** (1975), 327–337.

ON THE CURVATURE PROPERTIES OF REAL TIME-LIKE HYPERSURFACES OF KÄHLER MANIFOLDS WITH NORDEN METRIC

M. MANEV and M. TEOFILOVA

Department of Geometry, Faculty of Mathematics and Informatics, University of Plovdiv, 236 Bulgaria Blvd., Plovdiv 4003, Bulgaria
E-mail: mmanev@uni-plovdiv.bg, marta.teofilova@gmail.com

A type of almost contact hypersurfaces with Norden metric of a Kähler manifold with Norden metric is considered. The curvature tensor and the special sectional curvatures are characterized. The canonical connection on such manifolds is studied and the form of the corresponding Kähler curvature tensor is obtained. Some curvature properties of the manifolds belonging to the widest integrable main class of the considered type of hypersurfaces are given.

Keywords: Hypersurface; Curvature; Almost contact manifold; Norden metric.

Introduction

The Kähler manifolds with Norden metric have been introduced in [9]. These manifolds form the special class $\mathcal{W}_0$ in the decomposition of the almost complex manifolds with Norden metric, given in [2]. This most important class is contained in each of the basic classes in the mentioned classification.

The natural analogue of the almost complex manifolds with Norden metric in the odd dimensional case are the almost contact manifolds with Norden metric, classified in [4].

In [5] two types of hypersurfaces of an almost complex manifold with Norden metric are constructed as almost contact manifolds with Norden metric, and the class of these hypersurfaces of a $\mathcal{W}_0$-manifold is determined.

An important problem in the differential geometry of the Kähler manifolds with Norden metric is the studying of the manifolds of constant totally real sectional curvatures [3]. In this paper we study some curvature properties of the real time-like hypersurfaces of Kähler manifolds with Norden metric of constant totally real sectional curvatures and particularly curvature properties of nondegenerate special sections.

1. Preliminaries

1.1. *Almost complex manifolds with Norden metric*

Let (M', J, g') be a $2n'$-dimensional almost complex manifold with Norden metric, i.e. J is an almost complex structure and g' is a metric on M' such that:

$$J^2X = -X, \qquad g'(JX, JY) = -g'(X, Y)$$

for all vector fields $X, Y \in \mathfrak{X}(M')$ (the Lie algebra of the differentiable vector fields on M'). The associated metric $\widetilde{g'}$ of the manifold is given by $\widetilde{g'}(X,Y) = g'(X,JY)$. Both metrics are necessarily of signature (n',n').

Further, X, Y, Z, U will stand for arbitrary differentiable vector fields on the manifold, and x, y, z, u — arbitrary vectors in its tangent space at an arbitrary point.

The (0,3)-tensor F' on M' is defined by $F'(X,Y,Z) = g'((\nabla'_X J)Y,Z)$, where ∇' is the Levi-Civita connection of g'.

A decomposition to three basic classes of the considered manifolds with respect to F' is given in [2]. In this paper we shall consider only the class $\mathcal{W}_0 : F' = 0$ of the Kähler manifolds with Norden metric. The complex structure J is parallel on every $\mathcal{W}_0$-manifold, i.e. $\nabla' J = 0$.

The curvature tensor field R', defined by

$$R'(X, Y)Z = \nabla'_X \nabla'_Y Z - \nabla'_Y \nabla'_X Z - \nabla'_{[X,Y]} Z,$$

has the property $R'(X, Y, Z, U) = -R'(X, Y, JZ, JU)$ on a $\mathcal{W}_0$-manifold. Using the first Bianchi identity and the last property of R it follows $R'(X, JY, JZ, U) = -R'(X, Y, Z, U)$. Therefore, the tensor field $\widetilde{R'}$: $\widetilde{R'}(X, Y, Z, U) = R'(X, Y, Z, JU)$ has the properties of a Kähler curvature tensor and it is called *an associated curvature tensor.*

The essential curvature-like tensors are defined by:

$$\begin{aligned}
\pi'_1(x, y, z, u) &= g'(y, z)g'(x, u) - g'(x, z)g'(y, u),\\
\pi'_2(x, y, z, u) &= g'(y, Jz)g'(x, Ju) - g'(x, Jz)g'(y, Ju),\\
\pi'_3(x, y, z, u) &= -g'(y, z)g'(x, Ju) + g'(x, z)g'(y, Ju)\\
&\quad - g'(y, Jz)g'(x, u) + g'(x, Jz)g'(y, u).
\end{aligned}$$

For every nondegenerate section α' in $T_{p'}M'$, $p' \in M'$, with a basis $\{x, y\}$ there are known the following sectional curvatures [1]: $k'(\alpha'; p') =$

$k'(x,y) = R'(x,y,y,x)/\pi'_1(x,y,y,x)$ – the usual Riemannian sectional curvature; $\widetilde{k}'(\alpha';p') = \widetilde{k}'(x,y) = \widetilde{R}'(x,y,y,x)/\pi'_1(x,y,y,x)$ — an associated sectional curvature.

The sectional curvatures of an arbitrary holomorphic section α' (i.e. $J\alpha' = \alpha'$) is zero on a Kähler manifold with Norden metric [1].

For the totally real sections α' (i.e. $J\alpha' \perp \alpha'$) it is proved the following

Theorem 1.1 ([1]). *Let M' ($2n' \geq 4$) be a Kähler manifold with Norden metric. M' is of constant totally real sectional curvatures ν' and $\widetilde{\nu}'$, i.e. $k'(\alpha';p') = \nu'(p')$, $\widetilde{k}'(\alpha';p') = \widetilde{\nu}'(p')$ whenever α' is a nondegenerate totally real section in $T_{p'}M'$, $p' \in M'$, if and only if*

$$R' = \nu'\left[\pi'_1 - \pi'_2\right] + \widetilde{\nu}'\pi'_3.$$

Both functions ν' and $\widetilde{\nu}'$ are constant if M' is connected and $2n' \geq 6$.

1.2. *Almost contact manifolds with Norden metric*

Let (M,φ,ξ,η,g) be a $(2n+1)$-dimensional almost contact manifold with Norden metric, i.e. (φ,ξ,η) is an almost contact structure determined by a tensor field φ of type $(1,1)$, a vector field ξ and a 1-form η on M satisfying the conditions:

$$\varphi^2 X = -X + \eta(X)\xi, \qquad \eta(\xi) = 1,$$

and in addition the almost contact manifold (M,φ,ξ,η) admits a metric g such that [4]

$$g(\varphi X, \varphi Y) = -g(X,Y) + \eta(X)\eta(Y).$$

There are valid the following immediate corollaries: $\eta \circ \varphi = 0$, $\varphi\xi = 0$, $\eta(X) = g(X,\xi)$, $g(\varphi X,Y) = g(X,\varphi Y)$.

The associated metric $\widetilde{g}$ given by $\widetilde{g}(X,Y) = g(X,\varphi Y) + \eta(X)\eta(Y)$ is a Norden metric, too. Both metrics are indefinite of signature $(n, n+1)$.

The Levi-Civita connection of g will be denoted by ∇. The tensor field F of type (0,3) on M is defined by $F(X,Y,Z) = g((\nabla_X \varphi)Y,Z)$.

If $\{e_i,\xi\}$ $(i = 1,2,\dots,2n)$ is a basis of T_pM and (g^{ij}) is the inverse matrix of (g_{ij}), then the following 1-forms are associated with F:

$$\theta(\cdot) = g^{ij}F(e_i,e_j,\cdot), \quad \theta^*(\cdot) = g^{ij}F(e_i,\varphi e_j,\cdot), \quad \omega(\cdot) = F(\xi,\xi,\cdot).$$

A classification of the almost contact manifolds with Norden metric with respect to F is given in [4], where eleven basic classes $\mathcal{F}_i$ are defined. In the present paper we consider the following classes:

$$
\begin{aligned}
\mathcal{F}_4 &: F(x,y,z) = -\frac{\theta(\xi)}{2n}\{g(\varphi x,\varphi y)\eta(z)+g(\varphi x,\varphi z)\eta(y)\};\\
\mathcal{F}_5 &: F(x,y,z) = -\frac{\theta^*(\xi)}{2n}\{g(x,\varphi y)\eta(z)+g(x,\varphi z)\eta(y)\};\\
\mathcal{F}_6 &: F(x,y,z) = F(x,y,\xi)\eta(z)+F(x,\xi,z)\eta(y), \quad \theta(\xi)=\theta^*(\xi)=0,\\
&\quad F(x,y,\xi)=F(y,x,\xi), \quad F(\varphi x,\varphi y,\xi)=-F(x,y,\xi);\\
\mathcal{F}_{11} &: F(x,y,z) = \eta(x)\{\eta(y)\omega(z)+\eta(z)\omega(y)\}.
\end{aligned} \tag{1}
$$

The classes $\mathcal{F}_i \oplus \mathcal{F}_j$, etc., are defined in a natural way by the conditions of the basic classes. The special class $\mathcal{F}_0 : F = 0$ is contained in each of the defined classes. The $\mathcal{F}_i^0$-manifold is an $\mathcal{F}_i$ -manifold ($i = 1, 4, 5, 11$) with closed 1-forms θ, θ^* and $\omega \circ \varphi$.

The following tensors are essential curvature tensors on M:

$$
\begin{aligned}
\pi_1(x,y,z,u) &= g(y,z)g(x,u) - g(x,z)g(y,u),\\
\pi_2(x,y,z,u) &= g(y,\varphi z)g(x,\varphi u) - g(x,\varphi z)g(y,\varphi u),\\
\pi_3(x,y,z,u) &= -g(y,z)g(x,\varphi u) + g(x,z)g(y,\varphi u)\\
&\quad - g(y,\varphi z)g(x,u) + g(x,\varphi z)g(y,u),\\
\pi_4(x,y,z,u) &= \eta(y)\eta(z)g(x,u) - \eta(x)\eta(z)g(y,u)\\
&\quad + \eta(x)\eta(u)g(y,z) - \eta(y)\eta(u)g(x,z),\\
\pi_5(x,y,z,u) &= \eta(y)\eta(z)g(x,\varphi u) - \eta(x)\eta(z)g(y,\varphi u)\\
&\quad + \eta(x)\eta(u)g(y,\varphi z) - \eta(y)\eta(u)g(x,\varphi z).
\end{aligned}
$$

In [7] it is established that the tensors $\pi_1 - \pi_2 - \pi_4$ and $\pi_3 + \pi_5$ are Kählerian, i.e. they have the condition of a curvature-like tensor L: $L(X,Y,Z,U) = -L(X,Y,\varphi Z,\varphi U)$.

Let R be the curvature tensor of ∇. The tensors R and $\widetilde{R} : \widetilde{R}(x,y,z,u) = R(x,y,z,\varphi u)$ are Kählerian on any $\mathcal{F}_0$-manifold.

There are known the following sectional curvatures with respect to g and R for every nondegenerate section α in T_pM with a basis $\{x,y\}$:

$$
k(\alpha;p) = k(x,y) = \frac{R(x,y,y,x)}{\pi_1(x,y,y,x)}, \quad \widetilde{k}(\alpha;p) = \widetilde{k}(x,y) = \frac{\widetilde{R}(x,y,y,x)}{\pi_1(x,y,y,x)}.
$$

In [8] there are introduced the following special sections in T_pM: a ξ-section (e.g. $\{\xi,x\}$), a φ-holomorphic section (i.e. $\alpha = \varphi\alpha$) and a totally real section (i.e. $\alpha \perp \varphi\alpha$).

The canonical curvature tensor K is introduced in [7]. The tensor K is a curvature tensor with respect to the canonical connection D defined by

$$D_X Y = \nabla_X Y + \frac{1}{2}\{(\nabla_X \varphi)\varphi Y + (\nabla_X \eta)Y.\xi\} - \eta(Y)\nabla_X \xi. \tag{2}$$

The connection D is a natural connection, i.e. the structural tensors are parallel with respect to D. Let us note that the tensor K out of $\mathcal{F}_0$ has the properties of R in $\mathcal{F}_0$.

2. Curvatures on the real time-like hypersurfaces of a Kähler manifold with Norden metric

In [5] two types of real hypersurfaces of a complex manifold with Norden metric are introduced. The obtained submanifolds are almost contact manifolds with Norden metric. Let us recall the real time-like hypersurface with respect to the Norden metric.

The hypersurface M of an almost complex manifold with Norden metric (M', J, g'), determined by the condition the normal unit N to be time-like regarding g' (i.e. $g'(N,N) = -1$), equipped with the almost contact structure with Norden metric

$$\begin{aligned} &\varphi := J + \cos t.g'(\cdot, JN)\{\cos t.N - \sin t.JN\}, \\ &\xi := \sin t.N + \cos t.JN, \qquad \eta := \cos t.g'(\cdot, JN), \qquad g := g'|_M, \end{aligned} \tag{3}$$

where $t := \arctan\{g'(N, JN)\}$ for $t \in \left(-\frac{\pi}{2}; \frac{\pi}{2}\right)$, is called *a real time-like hypersurface* of (M', J, g').

In the case when (M', J, g') is a Kähler manifold with Norden metric (i.e. a $\mathcal{W}_0$-manifold), in [6] it is ascertained the following statement: The class $\mathcal{F}_4 \otimes \mathcal{F}_5 \otimes \mathcal{F}_6 \otimes \mathcal{F}_{11}$ is the class of the real time-like hypersurfaces of a Kähler manifold with Norden metric. There are 16 classes of these hypersurfaces in all. When $n = 1$ the class $\mathcal{F}_6$ is restricted to $\mathcal{F}_0$. Therefore, for a 4-dimensional Kähler manifold with Norden metric there are only 8 classes of the considered hypersurfaces.

The tensor F and the second fundamental tensor A of the considered type of hypersurfaces have the following form, respectively:

$$\begin{aligned} F(X,Y,Z) &= \sin t\,\{g(AX,\varphi Y)\eta(Z) + g(AX,\varphi Z)\eta(Y)\} \\ &\quad - \cos t\,\{g(AX,Y)\eta(Z) + g(AX,Z)\eta(Y) - 2\eta(AX)\eta(Y)\eta(Z)\}, \\ AX &= -\frac{dt(\xi)}{2\cos t}\eta(X)\xi - \sin t\{\nabla_X \xi + g(\nabla_\xi \xi, X)\xi\} \\ &\quad + \cos t\{\varphi\nabla_X \xi + g(\varphi\nabla_\xi \xi, X)\xi\}. \end{aligned} \tag{4}$$

The basic classes of the considered hypersurfaces are characterized in terms of the second fundamental tensor by the conditions [5]:

$$\mathcal{F}_0 \,:\, A = -\frac{dt(\xi)}{2\cos t}\eta\otimes\xi;$$

$$\mathcal{F}_4 \,:\, A = -\frac{dt(\xi)}{2\cos t}\eta\otimes\xi - \frac{\theta(\xi)}{2n}\{\sin t.\varphi - \cos t.\varphi^2\};$$

$$\mathcal{F}_5 \,:\, A = -\frac{dt(\xi)}{2\cos t}\eta\otimes\xi + \frac{\theta^*(\xi)}{2n}\{\cos t.\varphi + \sin t.\varphi^2\};$$

$$\mathcal{F}_6 \,:\, A\circ\varphi = \varphi\circ A, \qquad \operatorname{tr} A - \frac{dt(\xi)}{2\cos t} = \operatorname{tr}(A\circ\varphi) = 0;$$

$$\mathcal{F}_{11} \,:\, A = -\frac{dt(\xi)}{2\cos t}\eta\otimes\xi - \cos t\{\eta\otimes\Omega + \omega\otimes\xi\}$$
$$- \sin t\{\eta\otimes\varphi\Omega + (\omega\circ\varphi)\otimes\xi\}, \qquad \omega(\cdot) = g(\cdot,\Omega).$$

According to the formulas of Gauss and Weingarten in this case $\nabla'_X Y = \nabla_X Y - g(AX,Y)N$, $\nabla'_X N = -AX$, we get the relation between the curvature tensors R' and R of the $\mathcal{W}_0$-manifold (M', J, g') and its hypersurface (M,φ,ξ,η,g), respectively:

$$R'(x,y,z,u) = R(x,y,z,u) + \pi_1(Ax,Ay,z,u),$$
$$R'(x,y)N = -(\nabla_x A)\,y + (\nabla_y A)\,x.$$

Hence, having in mind Theorem 1.1, we obtain:

$$R(x,y,z,u) = \left\{\nu'\,[\pi'_1 - \pi'_2] + \widetilde{\nu'}\pi'_3\right\}(x,y,z,u) - \pi_1(Ax,Ay,z,u),$$

$$R(x,y,\varphi z,\varphi u) = -\left\{R - \nu'[\pi_4 - \tan t\pi_5] + \widetilde{\nu'}[\pi_5 + \tan t\pi_4]\right\}(x,y,z,u)$$
$$- [\pi_1 + \pi_2](Ax,Ay,z,u),$$

$$R(x,y)\xi = \left\{\nu'[\pi_4 - \tan t\pi_5] - \widetilde{\nu'}[\pi_5 + \tan t\pi_4]\right\}(x,y)\xi$$
$$- \pi_1(Ax,Ay)\xi,$$

$$R(x,y)N = -\frac{1}{\cos t}[\nu'\pi_5 + \widetilde{\nu'}\pi_4](x,y)\xi.$$

Therefore

$$(\nabla_x A)y - (\nabla_y A)x = \frac{1}{\cos t}[\nu'\pi_5 + \widetilde{\nu'}\pi_4](x,y)\xi. \tag{5}$$

Having in mind the equations:

$$g'(y,Jz) = g(y,\varphi z) + \tan t\;\eta(y)\eta(z), \qquad \pi'_1 = \pi_1,$$
$$\pi'_2 = \pi_2 + \tan t\;\pi_5, \qquad \pi'_3 = \pi_3 - \tan t\;\pi_4,$$

which are valid for real time-like hypersurfaces, we obtain

Proposition 2.1. *A real time-like hypersurface of a Kähler manifold with Norden metric of constant totally real sectional curvatures ν' and $\widetilde{\nu'}$ has the following curvature properties:*

$$\begin{aligned} R(x,y,z,u) = &\left\{\nu'[\pi_1 - \pi_2 - \tan t\pi_5] + \widetilde{\nu'}[\pi_3 - \tan t\pi_4]\right\}(x,y,z,u) \\ &- \pi_1(Ax,Ay,z,u), \\ \tau = &\, 4n^2\nu' - 4n\widetilde{\nu'}\tan t - (\operatorname{tr} A)^2 + \operatorname{tr} A^2, \\ \widetilde{\tau} = &-2n\nu'\tan t + 2n(2n-1)\widetilde{\nu'} - \operatorname{tr} A \operatorname{tr}(A\circ\varphi) + \operatorname{tr}(A^2\circ\varphi); \end{aligned}$$

for a ξ-section $\{\xi, x\}$

$$k(\xi,x) = \nu' - \widetilde{\nu'}\tan t - [\nu'\tan t + \widetilde{\nu'}]\frac{g(x,\varphi x)}{g(x,x)-\eta(x)^2} - \frac{\pi_1(A\xi,Ax,x,\xi)}{g(x,x)-\eta(x)^2};$$

for a φ-holomorphic section $\{\varphi x, \varphi^2 x\}$ and for a totally real section $\{x,y\}$, orthogonal to ξ, respectively:

$$k(\varphi x,\varphi^2 x) = -\frac{\pi_1(A\varphi x, A\varphi^2 x, \varphi^2 x, \varphi x)}{\pi_1(\varphi x, \varphi^2 x, \varphi^2 x, \varphi x)}, \quad k(x,y) = \nu' - \frac{\pi_1(Ax,Ay,y,x)}{\pi_1(x,y,y,x)}.$$

If (M,φ,ξ,η,g) is a real time-like hypersurface of $\mathcal{W}_0$-manifold, then (2), (4) and (5) imply that the canonical curvature tensor has the form

$$\begin{aligned} K(x,y,z,u) = &\, R(x,y,\varphi^2 z,\varphi^2 u) + \pi_1(Ax,Ay,\varphi z,\varphi u) \\ &+ \sin t\left\{\sin t[\pi_1 - \pi_2 - \pi_4] - \cos t[\pi_3 + \pi_5]\right\}(Ax,Ay,z,u). \end{aligned}$$

Then, because of the last equation and Proposition 2.1 we have

Proposition 2.2. *Let (M,φ,ξ,η,g) be a real time-like hypersurface of a $\mathcal{W}_0$-manifold (M',J,g') of constant totally real sectional curvatures. Then K of M is Kählerian and*

$$\begin{aligned} K(x,y,z,u) = &\left\{\nu'[\pi_1 - \pi_2 - \pi_4] + \widetilde{\nu'}[\pi_3 + \pi_5]\right\}(x,y,z,u) \\ &- \cos t\{\cos t[\pi_1 - \pi_2 - \pi_4] + \sin t[\pi_3 + \pi_5]\}(Ax,Ay,z,u), \\ \tau(K) = &\, 4n(n-1)\nu' - \cos t(a\cos t + 2b\sin t), \\ \widetilde{\tau}(K) = &\, 4n(n-1)\widetilde{\nu'} - \cos t(a\sin t - 2b\cos t), \\ a = &\, (\operatorname{tr} A)^2 - \operatorname{tr} A^2 - [\operatorname{tr}(A\circ\varphi)]^2 + \operatorname{tr}(A\circ\varphi)^2 - 2\eta(A\xi)\operatorname{tr} A + 2g(A\xi,A\xi), \\ b = &\, \operatorname{tr}(A^2\circ\varphi) - \operatorname{tr} A \operatorname{tr}(A\circ\varphi) + \eta(A\xi)\operatorname{tr}(A\circ\varphi) - g(\varphi A\xi, A\xi). \end{aligned}$$

3. Curvatures on $\mathcal{W}_0$'s real time-like hypersurfaces, belonging to the main classes

Now, let (M,φ,ξ,η,g) belong to the widest integrable main class $\mathcal{F}_4 \oplus \mathcal{F}_5$ of the real time-like hypersurfaces. Let us recall that a class of almost contact manifolds with Norden metric is said to be main if the tensor F is expressed explicitly by the structural tensors φ,ξ,η,g. In this case for the second fundamental tensor we have [5]:

$$\begin{aligned}
A &= -\frac{dt(\xi)}{2\cos t}\eta\otimes\xi - \frac{1}{2n}\{[\theta(\xi)\sin t - \theta^*(\xi)\cos t]\varphi \\
&\quad - [\theta(\xi)\cos t + \theta^*(\xi)\sin t]\,\varphi^2\}, \\
\operatorname{tr} A &= -\frac{dt(\xi)}{2\cos t} - \theta(\xi)\cos t - \theta^*(\xi)\sin t, \\
\operatorname{tr}(A\circ\varphi) &= \theta(\xi)\sin t - \theta^*(\xi)\cos t.
\end{aligned} \tag{6}$$

Then, having in mind the last identities and Proposition 2.1, we obtain

Corollary 3.1. *If a real time-like hypersurface of a Kähler manifold with Norden metric of constant totally real sectional curvatures is an $\mathcal{F}_4\oplus\mathcal{F}_5$-manifold, then it has the following curvature properties:*

$$\begin{aligned}
R &= \nu'[\pi_1 - \pi_2 - \tan t\pi_5] + \widetilde{\nu'}[\pi_3 - \tan t\pi_4] \\
&\quad - \frac{dt(\xi)}{4n\cos t}\{\theta(\xi)[\sin t\pi_5 + \cos t\pi_4] + \theta^*(\xi)[\sin t\pi_4 - \cos t\pi_5]\} \\
&\quad - \frac{\theta(\xi)^2 + \theta^*(\xi)^2}{4n^2}\pi_2 - \frac{\left(\theta(\xi)\cos t + \theta^*(\xi)\sin t\right)^2}{4n^2}[\pi_1 - \pi_2 - \pi_4] \\
&\quad + \frac{\left(\theta(\xi)\cos t + \theta^*(\xi)\sin t\right)\left(\theta(\xi)\sin t - \theta^*(\xi)\cos t\right)}{4n^2}[\pi_3 + \pi_5], \\
\tau &= 4n(n\nu' - \widetilde{\nu'}\tan t) - dt(\xi)\theta(\xi) - dt(\xi)\theta^*(\xi)\tan t \\
&\quad - \frac{n-1}{n}(\theta(\xi)\cos t + \theta^*(\xi)\sin t)^2 - \frac{\theta(\xi)^2 + \theta^*(\xi)^2}{2n}, \\
\widetilde{\tau} &= 2n(n-1)\widetilde{\nu'} + 2n\nu'\tan t + \frac{dt(\xi)\theta(\xi)}{2}\tan t - \frac{dt(\xi)\theta^*(\xi)}{2} \\
&\quad + \frac{n-1}{n}(\theta(\xi)\sin t - \theta^*(\xi)\cos t)(\theta(\xi)\cos t + \theta^*(\xi)\sin t),
\end{aligned}$$

$$k(\varphi x,\varphi^2 x) = -\frac{\theta(\xi)^2 + \theta^*(\xi)^2}{4n^2}, \quad k(x,y) = \nu' - \frac{\left(\theta(\xi)\cos t + \theta^*(\xi)\sin t\right)^2}{4n^2}.$$

Let us remark that we can obtain the corresponding properties for the classes $\mathcal{F}_4$, $\mathcal{F}_5$ and $\mathcal{F}_0$, if we substitute $\theta^*(\xi) = 0$, $\theta(\xi) = 0$ and $\theta(\xi) = \theta^*(\xi) = 0$, respectively.

Using the equations (2) and (1), we express the canonical connection explicitly for the class $\mathcal{F}_4 \oplus \mathcal{F}_5$ as follows

$$D_X Y = \nabla_X Y + \frac{\theta(\xi)}{2n}\left\{g(x,\varphi y)\xi - \eta(y)\varphi x\right\} - \frac{\theta^*(\xi)}{2n}\left\{g(\varphi x,\varphi y)\xi - \eta(y)\varphi^2 x\right\}.$$

Let $(M,\varphi,\xi,\eta,g) \in \mathcal{F}_4^0 \oplus \mathcal{F}_5^0$, i.e. (M,φ,ξ,η,g) is an $(\mathcal{F}_4 \oplus \mathcal{F}_5)$-manifold with closed 1-forms θ and θ^*. The canonical curvature tensor K of any $(\mathcal{F}_4^0 \oplus \mathcal{F}_5^0)$-manifold is Kählerian and it has the form

$$\begin{aligned} K = R &+ \frac{\xi\theta(\xi)}{2n}\pi_5 + \frac{\xi\theta^*(\xi)}{2n}\pi_4 \\ &+ \frac{\theta^2(\xi)}{4n^2}\left[\pi_2 - \pi_4\right] + \frac{\theta^{*2}(\xi)}{4n^2}\pi_1 - \frac{\theta(\xi)\theta^*(\xi)}{4n^2}\left[\pi_3 - \pi_5\right]. \end{aligned}$$

Then, using Corollary 3.1, we ascertain the truthfulness of the following

Corollary 3.2. *If a real time-like hypersurface of a Kähler manifold with Norden metric of constant totally real sectional curvatures is an* $(\mathcal{F}_4^0 \oplus \mathcal{F}_5^0)$*-manifold, then* K *is expressed in the following way:*

$$\begin{aligned} K = &\left(\nu' + \frac{\theta^*(\xi)}{4n^2}\right)\left[\pi_1 - \pi_2\right] + \left(\widetilde{\nu'} - \frac{\theta(\xi)\theta^*(\xi)}{4n^2}\right)\pi_3 \\ &- \left(\widetilde{\nu'}\tan t + \frac{dt(\xi)\theta(\xi)}{4n} + \frac{dt(\xi)\theta^*(\xi)}{4n}\tan t - \frac{\xi\theta^*(\xi)}{2n} + \frac{\theta(\xi)^2}{4n^2}\right)\pi_4 \\ &- \left(\nu'\tan t + \frac{dt(\xi)\theta(\xi)}{4n}\tan t - \frac{dt(\xi)\theta^*(\xi)}{4n} - \frac{\xi\theta(\xi)}{2n} - \frac{\theta(\xi)\theta^*(\xi)}{4n^2}\right)\pi_5 \\ &- \frac{\left(\theta(\xi)\cos t + \theta^*(\xi)\sin t\right)^2}{4n^2}\left[\pi_1 - \pi_2 - \pi_4\right] \\ &+ \frac{\left(\theta(\xi)\sin t - \theta^*(\xi)\cos t\right)\left(\theta(\xi)\cos t + \theta^*(\xi)\sin t\right)}{4n^2}\left[\pi_3 + \pi_5\right]. \end{aligned}$$

We compute the expression $(\nabla_x A)\,y - (\nabla_y A)\,x$ using (6) and we compare the result with (5). Thus, we get the relations

$$\begin{aligned} \nu' = &-\frac{dt(\xi)\theta(\xi)}{4n} + \frac{\cos t}{2n}\left[\xi\theta(\xi)\sin t - \xi\theta^*(\xi)\cos t\right] \\ &+ \frac{\cos^2 t}{4n^2}\left[\theta(\xi)^2 - \theta^*(\xi)^2\right] + \frac{\sin t\cos t}{2n^2}\theta(\xi)\theta^*(\xi) \\ &+ \frac{dt(\xi)}{2n}\cos t\left[\theta(\xi)\cos t + \theta^*(\xi)\sin t\right], \end{aligned} \tag{7}$$

$$\begin{aligned}\widetilde{\nu}' = &-\frac{dt(\xi)\theta^*(\xi)}{4n} + \frac{\cos t}{2n}\left[\xi\theta(\xi)\cos t + \xi\theta^*(\xi)\sin t\right] \\ &+ \frac{\sin t\cos t}{4n^2}\left[\theta^*(\xi)^2 - \theta(\xi)^2\right] + \frac{\cos^2 t}{2n^2}\theta(\xi)\theta^*(\xi) \\ &- \frac{dt(\xi)}{2n}\cos t\left[\theta(\xi)\sin t - \theta^*(\xi)\cos t\right].\end{aligned} \tag{8}$$

Hence, we have

$$\begin{aligned}K &= \lambda[\pi_1 - \pi_2 - \pi_4] + \mu[\pi_3 + \pi_5], \\ R &= \lambda[\pi_1 - \pi_2 - \pi_4] + \mu[\pi_3 + \pi_5] - \frac{\xi\theta^*(\xi)}{2n}\pi_4 - \frac{\xi\theta(\xi)}{2n}\pi_5 \\ &\quad - \frac{\theta^*(\xi)^2}{4n^2}\pi_1 - \frac{\theta(\xi)^2}{4n^2}[\pi_2 - \pi_4] + \frac{\theta(\xi)\theta^*(\xi)}{4n^2}[\pi_3 - \pi_5], \\ \lambda &= -\frac{dt(\xi)\theta(\xi)}{4n} + \frac{dt(\xi)}{2n}\cos t\left[\theta(\xi)\cos t + \theta^*(\xi)\sin t\right] \\ &\quad + \frac{\cos t}{2n}\left[\xi\theta(\xi)\sin t - \xi\theta^*(\xi)\cos t\right], \\ \mu &= -\frac{dt(\xi)\theta^*(\xi)}{4n} - \frac{dt(\xi)}{2n}\cos t\left[\theta(\xi)\sin t - \theta^*(\xi)\cos t\right] \\ &\quad + \frac{\cos t}{2n}\left[\xi\theta(\xi)\cos t + \xi\theta^*(\xi)\sin t\right].\end{aligned}$$

We solve the system (7), (8) with respect to the functions $\theta(\xi)$ and $\theta^*(\xi)$ for $t = \mathrm{const}$ and get

$$\theta(\xi) = 2\varepsilon n\sqrt{\frac{\nu'\cos t - \widetilde{\nu}'\sin t + \sqrt{\nu'^2 + \widetilde{\nu}'^2}}{2\cos t}},$$

$$\theta^*(\xi) = 2\varepsilon n\frac{\sqrt{\cos t}(\nu'\tan t + \widetilde{\nu}')}{\sqrt{2(\nu'\cos t - \widetilde{\nu}'\sin t + \sqrt{\nu'^2 + \widetilde{\nu}'^2})}},$$

where $\varepsilon = \pm 1$. Since ν' and $\widetilde{\nu}'$ are point-wise constant for M'^4 $(n = 1)$ and they are absolute constants for M'^{2n+2} $(n \geq 2)$ (Theorem 1.1), then the functions $\theta(\xi)$ and $\theta^*(\xi)$, which determine the real time-like hypersurface as an almost contact manifold with Norden metric, are also pointwise constant on M^3 and absolute constants on M^{2n+1} $(n \geq 2)$. Hence, we have

Theorem 3.1. *Let (M', J, g') be a Kähler manifold with Norden metric of constant totally real sectional curvatures. Let the $(\mathcal{F}_4^0 \oplus \mathcal{F}_5^0)$-manifold $(M, \varphi, \xi, \eta, g)$, $\dim M \geq 5$, be its real time-like hypersurface, defined by (3).*

If $t = \text{const}$, *then* $K = 0$ *on* M *and*

$$\begin{aligned}
R &= -\frac{\theta(\xi)^2}{4n^2}\left[\pi_2 - \pi_4\right] - \frac{\theta^*(\xi)^2}{4n^2}\pi_1 + \frac{\theta(\xi)\theta^*(\xi)}{4n^2}\left[\pi_3 - \pi_5\right], \\
\tau &= \frac{\theta(\xi)^2}{2n} - (2n+1)\frac{\theta^*(\xi)^2}{2n}, \qquad \widetilde{\tau} = \frac{\theta(\xi)\theta^*(\xi)}{2n}, \\
k(\xi, x) &= \frac{\theta(\xi)^2 - \theta^*(\xi)^2}{4n^2} + \frac{2\theta(\xi)\theta^*(\xi)}{4n^2}\frac{g(x,\varphi x)}{g(\varphi x,\varphi x)}, \\
k\left(\varphi x, \varphi^2 x\right) &= -\frac{\theta(\xi)^2 + \theta^*(\xi)^2}{4n^2}, \qquad k(x,y) = -\frac{\theta^*(\xi)^2}{4n^2}.
\end{aligned}$$

Acknowledgement

A support by the NIMP Fund of Plovdiv University is acknowledged.

References

1. A. Borisov and G. Ganchev, *Curvature properties of Kaehlerian manifolds with B-metric*, Math. Educ. Math., Proc. of the 14th Spring Conference of UBM, Sunny Beach, 1985, 220–226.
2. G. Ganchev and A. Borisov, *Note on the almost complex manifolds with Norden metric*, C. R. Acad. Bulgare Sci. **39** (5), 1986, 31–34.
3. G. Ganchev, K. Gribachev and V. Mihova, *Holomorphic hypersurfaces of Kaehlerian manifolds with Norden metric*, Plovdiv Univ. Sci. Works — Math. **23** (2), 1985, 221–236.
4. G. Ganchev, V. Mihova and K. Gribachev, *Almost contact manifolds with B-metric*, Math. Balkanica (N.S.) **7** (3–4), 1993, 261–276.
5. M. Manev, *Almost contact B-metric hypersurfaces of Kaehler manifolds with B-metric*, In: Perspectives of Complex Analysis, Differential Geometry and Mathematical Physics, eds. S. Dimiev and K. Sekigawa, World Scientific, Singapore, 2001, 159–170.
6. M. Manev, *Classes of real time-like hypersurfaces of a Kaehler manifold with B-metric*, J. Geom. **75**, 2002, 113–122.
7. M. Manev and K. Gribachev, *Conformally invariant tensors on almost contact manifolds with B-metric*, Serdica - Bulg. Math. Publ. **20**, 1994, 133–147.
8. G. Nakova and K. Gribachev, *Submanifolds of some almost contact manifolds with B-metric with codimension two. I*, Math. Balkanica (N.S.) **12** (1–2), 1998, 93–108.
9. A. P. Norden, *On a class of four-dimensional A-spaces* (in Russian), Izv. Vyssh. Uchebn. Zaved. Mat. **17** (4), 1960, 145–157.

SOME SUBMANIFOLDS OF ALMOST CONTACT MANIFOLDS WITH NORDEN METRIC

G. NAKOVA

Department of Algebra and Geometry,
Faculty of Pedagogics, University of Veliko Tarnovo,
1 Theodosij Tirnovsky Str.,
5000 Veliko Tarnovo, Bulgaria
E-mail: gnakova@yahoo.com

In this paper we study submanifolds of almost contact manifolds with Norden metric of codimension two with totally real normal spaces. Examples of such submanifolds as a Lie subgroups are constructed.

Keywords: Norden metric; Almost contact manifold; Submanifold; Lie group.

1. Introduction

Let $(M, \varphi, \xi, \eta, g)$ be a $(2n+1)$-dimensional almost contact manifold with Norden metric, i. e. (φ, ξ, η) is an almost contact structure [1] and g is a metric [3] on M such that

$$\begin{aligned} \varphi^2 X = -\mathrm{id} + \eta \otimes \xi, \qquad \eta(\xi) = 1, \\ g(\varphi X, \varphi Y) = -g(X, Y) + \eta(X)\eta(Y), \end{aligned} \tag{1}$$

where id denotes the identity transformation and X, Y are differentiable vector fields on M, i. e., $X, Y \in \mathfrak{X}(M)$. The tensor $\tilde{g}$ given by

$$\tilde{g}(X, Y) = g(X, \varphi Y) + \eta(X)\eta(Y)$$

is a Norden metric, too. Both metrics g and $\tilde{g}$ are indefinite of signature $(n+1, n)$.

Let ∇ be the Levi-Civita connection of the metric g. The tensor field F of type $(0,3)$ on M is defined by

$$F(X, Y, Z) = g((\nabla_X \varphi)Y, Z).$$

A classification of the almost contact manifolds with Norden metric with respect to the tensor F is given in [3] and eleven basic classes $\mathcal{F}_i$, $(i = 1, 2, \ldots, 11)$ are obtained.

Let R be the curvature tensor field of ∇ defined by

$$R(X,Y)Z = \nabla_X \nabla_Y Z - \nabla_Y \nabla_X Z - \nabla_{[X,Y]} Z.$$

The corresponding tensor field of type $(0,4)$ is determined as follows

$$R(X,Y,Z,W) = g(R(X,Y)Z,W).$$

Let $(\overline{M}, \overline{\varphi}, \overline{\xi}, \overline{\eta}, g)$, $(\dim \overline{M} = 2n+3)$ be an almost contact manifold with Norden metric and let M be a submanifold of $\overline{M}$. Then for each point $p \in \overline{M}$ we have

$$T_p\overline{M} = T_pM \oplus (T_pM)^{\perp},$$

where T_pM and $(T_pM)^{\perp}$ are the tangent space and the normal space of $\overline{M}$ at p respectively. When the submanifold M of $\overline{M}$ is of codimension 2 we denote $(T_pM)^{\perp}$ by $\alpha = \{N_1, N_2\}$, i.e. α is a normal section of M. Let α be a 2-dimensional section in $T_p\overline{M}$. Let us recall a section α is said to be

- *non-degenerate*, *weakly isotropic* or *strongly isotropic* if the rank of the restriction of the metric g on α is 2, 1 or 0 respectively;
- of *pure* or *hybrid* type if the restriction of g on α has a signature $(2,0)$, $(0,2)$ or $(1,1)$ respectively;
- *holomorphic* if $\overline{\varphi}\alpha = \alpha$;
- $\overline{\xi}$- section if $\overline{\xi} \in \alpha$;
- *totally real* if $\overline{\varphi}\alpha \perp \alpha$.

Submanifolds M of $\overline{M}$ of codimension 2 with a non-degenerate of hybrid type normal section α are studied. In [4] two basic types of such submanifolds are considered: α is a holomorphic section and α is a $\overline{\xi}$-section. In [5] the normal section $\alpha = \{N_1, N_2\}$ is such that $\overline{\varphi}N_1 \notin \alpha$, $\overline{\varphi}N_2 \in \alpha$. In this paper we consider submanifolds M of $\overline{M}$ of codimension 2 in the case when the normal section α is a non-degenerate of hybrid type and α is a totally real. The totally real sections α are two types: α is non-orthogonal to $\overline{\xi}$ and α is orthogonal to $\overline{\xi}$.

2. Submanifolds of codimension 2 of almost contact manifolds with Norden metric with totally real non-orthogonal to $\overline{\xi}$ normal spaces

Let $(\overline{M}, \overline{\varphi}, \overline{\xi}, \overline{\eta}, g)$ $(\dim \overline{M} = 2n+3)$ be an almost contact manifold with Norden metric and let M be a submanifold of codimension 2 of $\overline{M}$. We assume that there exists a normal section $\alpha = \{N_1, N_2\}$ defined globally over the submanifold M such that

- α is a non-degenerate of hybrid type, i.e.

$$g(N_1, N_1) = -g(N_2, N_2) = 1, \quad g(N_1, N_2) = 0; \tag{2}$$

- α is a totally real, i.e.

$$g(N_1, \overline{\varphi} N_1) = g(N_2, \overline{\varphi} N_2) = g(N_1, \overline{\varphi} N_2) = g(N_2, \overline{\varphi} N_1) = 0; \tag{3}$$

- α is a non-orthogonal to $\overline{\xi}$ ($\overline{\xi} \notin T_pM$) and $\overline{\xi} \notin \alpha$.

Then we obtain the following decomposition for $\overline{\xi}$, $\overline{\varphi}X$, $\overline{\varphi}N_1$, $\overline{\varphi}N_2$ with respect to $\{N_1, N_2\}$ and T_pM

$$\begin{aligned} \overline{\xi} &= \xi_0 + aN_1 + bN_2; \\ \overline{\varphi}X &= \varphi X + \eta^1(X)N_1 + \eta^2(X)N_2, \qquad X \in \chi(M); \\ \overline{\varphi}N_1 &= \xi_1; \qquad \overline{\varphi}N_2 = -\xi_2; \end{aligned} \tag{4}$$

where φ denotes a tensor field of type $(1,1)$ on M; $\xi_0, \xi_1, \xi_2 \in \chi(M)$; η^1 and η^2 are 1-forms on M; a, b are functions on M such that $(a, b) \neq (0, 0)$. We denote the restriction of g on M by the same letter.

Let $a \neq 0$, $|a| > b$ and $a^2 - b^2 = k^2$. Taking into account the equalities (1)–(4) we compute

$$\eta^i(X) = g(X, \xi_i), \qquad (i = 0, 1, 2); \tag{5}$$

$$g(\varphi X, \varphi Y) = -g(X, Y) + \eta^0(X)\eta^0(Y) - \eta^1(X)\eta^1(Y) + \eta^2(X)\eta^2(Y); \tag{6}$$

$$\begin{aligned} \varphi^2 X &= -X + \eta^0(X)\xi_0 - \eta^1(X)\xi_1 + \eta^2(X)\xi_2; \\ \eta^0(\varphi X) &= -a\eta^1(X) + b\eta^2(X); \\ \eta^1(\varphi X) &= a\eta^0(X); \qquad \eta^2(\varphi X) = b\eta^0(X); \end{aligned} \tag{7}$$

$$\varphi\xi_0 = -a\xi_1 + b\xi_2; \qquad \varphi\xi_1 = a\xi_0; \qquad \varphi\xi_2 = b\xi_0; \tag{8}$$

$$\begin{aligned} &g(\xi_0, \xi_0) = 1 - a^2 + b^2; && g(\xi_1, \xi_1) = a^2 - 1; \\ &g(\xi_2, \xi_2) = 1 + b^2; && g(\xi_0, \xi_1) = g(\xi_0, \xi_2) = 0; \\ &g(\xi_1, \xi_2) = ab; \end{aligned} \tag{9}$$

for arbitrary $X, Y \in \chi(M)$.

Now we define a vector field ξ, an 1-form η and a tensor field ϕ of type $(1,1)$ on M by

$$\begin{aligned} \xi &= -\frac{b}{k}.\xi_1 + \frac{a}{k}.\xi_2; \\ \eta(X) &= -\frac{b}{k}.\eta^1(X) + \frac{a}{k}.\eta^2(X), \qquad X \in \chi(M); \\ \phi X &= \lambda.\varphi^3 X + \mu.\varphi X, \qquad X \in \chi(M); \end{aligned} \tag{10}$$

where

$$\lambda_1 = \frac{\epsilon}{k(k+1)}, \qquad \mu_1 = \frac{\epsilon(1+k^2+k)}{k(k+1)};$$
$$\lambda_2 = \frac{\epsilon}{k(k-1)}, \qquad \mu_2 = \frac{\epsilon(1+k^2-k)}{k(k-1)}; \qquad \epsilon = \pm 1.$$

Further we consider the following cases for k:

1) $k^2 \neq 1 \iff k \neq \pm 1$. In this case ϕ, ξ, η are given by (10) and $\lambda = \lambda_1$, $\mu = \mu_1$ or $\lambda = \lambda_2$, $\mu = \mu_2$.
2) $k = -1$. We obtain ϕ, ξ, η from (10) by $k = -1$ and $\lambda = \lambda_2 = \epsilon/2$, $\mu = \mu_2 = 3\epsilon/2$.
3) $k = 1$. We obtain ϕ, ξ, η from (10) by $k = 1$ and $\lambda = \lambda_1 = \epsilon/2$, $\mu = \mu_1 = 3\epsilon/2$.

Using (5)–(10) we verify that (ϕ, ξ, η) is an almost contact structure on M and the restriction of g on M is Norden metric. Thus, the submanifolds (M, ϕ, ξ, η, g) of $\overline{M}$ considered in 1), 2), 3) are $(2n+1)$-dimensional almost contact manifolds with Norden metric.

Denoting by $\overline{\nabla}$ and ∇ the Levi-Civita connections of the metric g in $\overline{M}$ and M respectively, the formulas of Gauss and Weingarten are

$$\begin{aligned}
\overline{\nabla}_X Y &= \nabla_X Y + g(A_{N_1}X, Y)N_1 - g(A_{N_2}X, Y)N_2; \\
\overline{\nabla}_X N_1 &= -A_{N_1}X + \gamma(X)N_2; \\
\overline{\nabla}_X N_2 &= -A_{N_2}X + \gamma(X)N_1, \qquad X, Y \in \chi(M);
\end{aligned} \tag{11}$$

where A_{N_i} $(i = 1, 2)$ are the second fundamental tensors and γ is a 1-form on M.

3. Submanifolds of codimension 2 of almost contact manifolds with Norden metric with totally real orthogonal to $\overline{\xi}$ normal spaces

Let $(\overline{M}, \overline{\varphi}, \overline{\xi}, \overline{\eta}, g)$ $(\dim \overline{M} = 2n+3)$ be an almost contact manifold with Norden metric and let M be a submanifold of codimension 2 of $\overline{M}$. We assume that there exists a normal section $\alpha = \{N_1, N_2\}$ defined globally over the submanifold M such that

- α is a non-degenerate of hybrid type, i.e. the equality (2) holds;
- α is a totally real;
- α is orthogonal to $\overline{\xi}$, i.e. $\overline{\xi} \in T_pM$.

Then from (4) by $a = b = 0$ we obtain the following decomposition with respect to $\{N_1, N_2\}$ and T_pM

$$\begin{aligned} \overline{\xi} &= \xi_0; \\ \overline{\varphi}X &= \varphi X + \eta^1(X)N_1 + \eta^2(X)N_2, \qquad X \in \chi(M); \\ \overline{\varphi}N_1 &= \xi_1; \qquad \overline{\varphi}N_2 = -\xi_2. \end{aligned} \tag{12}$$

Substituting $a = b = 0$ in (7), (8), (9) we have

$$\begin{aligned} \eta^0(\varphi X) &= \eta^1(\varphi X) = \eta^2(\varphi X) = 0; \\ \varphi\xi_0 &= \varphi\xi_1 = \varphi\xi_2 = 0; \\ g(\xi_0, \xi_0) &= g(\xi_2, \xi_2) = 1; \qquad g(\xi_1, \xi_1) = -1; \\ g(\xi_0, \xi_1) &= g(\xi_0, \xi_2) = g(\xi_1, \xi_2) = 0. \end{aligned} \tag{13}$$

Now we define a vector field ξ, an 1-form η and a tensor field ϕ of type $(1, 1)$ on M by

$$\begin{aligned} \xi &= t_0\xi_0 - t_2\xi_2; \\ \eta(X) &= t_0\eta^0(X) - t_2\eta^2(X), \qquad X \in \chi(M); \\ \phi X &= \varphi X + t_0\{\eta^1(X).\xi_2 + \eta^2(X).\xi_1\} + t_2\{\eta^0(X).\xi_1 + \eta^1(X).\xi_0\}; \end{aligned} \tag{14}$$

where t_0, t_2 are functions on M and $t_0^2 + t_2^2 = 1$.

Using equations (5), (6), (13), (14) we verify that (ϕ, ξ, η) is an almost contact structure on M and the restriction of g on M is Norden metric. So, the submanifolds (M, ϕ, ξ, η, g) of $\overline{M}$ are $(2n+1)$-dimensional almost contact manifolds with Norden metric. The formulas of Gauss and Weingarten are the same as those in section 2.

4. Examples of submanifolds of codimension 2 of almost contact manifolds with Norden metric with totally real normal spaces

In [6] a Lie group as a 5-dimensional almost contact manifold with Norden metric of the class $\mathcal{F}_9$ is constructed. We will use this Lie group to obtain examples of submanifolds considered in sections 2 and 3.

First we recall some facts from [6] which we need. Let $\mathfrak{g}$ be a real Lie algebra with a global basis of left invariant vector fields $\{X_1, X_2, X_3, X_4, X_5\}$ and G be the associated with $\mathfrak{g}$ real connected Lie group. The almost contact structure $(\overline{\varphi}, \overline{\xi}, \overline{\eta})$ and the Norden metric g on G are defined by:

$$\begin{aligned}
&\varphi X_i = X_{2+i}, \qquad \varphi X_{2+i} = -X_i, \qquad \varphi X_5 = 0, && (i = 1, 2);\\
&g(X_i, X_i) = -g(X_{2+i}, X_{2+i}) = g(X_5, X_5) = 1, && (i = 1, 2);\\
&g(X_j, X_k) = 0, && (j \neq k,\ j, k = 1, \ldots, 5);\\
&\overline{\xi} = X_5, \qquad \overline{\eta}(X_i) = g(X_i, X_5), && (i = 1, \ldots, 5).
\end{aligned} \tag{15}$$

The commutators of the basis vector fields are given by:

$$\begin{aligned}
&[X_1, X_2] = -[X_1, X_3] = aX_4, && [X_2, X_3] = aX_2 + aX_3,\\
&[X_3, X_4] = -[X_2, X_4] = aX_1, && [X_2, \overline{\xi}] = 2mX_1,\\
&[X_3, \overline{\xi}] = -2mX_4, && [X_1, X_4] = [X_1, \overline{\xi}] = [X_4, \overline{\xi}] = 0,
\end{aligned} \tag{16}$$

where $a, m \in \mathbb{R}$. So, the manifold $(G, \overline{\varphi}, \overline{\xi}, \overline{\eta}, g)$ is an almost contact manifold with Norden metric in the class $\mathcal{F}_9$.

Theorem 4.1 **([2]).** *Let G be a Lie group with a Lie algebra $\mathfrak{g}$ and $\widetilde{b}$ be a subalgebra of $\mathfrak{g}$. There exists a unique connected Lie subgroup H of G such that the Lie algebra b of H coincides with $\widetilde{b}$.*

From the equalities for the commutators of the basis vector fields $\{X_1, X_2, X_3, X_4, X_5\}$ it follows that the 3-dimensional subspaces of $\mathfrak{g}$ $\mathbf{b_1}$ with a basis $\{X_1, X_2, X_3\}$, $\mathbf{b_2}$ with a basis $\{X_1, X_3, X_4\}$ and $\mathbf{b_3}$ with a basis $\{X_1, X_4, \overline{\xi}\,\}$ are closed under the bracket operation. Hence b_i $(i = 1, 2, 3)$ are real subalgebras of $\mathfrak{g}$. Taking into account Theorem 4.1 we have there exist Lie subgroups H_i $(i = 1, 2, 3)$ of the Lie group G with Lie algebras b_i $(i = 1, 2, 3)$ respectively. The normal spaces α_i $(i = 1, 2, 3)$ of the submanifolds H_i $(i = 1, 2, 3)$ of G are: $\alpha_1 = \{X_4, \overline{\xi}\}$, $\alpha_2 = \{X_2, \overline{\xi}\}$, $\alpha_3 = \{X_2, X_3\}$. Because of (15) we have α_1 is $\overline{\xi}$-section of hybrid type, α_2 is $\overline{\xi}$-section of pure type and α_3 is a totally real orthogonal to $\overline{\xi}$ section of hybrid type. So, the submanifold H_3 of G is of the same type submanifolds considered in section 3.

We choose the unit normal fields of H_3 $N_1 = X_2$ and $N_2 = X_3$. For an arbitrary $X \in \chi(H_3)$ we have $X = x^1 X_1 + x^4 X_4 + \overline{\eta}(X)\overline{\xi}$. Taking into account (15) we compute

$$\begin{aligned}
&\overline{\xi} = \xi_0; && \overline{\varphi} X = -x^4 X_2 + x^1 X_3;\\
&\overline{\varphi} X_2 = X_4; && \overline{\varphi} X_3 = -X_1.
\end{aligned} \tag{17}$$

From (12), (17) it follows

$$\begin{aligned} &\eta^0(X) = \overline{\eta}(X); \\ &\varphi X = 0; \qquad \eta^1(X) = -x^4; \qquad \eta^2(X) = x^1; \\ &\xi_1 = X_4; \qquad \xi_2 = X_1. \end{aligned} \tag{18}$$

Substituting (18) in (14) for the almost contact structure on H_3 we obtain

$$\begin{aligned} \xi &= t_0\overline{\xi} - t_2X_1; \\ \eta(X) &= t_0\overline{\eta}(X) - t_2x^1; \\ \phi X &= t_0\{-x^4X_1 + x^1X_4\} + t_2\{\overline{\eta}(X)X_4 - x^4\overline{\xi}\}; \end{aligned} \tag{19}$$

where t_0, $t_2 \in \mathbb{R}$ and $t_0^2 + t_2^2 = 1$.

Using the well known condition for the Levi-Civita connection ∇ of g

$$\begin{aligned} 2g(\nabla_X Y, Z) = Xg(Y,Z) + Yg(X,Z) - Zg(X,Y) + g([X,Y],Z) \\ + g([Z,X],Y) + g([Z,Y],X) \end{aligned} \tag{20}$$

we get the following equation for the tensor F of H_3

$$\begin{aligned} F(X,Y,Z) = \frac{1}{2}\{g([X,\phi Y] - \phi[X,Y],Z) + g(\phi[Z,X] - [\phi Z,X],Y) \\ +g([Z,\phi Y] - [\phi Z,Y],X)\}, \qquad X,Y,Z \in \chi(H_3). \end{aligned} \tag{21}$$

From (16) we have $[X_1, X_4] = [X_1, \overline{\xi}] = [X_4, \overline{\xi}] = 0$. Having in mind the last equalities, (19) and (21) for the tensor F of H_3 we obtain $F = 0$. Thus, the submanifold $(H_3, \phi, \xi, \eta, g)$ of G, where (ϕ, ξ, η) is defined by (19) is an almost contact manifold with Norden metric in the class $\mathcal{F}_0$.

In order to construct an example for a submanifold from section 2 we make the following change of the basis of $\mathfrak{g}$

$$\begin{pmatrix} E_1 \\ E_2 \\ E_3 \\ E_4 \\ E_5 \end{pmatrix} = T^T \begin{pmatrix} X_1 \\ X_2 \\ \overline{\xi} \\ X_3 \\ X_4 \end{pmatrix}, \qquad T = \begin{pmatrix} 1 & 0 & 0 & 0 & 0 \\ 0 & \frac{\sqrt{3}}{2} & \frac{1}{2} & 0 & 0 \\ 0 & -\frac{1}{2} & \frac{\sqrt{3}}{2} & 0 & 0 \\ 0 & 0 & 0 & 1 & 0 \\ 0 & 0 & 0 & 0 & 1 \end{pmatrix} \in O(3,2). \tag{22}$$

Taking into account (16) and (22) we compute the commutators of the basis vector fields $\{E_1, E_2, E_3, E_4, E_5\}$ of $\mathfrak{g}$

$$
\begin{aligned}
&[E_1, E_2] = \frac{\sqrt{3}}{2} a E_5, \qquad [E_1, E_3] = \frac{1}{2} a E_5, \qquad [E_3, E_5] = -\frac{1}{2} a E_1, \\
&[E_2, E_5] = -\frac{\sqrt{3}}{2} a E_1, \qquad [E_1, E_4] = -a E_5, \\
&[E_2, E_4] = \frac{3}{4} a E_2 + \frac{\sqrt{3}}{4} a E_3 + \frac{\sqrt{3}}{2} a E_4 - m E_5, \\
&[E_3, E_4] = \frac{\sqrt{3}}{4} a E_2 + \frac{1}{4} a E_3 + \frac{1}{2} a E_4 + \sqrt{3} m E_5, \\
&[E_4, E_5] = a E_1, \qquad [E_2, E_3] = 2 m E_1, \qquad [E_1, E_5] = 0.
\end{aligned} \tag{23}
$$

Because of the elements of the matrix T are constants the Jacobi identity for the vector fields $\{E_1, E_2, E_3, E_4, E_5\}$ is valid. Now, we compute the matrix B of $\overline{\varphi}$ and the coordinates of $\overline{\xi}$ with respect to the basis $\{E_1, E_2, E_3, E_4, E_5\}$

$$
B = \begin{pmatrix} 0 & 0 & 0 & -1 & 0 \\ 0 & 0 & 0 & 0 & -\frac{\sqrt{3}}{2} \\ 0 & 0 & 0 & 0 & -\frac{1}{2} \\ 1 & 0 & 0 & 0 & 0 \\ 0 & \frac{\sqrt{3}}{2} & \frac{1}{2} & 0 & 0 \end{pmatrix}; \qquad \overline{\xi} = \left(0, -\frac{1}{2}, \frac{\sqrt{3}}{2}, 0, 0\right). \tag{24}
$$

From $T \in O(3,2)$ it follows the matrix of the metric g with respect to the basis $\{E_1, E_2, E_3, E_4, E_5\}$ is the same as the matrix

$$
C = \begin{pmatrix} 1 & 0 & 0 & 0 & 0 \\ 0 & 1 & 0 & 0 & 0 \\ 0 & 0 & 1 & 0 & 0 \\ 0 & 0 & 0 & -1 & 0 \\ 0 & 0 & 0 & 0 & -1 \end{pmatrix} \tag{25}
$$

of g with respect to the basis $\{X_1, X_2, \overline{\xi}, X_3, X_4\}$.

Using (23) we have that the 3-dimensional subspace b of $\mathfrak{g}$ with a basis $\{E_1, E_2, E_5\}$ is a subalgebra of $\mathfrak{g}$. Let H be the Lie subgroup of G with a Lie algebra b. Having in mind (24), (25) we obtain that the section $\alpha = \{E_3, E_4\}$ is a normal to the submanifold H, α is a totally real non-orthogonal to $\overline{\xi}$ section of hybrid type and $\overline{\xi} \notin \alpha$, i.e. H is of the same type submanifolds considered in section 2. We have the following decomposition of $\overline{\xi}$, $\overline{\varphi}X$, $\overline{\varphi}E_3$, $\overline{\varphi}E_4$ with respect to $\{E_3, E_4\}$ and T_pH

$$\overline{\xi} = -\frac{1}{2}E_2 + \frac{\sqrt{3}}{2}E_3;$$
$$\overline{\varphi}X = -\frac{\sqrt{3}}{2}\overline{x}^5 E_2 + \frac{\sqrt{3}}{2}\overline{x}^2 E_5 - \frac{1}{2}\overline{x}^5 E_3 + \overline{x}^1 E_4; \tag{26}$$
$$\overline{\varphi}E_3 = \frac{1}{2}E_5; \qquad \overline{\varphi}E_4 = -E_1;$$

where $X \in \chi(H)$ and $X = \overline{x}^1 E_1 + \overline{x}^2 E_2 + \overline{x}^5 E_5$. We substitute

$$a = \frac{\sqrt{3}}{2}, \qquad b = 0, \qquad k = \frac{\sqrt{3}}{2}, \qquad \lambda = \frac{4\sqrt{3}(2-\sqrt{3})}{3}, \qquad \mu = \lambda + 1,$$
$$\xi_2 = E_1, \qquad \eta^2(X) = \overline{x}^1, \qquad \varphi X = -\frac{\sqrt{3}}{2}\overline{x}^5 E_2 + \frac{\sqrt{3}}{2}\overline{x}^2 E_5$$

in (10) and obtain an almost contact structure (ϕ, ξ, η)

$$\xi = E_1; \qquad \eta(X) = \overline{x}^1;$$
$$\phi X = \frac{2\sqrt{3}}{3}\varphi X = -\overline{x}^5 E_2 + \overline{x}^2 E_5; \tag{27}$$

on the submanifold H.

Using (11), (16) and (20) we get

$$A_{E_3}X = -m\overline{x}^2 E_1 - m\overline{x}^1 E_2;$$
$$A_{E_4}X = -\frac{1}{2}\left(\frac{3}{2}a\overline{x}^2 + m\overline{x}^5\right)E_2 + \frac{1}{2}m\overline{x}^2 E_5;$$
$$\gamma(X) = \frac{\sqrt{3}}{2}\left(a\overline{x}^2 - m\overline{x}^5\right).$$

Then the formulae of Gauss and Weingarten (11) become

$$\overline{\nabla}_X Y = \nabla_X Y - m\left(\overline{x}^2\overline{y}^1 + \overline{x}^1\overline{y}^2\right)E_3 + \frac{1}{2}\left(\left(\frac{3}{2}a\overline{x}^2 + m\overline{x}^5\right)\overline{y}^2 + m\overline{x}^2\overline{y}^5\right)E_4;$$
$$\overline{\nabla}_X E_3 = m\overline{x}^2 E_1 + m\overline{x}^1 E_2 + \frac{\sqrt{3}}{2}\left(a\overline{x}^2 - m\overline{x}^5\right)E_4;$$
$$\overline{\nabla}_X E_4 = \frac{1}{2}\left(\frac{3}{2}a\overline{x}^2 + m\overline{x}^5\right)E_2 - \frac{1}{2}m\overline{x}^2 E_5 + \frac{\sqrt{3}}{2}\left(a\overline{x}^2 - m\overline{x}^5\right)E_3.$$

Having in mind the last formulas, (26) and (27) we compute the tensor F of H

$$F(X,Y,Z) = -\frac{\sqrt{3}}{2}a\overline{x}^2(\overline{y}^1\overline{z}^2 + \overline{y}^2\overline{z}^1), \quad X, Y, Z \in \chi(H)$$

and verify that the submanifold (H, ϕ, ξ, η, g) of G, where (ϕ, ξ, η) is defined by (27) is an almost contact manifold with Norden metric in the class $\mathcal{F}_4 \oplus \mathcal{F}_8$.

Acknowledgement

Partially supported by Scientific researches fund of "St. Cyril and St. Methodius" University of Veliko Tarnovo under contract RD-491-08 from 27. 06. 2008.

References

1. D. Blair, *Contact manifolds in Riemannian geometry*, Lecture Notes in Math., vol. 509, Springer Verlag, Berlin, 1976.
2. F. Warner, *Foundations of differentiable manifolds and Lie groups*, Springer Verlag, New York Berlin Heidelberg Tokyo, 1983.
3. G. Ganchev, V. Mihova and K. Gribachev, *Almost contact manifolds with B-metric*, Math. Balkanica, vol. 7, 1993, 262–276.
4. G. Nakova, K. Gribachev, *Submanifolds of some almost contact manifolds with B-metric with codimension two, I*, Math. Balkanica, vol. 11, 1997, 255–267.
5. G. Nakova, K. Gribachev, *Submanifolds of some almost contact manifolds with B-metric with codimension two, II*, Math. Balkanica, vol. 12, 1998, 93–108.
6. G. Nakova, *On some non-integrable almost contact manifolds with Norden metric of dimension 5*, Topics in contemporary differential geometry, complex analysis and mathematical physics, Proceedings of the 8th International Workshop on Complex Structures and Vector Fields, Singapore, 2007, 252–260.

A SHORT NOTE ON THE DOUBLE-COMPLEX LAPLACE OPERATOR

P. POPIVANOV

Institute of Mathematics and Informatics,
of the Bulgarian Academy of Sciences,
Sofia 1113, Bulgaria
E-mail: popivano@math.bas.bg

This short communication deals with several properties of the double-complex Laplace operator, namely it is constructed a fundamental distribution solution, it is proved the partial C^∞ hypoellipticity and it is proved a theorem on microlocal propagation of singularities of the solutions along the leaves (two dimensional) of the real and imaginary parts of the Hamiltonian vector field of the symbol.

Keywords: Fundamental solution; Partial hypoellipticity; Microlocal propagation of singularities.

1. Introduction and formulation of the main results

1. The double-complex Laplace operator was introduced by S. Dimiev in [1]. The Cauchy problem for it was studied by L. Apostolova in [2]. Stimulated by the talk of these two authors on the 9th International workshop on complex structures, integrability and vector fields, held in Sofia, August 2008, we propose a look at the same operator from the point of view of the general theory of linear partial differential operators and microlocal analysis too. Therefore, we have constructed its fundamental solution in the sense of L. Schwartz, we have shown its non-hypoellipticity and its partial C^∞ hypoellipticity, we have proved a result on propagation of microlocal singularities along canonical two-dimensional leaves etc.

2. This is the definition of the double-complex Laplace operator:

$$P(\mathcal{D}) = \partial^2/\partial u^2 + i\partial^2/\partial v^2, \tag{1}$$

where $u = x_1 + ix_2 \in \mathbf{C^1}$, $v = x_3 + ix_4 \in \mathbf{C}^1$ and as usual $\dfrac{\partial}{\partial u} = 1/2(\dfrac{\partial}{\partial x_1} +$

$\frac{1}{i}\frac{\partial}{\partial x_2}$), $x' = (x_1, x_2), x'' = (x_3, x_4) \in \mathbf{R}^2$.

Evidently, if $f(.)$ is analytic, then $\frac{\partial}{\partial u} f(\overline{u}) = 0$, while $\frac{\partial}{\partial u} f(u) = f'(u)$.

We remind of the reader that a distribution $\mathcal{E}(x) \in \mathcal{D}'$ is called a fundamental solution of (1) if $P(\mathcal{D})\mathcal{E}(x) = \delta(x)$ and $\delta(x)$, $x \in \mathbf{R}^4$ is the Dirac delta function. The characteristic set of (1) is given by:

$$\text{Char } P = \{(x, \xi) \in T^*(\mathbf{R}^4) \setminus 0 : \ p(\xi) = 0\}$$

where the symbol p of the operator P is the following one:

$$p(\xi) := 1/4[(i\xi_1 + \xi_2)^2 + i(i\xi_3 + \xi_4)^2] = -1/4[(\xi_1 - i\xi_2)^2 + i(\xi_3 - i\xi_4)^2].$$

Let us put $\lambda = \xi_1 + i\xi_2, \ \ \mu = \xi_3 + i\xi_4$. Thus, $-4p(\xi) = \overline{\lambda}^2 + i\overline{\mu}^2$.

Evidently,

$$p(\xi) = 0 \iff \overline{p(\xi)} = 0 \Rightarrow p(\xi) = 0 \iff \lambda^2 - i\mu^2 = 0 \iff \lambda = \pm e^{i\frac{\pi}{4}}\mu,$$

i.e. $\lambda\mu \neq 0$. This way we conclude that Char P can be written in real and complex forms:

$$\text{Char } P = \mathbf{R}^4_x \times \text{Char}_\xi P, \tag{2}$$

$$\text{Char}_\xi P = \{\xi \in \mathbf{R}^4 \setminus 0 : \xi_1^2 - \xi_2^2 + 2\xi_3\xi_4 = 0, \ \xi_3^2 - \xi_4^2 - 2\xi_1\xi_2 = 0\}$$
$$= \{(\lambda, \mu) \in \mathbf{C}^2 : \lambda = \pm e^{i\pi/4}\mu\} \ \text{ and } \ \lambda \neq 0, \ \mu \neq 0.$$

Evidently, $\nabla_\xi \text{Re}\, p(\xi) \neq 0, \ \ \nabla_\xi \text{Im}\, p \neq 0$ and $\nabla_\xi \text{Re} p(\xi) \not\parallel \nabla_\xi \text{Im}\, p(\xi)$. Consequently, $\text{Char}_\xi P$ is 2 dimensional conical submanifold of $\mathbf{R}^4_\xi \setminus 0$.

We shall denote $\xi' = (\xi_1, \xi_2), \ \ \xi'' = (\xi_3, \xi_4)$ and $\mathcal{D}^1_{\varepsilon, e^{i\pi/4}} = \{z \in \mathbf{C}^1 : |z - e^{i\pi/4}| \leq \varepsilon\}$, $\mathcal{D}^2_{\varepsilon, -e^{i\pi/4}} = \{z \in \mathbf{C}^1 : |z + e^{i\pi/4}| \leq \varepsilon\}$, where $0 < \varepsilon < 1$ is arbitrary small. Put $M^1 = \mathcal{D}^1 \setminus (\mathcal{D}^1_{\varepsilon, e^{i\pi/4}} \cup \mathcal{D}^2_{\varepsilon, -e^{i\pi/4}})$. In a similar way we define $\mathcal{D}^1_{\varepsilon, e^{-i\pi/4}}, \mathcal{D}^2_{\varepsilon, -e^{-i\pi/4}}, \ M^2$; $\mathcal{D}^1 = \{z : |z| \leq 1\}$.

One can easily see that $|P(\lambda, \mu)| \geq c_1|\mu|^2$, $c_1 > 0$ if $\lambda/\mu \in M^1$, $\mu \neq 0$ and $|P(\lambda, \mu)| \geq c_2|\lambda|^2$, $c_2 > 0$ if $\mu/\lambda \in M^2$, $\lambda \neq 0$. Certainly, $c_{1,2} = c_{1,2}(\varepsilon)$.

We point out that $\text{Char} P \subset T^*(\mathbf{R}^4) \setminus 0$ is a conic manifold of codimension 2 and the Poisson bracket $\{\text{Re}\, p(\xi), \text{Im}\, p(\xi)\} \equiv 0$. Therefore, Char P is involutive manifold of codimension 2. If

$$\text{H}_{\text{Re}p} = \langle \nabla_\xi \text{Re}\, p, \frac{\partial}{\partial x} \rangle, \quad \text{H}_{\text{Im}p} = \langle \nabla_\xi \text{Im}\, p, \frac{\partial}{\partial x} \rangle, \quad ([\text{H}_{\text{Re}p}, \text{H}_{\text{Im}p}] = 0)$$

are the corresponding Hamilton vector fields of $\text{Re}\, p$ and $\text{Im}\, p$ we obtain according to Frobenius theorem that $\text{Char} P$ foliates by 2 dimensional leaves Γ. We shall call them bicharacteristics of P. This is the complex involutive case considered in section 26.2 of [3].

We are ready now to formulate the main result of this short communication.

Theorem 1.1. *The following results hold for the double-complex Laplace operator P:*

1.) *P is neither C^∞, nor analytic hypoelliptic.*
2.) *P has the fundamental solution $\mathcal{E}(x) = \dfrac{c_3}{\overline{u}^2 + i\overline{v}^2} \in \mathcal{D}'(\mathbf{R}^4)$, $c_3 =$ const. $\neq 0$.*
3.) *P is partially C^∞ hypoelliptic, i.e. if $Pw \in C^\infty(\Omega)$, $w \in \mathcal{D}'(\Omega)$ and for each multiindex $\alpha' = (\alpha_1, \alpha_2) \in \mathbf{Z}^2_+$ we have that $\mathcal{D}^{\alpha'}_{x'} w \in C^0(\Omega)$ then $w \in C^\infty(\Omega)$, (respectively, if $\mathcal{D}^{\alpha''}_{x''} w \in C^0(\Omega)$ then again $w \in C^\infty(\Omega)$).*
4.) *P is operator of principal type, i.e. $p(\xi) = 0 \Rightarrow \nabla_\xi p(\xi) \neq 0$. If $w \in \mathcal{D}'(\Omega)$ and $Pw = f$ then $(Char\, P \cap WF(w)) \setminus WF(f)$ is invariant under the bicharacteristic foliation in* Char$P \setminus WF(f)$.
5.) *Let $Pw \in C^\infty$, $w \in \mathcal{D}'(\Omega)$. Then for each $\varphi \in C_0^\infty(\Omega)$ and each $N \in \mathbf{N}$ there exists a constant $C_N > 0$ and such that $|\hat{\varphi w}(\xi)| \leq C_N(1 + |\xi''|)^{-N}$, when $\dfrac{\xi_1 + i\xi_2}{\xi_3 + i\xi_4} \in M^1$ (respectively, $|\hat{\varphi w}(\xi)| \leq D_N(1 + |\xi'|)^{-N}$, when $\dfrac{\xi_3 + i\xi_4}{\xi_1 + i\xi_2} \in M^2$).*

If $g \in \mathcal{D}'$ then $\hat{\varphi g}(\xi)$ stands for its Fourier transformation, while $WF(f)$ is the wave front set of the Schwartz distribution $f \in \mathcal{D}'$ (see [3], vol. I]).

The result, formulated in p. 4 of Theorem 1 can be given in the form: Singularities of $Pw = f \in C^\infty$ propagate along the bicharacteristics Γ of p.

Proofs of the results of Theorem 1

1. P.1. follows from p.2 as sing supp $\mathcal{E} \neq \{0\}$. So we shall prove at first p.2.

As it is known (see [4]) $\mathcal{E}_1(u) \equiv \mathcal{E}_1(x_1, x_2) = \dfrac{1}{\pi \overline{u}} = \dfrac{1}{\pi(x_1 - ix_2)}$ is a fundamental solution of $\dfrac{\partial}{\partial u}$, i. e.

$$\mathcal{E}_1'(u) = \frac{\partial}{\partial u}\mathcal{E}_1(x_1, x_2) = \delta(x_1, x_2) = \delta(\mathrm{Re}\, u, \mathrm{Im}\, u).$$

Let $a \in \mathbf{C}^1$. Then $\dfrac{\partial}{\partial u}\mathcal{E}_1(u + a) = \mathcal{E}_1'(u + a) = \delta((x_1, x_2) + (\mathrm{Re}\, a, \mathrm{Im}\, a))$.

We are looking for the fundamental solution of the form

$$\mathcal{E}(x) = \mathcal{E}(u, v) = \mathcal{E}_1(u + e^{i\pi/4v})\mathcal{E}_1(u - e^{i\pi/4v}).$$

Therefore,

$$\begin{aligned}P\mathcal{E} &= 4\mathcal{E}_1'(u + e^{i\pi/4}v)\mathcal{E}_1'(u - e^{i\pi/4}v)\\ &= 4\delta(x' + (\operatorname{Re} e^{i\pi/4}v, \operatorname{Im} e^{i\pi/4}v))\delta(x' - (\operatorname{Re} e^{i\pi/4}v, \operatorname{Im} e^{i\pi/4}v)).\end{aligned}$$

Put $y_1 = \operatorname{Re}(e^{i\pi/4}v)$, $y_2 = \operatorname{Im}(e^{i\pi/4}v)$. Then the identity $\delta(x' + y')\delta(x' - y') = const.\delta(x', y')$ completes the proof. (See ([3], Chapter 8)).

In fact, let $\varphi(x', y') \in C_0^\infty(\mathbf{R}^4)$. Then the linear change $x' - y' = z$, $x' + y' = w$ leads to $\langle \delta(x' - y')\delta(x' + y'), \varphi(x', y')\rangle = const.\langle \delta(z) \otimes \delta(w), \varphi\left(\frac{z+w}{2}, \frac{w-z}{2}\right)\rangle = const.\varphi(0,0) = const.\langle \delta(x', y'), \varphi(x', y')\rangle$; $const. \neq 0$, $y' = 0 \iff e^{i\pi/4}v = 0 \iff v = 0 \iff x_3 = x_4 = 0$.

P.3. of Theorem 1 follows immediately from Theorem 11.2.3 and 11.2.5 from [3].

P.5 follows from the fact that P is microelliptic outside Char P (see also [3], vol. 2, page 72).

P.4. of Theorem 1 is a special case of Corollary 26.2.2. of [3].

As we have seen, the double complex Laplace operator has very interesting properties from the point of view of the linear partial differential operators.

References

1. S. Dimiev, Double-complex functions, Ed. R.K. Kovacheva, J. Lawrynowicz, and S. Marchiafava, in: *"Applied Complex and Quaternionic Approximation"*, World Scientific, Singarore (2009), 58–75, also talk at the 9th International conference on complex structures, integrability and vector fields.
2. L. Apostolova, Initial value problem for the double-complex Laplace equation, *C. R. acad. Bulg. Sci.* t. 62, no. 4, (2009), (in press).
3. L. Hörmander, *The Analysis of Linear Partial Differential Operators, vol. I, II, IV*, Springer Verlag, Berlin, 1985.
4. V. Vladimirov, *Equations of Mathematical Physics*, Nauka, Moskow, 1967 (in Russian).

MONOGENIC, HYPERMONOGENIC AND HOLOMORPHIC CLIFFORDIAN FUNCTIONS — A SURVEY

IVAN P. RAMADANOFF

Université de Caen, CNRS UMR 6139,
Laboratoire de Mathématiques, Nicolas ORESME,
14032 Caen, France
E-mail: rama@math.unicaen.fr

The aim of this paper is to make an overview on three generalizations to higher dimensions of the function's theory of a complex variable. The first one concerns the so-called monogenic functions which were introduced by F. Brackx, R. Delanghe and F. Sommen, the second is the theory of hyper-monogenic functions developed by H. Leutwiler, and the last one studies the theory of holomorphic Cliffordian functions due to G. Laville and I. Ramadanoff. The basic notions in this three theories will be given. Their links and differences will also be commented.

Keywords: Clifford analysis; Monogenic; Hypermonogenic and holomorphic Cliffordian functions; Special functions.

1. Clifford algebras

Denote by $\mathbf{R}^{p+q}$ a real vector space of dimension $d = p + q$ provided with a non-degenerate quadratic form Q of signature (p, q).

Main definition: The Clifford algebra, we will denote by $\mathbf{R}_{p,q}$, of the quadratic form Q on the vector space $\mathbf{R}^{p+q}$ is an associative algebra over $\mathbf{R}$, generated by $\mathbf{R}^{p+q}$, with unit 1, if it contains $\mathbf{R}$ and $\mathbf{R}^{p+q}$ as distinct subspaces and

(1) $\forall v \in \mathbf{R}^{p+q}, v^2 = Q(v)$,
(2) the algebra is not generated by any proper subspace of $\mathbf{R}^{p+q}$.

Actually, if we consider the Clifford algebra $\mathbf{R}_{p,q}$ as a vector space, it has the splitting:

$$\mathbf{R}_{p,q} = \mathbf{R}^0_{p,q} \oplus \mathbf{R}^1_{p,q} \oplus \cdots \oplus \mathbf{R}^k_{p,q} \oplus \cdots \oplus \mathbf{R}^d_{p,q},$$

where $\mathbf{R}^0_{p,q} = \mathbf{R}$ are the scalars, $\mathbf{R}^1_{p,q} = \mathbf{R}^{p+q}$ is the vector space, $\mathbf{R}^2_{p,q}$ is the vector space of the so-called bivectors corresponding to the planes in $\mathbf{R}^{p+q}$, and so on. Finally, $\mathbf{R}^d_{p,q}$ contains what we call the pseudoscalars. Moreover, $\dim_{\mathbf{R}} \mathbf{R}^k_{p,q} = C^k_d$, $\dim_{\mathbf{R}} \mathbf{R}_{p,q} = 2^d$. Now, set $e_0 = 1$ as basis of $\mathbf{R}^0_{p,q} = \mathbf{R}$ and suppose $\{e_1, e_2, \ldots, e_d\}$ be an orthonormal basis for $\mathbf{R}^1_{p,q} = \mathbf{R}^{p+q}$. Thus, the corresponding vector spaces of the splitting will be provided with respective basis

$$\{e_0 = 1\}, \qquad \{e_1, e_2, \ldots, e_d\}, \qquad \{e_{ij} = e_i e_j, \quad 1 \le i < j \le d\}, \quad \ldots,$$
$$\ldots, \quad \{e_{i_1 \ldots i_k} = e_{i_1} e_{i_2} \ldots e_{i_k}, \quad 1 \le i_1 < i_2 < \cdots < i_k \le d\}, \quad \ldots,$$
$$\ldots, \quad \{e_{12\ldots d} = e_1 e_2 \ldots e_d\},$$

and the algebra will obey to the laws:

$$e_i^2 = 1, \quad i = 1, \ldots, p, \qquad e_i^2 = -1, \quad i = p+1, \ldots, d = p+q,$$
$$e_i e_j = -e_j e_i, \qquad i \neq j.$$

This allows us to write down any Clifford number $a \in \mathbf{R}_{p,q}$ as a sum of its scalar part $\langle a \rangle_0$, its vector part $\langle a \rangle_1 \in \mathbf{R}^1_{p,q}$, its bivector part $\langle a \rangle_2 \in \mathbf{R}^2_{p,q}$, up to its pseudoscalar part $\langle a \rangle_d \in \mathbf{R}^d_{p,q}$, namely

$$a = \langle a \rangle_0 + \langle a \rangle_1 + \cdots + \langle a \rangle_d,$$

where $\langle a \rangle_k = \sum_{|J|=k} a_J e_J$, with $J = (j_1, \ldots, j_k)$ is a strictly increasing multiindex of length k and $e_J = e_{j_1} e_{j_2} \ldots e_{j_k}$, while $a_J \in \mathbf{R}$.

Some examples. The Clifford algebra $\mathbf{R}_{0,1}$ can be identified with the complex numbers $\mathbf{C}$. The algebra $\mathbf{R}_{0,2}$ is nothing else than the set of quaternions $\mathbf{H}$ if we identify $e_1 = i, e_2 = j, e_{12} = k$ using the traditional notations. Physicists are working very often with the algebras $\mathbf{R}_{1,3}$ or $\mathbf{R}_{3,1}$. It suffices to note the nature of the corresponding signatures $(+,-,-,-)$ and $(+,+,+,-)$, respectively. However, $\mathbf{R}_{1,3}$ and $\mathbf{R}_{3,1}$ are not isomorphic as algebras.

Some models. Let $M_2(\mathbf{R})$ be the algebra of all 2×2 matrices with real entries. Put:

$$\varepsilon_0 = \begin{pmatrix} 1 & 0 \\ 0 & 1 \end{pmatrix}, \quad \varepsilon_1 = \begin{pmatrix} 1 & 0 \\ 0 & -1 \end{pmatrix}, \quad \varepsilon_2 = \begin{pmatrix} 0 & 1 \\ 1 & 0 \end{pmatrix}, \quad \varepsilon_3 = \begin{pmatrix} 0 & 1 \\ -1 & 0 \end{pmatrix}.$$

In a first case, let us identify $e_0 = \varepsilon_0, e_1 = \varepsilon_1, e_2 = \varepsilon_2, e_{12} = \varepsilon_3$. Thus we get a model for the algebra $\mathbf{R}_{2,0}$. On the other hand, if we identify $e_0 = \varepsilon_0, e_1 = \varepsilon_1, e_2 = \varepsilon_3, e_{12} = \varepsilon_2$, then we will obtain a model for $\mathbf{R}_{1,1}$.

Recall that the main involution *, the reversion $^\sim$ and the conjugation $^-$ act on $a \in \mathbf{R}_{p,q}$ as follows:

$$a^* = \sum_{k=0}^{d} (-1)^k \langle a \rangle_k, \; a^\sim = \sum_{k=0}^{d} (-1)^{k(k-1)/2} \langle a \rangle_k, \; \bar{a} = \sum_{k=0}^{d} (-1)^{k(k+1)/2} \langle a \rangle_k.$$

These operations are not of algebraic type, they are of geometric type.

Remark 1.1. For $\mathbf{C} = \mathbf{R}_{0,1}$, the usual complex conjugation is the main involution, as well as the conjugation. The reversion is useless.

Remark 1.2. For $\mathbf{H} = \mathbf{R}_{0,2}$, with the classical notations, we have:

$$a = \alpha + \beta i + \gamma j + \delta k, \qquad a^* = \alpha - \beta i - \gamma j + \delta k,$$
$$a^\sim = \alpha + \beta i + \gamma j - \delta k, \qquad \bar{a} = \alpha - \beta i - \gamma j - \delta k.$$

Periodicity properties. Let us turn to the problem how to classify Clifford algebras. There is an algebraic point of view 13,14. Here we will illustrate the pseudoscalar computational point of view. Actually, they are four possibilities for the behavior of the pseudoscalar basis element $e_{12\ldots d}$:

(1) $d = p + q$ even $\iff \forall a \in \mathbf{R}^{p+q}$, $ae_{12\ldots d} = -e_{12\ldots d}a$.
(2) $d = p + q$ odd $\iff \forall a \in \mathbf{R}^{p+q}$, $ae_{12\ldots d} = e_{12\ldots d}a$.
(3) $p - q \equiv 0 \pmod 4$ or $p - q \equiv 1 \pmod 4$ $\iff e_{12\ldots d}^2 = 1$.
(4) $p - q \equiv 2 \pmod 4$ or $p - q \equiv 3 \pmod 4$ $\iff e_{12\ldots d}^2 = -1$.

In [5] Guy Laville introduced the notations i_+, i_-, e_+, e_- for the pseudoscalar $e_{12\ldots d}$ having in mind that i commutes, e does not, the index $+$ corresponds to a square equal to $+1$, the index $-$ to -1.

2. Generalizations of the one complex variable theory

Henceforth, we will consider Clifford algebras of antieuclidean type [1,2], namely $\mathbf{R}_{0,d}$. Note the first three, for $d = 0, 1, 2 : \mathbf{R}, \mathbf{C}$ and $\mathbf{H}$, are division algebras by the well known theorem of Frobenius.

Our aim is to survey different generalizations of the function theory of a complex variable which can be viewed as the study of those functions defined in a domain of $\mathbf{R}^2$ and taking their values in the Clifford algebra $\mathbf{R}_{0,1} = \mathbf{C}$.

The first key of the theory of holomorphic functions is, of course, the Cauchy-Riemann operator ($\partial/\partial\bar{z}$ in the classical notations), which can be written now as:

$$D = \frac{\partial}{\partial x_0} + e_1 \frac{\partial}{\partial x_1},$$

omitting the famous normalization constant 1/2. It should be noted that the definition domain of f lies in a space whose elements are couples of a scalar and a vector, so that $\mathbf{R}^2$ should be identified to $\mathbf{R} \oplus i\mathbf{R}$.

Monogenic functions. Let $\mathbf{R}_{0,n}$ be the Clifford algebra of the real vector space V of dimension n provided with a quadratic form of negative signature, $n \in \mathbf{N}$. denote by S the set of the scalars in $\mathbf{R}_{0,n}$, identified with $\mathbf{R}$. Let $\{e_i\}_{i=1}^n$ be an orthonormal basis of V and set also $e_0 = 1$.

A point $x = (x_0, x_1, \ldots, x_n)$ of $\mathbf{R}^{n+1}$ will be considered as an element of $S \oplus V$, namely $x = \sum_{i=0}^{n} x_i e_i$. Such an element will be called a *paravector*. Obviously, it belongs to $\mathbf{R}_{0,n}$ and we can act on him with the main involution: $x^* = x_0 - \sum_{i=1}^n x_i e_i$. It is remarkable that:

$$xx^* = x^*x = |x|^2,$$

where $\mid x \mid$ denotes the usual euclidean norm of x in $\mathbf{R}^{n+1}$ and it shows that every non-zero paravector is invertible. Sometime, if necessary, we will resort to the notation $x = x_0 + \overrightarrow{x}$, , where $\overrightarrow{x}$ is the vector part of x, i.e. $\overrightarrow{x} = \sum_{i=1}^n x_i e_i$.

Let $f : \Omega \to \mathbf{R}_{0,n}$, where Ω is an open subset of $S \oplus V$. Introduce the Cauchy-Dirac-Fueter operator:

$$D = \sum_{i=0}^{n} e_i \frac{\partial}{\partial x_i}.$$

Note D possesses a conjugate operator

$$D^* = \frac{\partial}{\partial x_0} - \sum_{i=1}^{n} e_i \frac{\partial}{\partial x_i}$$

and that $DD^* = D^*D = \Delta$, where Δ is the usual Laplacian.

Definition 2.1. A function $f : \Omega \to \mathbf{R}_{0,n}$, of class C^1, is said to be (left) *monogenic* in Ω if and only if $Df(x) = 0$ for each $x \in \Omega$.

Obviously, in the case $n = 1$, we get the holomorphic functions of one complex variable.

Remark 2.1. If $n > 1$, then the functions $x \mapsto x$ and $x \mapsto x^m, x \in S \oplus V, m \in \mathbf{N}$ are not monogenic.

Following R. Brackx, R. Delanghe and F. Sommen [1], recall that there exists a Cauchy kernel: $E(x) = (\omega_n^{-1})(x^*/|x|^{n+1})$ for $x \in S \oplus V - \{0\}$, where ω_n is the area of the unit sphere in $\mathbf{R}^{n+1}$. This kernel is really well adapted to the monogenic functions because it is himself a monogenic function with a singularity at the origin, i.e. $DE(x) = \delta$ for $x \in S \oplus V$.

Then, put $\omega(y) = dy_0 \wedge \cdots \wedge dy_n$ and

$$\gamma(y) = \sum_{i=0}^{n} e_i dy_0 \wedge \cdots \wedge \hat{dy_i} \wedge \cdots \wedge dy_n.$$

Thus, we have:

Integral representation formula. If f is monogenic in Ω and U is an oriented compact differentiable variety of dimension $n+1$ with boundary ∂U and $U \subset \Omega$, then

$$\int_{\partial U} E(y-x)\gamma(y)f(y) = f(x), \quad x \in \dot{U}.$$

Thanks to this, the analogous of the mean value theorem, the maximum modulus principle, Morera's theorem follow easily.

The depth and the wealth of the one complex variable theory come also thanks to the "duality": Cauchy-Riemann and Weierstrass, i.e. every holomorphic function is analytic and the reciproque. How to understand what is the generalization of a power series?

In the frame of monogenic functions, an answer exists because, fortunately, the functions $w_k = x_k e_0 - x_0 e_k, k = 1, \ldots, n$ are monogenic. Omitting the details and roughly speaking one can expand every monogenic function in a series of polynomials which elementary monomials are the w_k and their powers. Just for an illustration let us show this phenomena is somehow natural: suppose $f : S \oplus V \to \mathbf{R}_{0,n}$ is real analytic on a neighborhood of the origin, so

$$f(h) = \sum_{k=0}^{\infty} \left(h_0 \frac{\partial}{\partial x_0} + \cdots + h_n \frac{\partial}{\partial x_n} \right)^k f(0).$$

But at the same time f is monogenic, i. e.

$$\frac{\partial f}{\partial x_0} = -\sum_{i=1}^{n} e_i \frac{\partial f}{\partial x_i}.$$

Hence

$$f(h) = \sum_{k=0}^{\infty} \left(\sum_{i=1}^{n} (h_i - e_i h_0) \frac{\partial}{\partial x_i} \right)^k f(0).$$

Holomorphic Cliffordian functions. Here, consider functions $f : \Omega \to \mathbf{R}_{0,2m+1}$, where Ω is an open subset of $S \oplus V = \mathbf{R} \oplus \mathbf{R}^{2m+1} = \mathbf{R}^{2m+2}$. The paravectors of $S \oplus V$ will be written as

$$x = x_0 + \overrightarrow{x}, \qquad x_0 \in \mathbf{R}, \qquad \overrightarrow{x} = \sum_{i=1}^{2m+1} x_i e_i.$$

Definition 2.2. A function $f : \Omega \to \mathbf{R}_{0,2m+1}$ is said to be (left) *holomorphic Cliffordian* in Ω if and only if:

$$D\Delta^m f(x) = 0$$

for each $x \in \Omega$. Here Δ^m means the iterated Laplacian [7].

The set of holomorphic Cliffordian functions is wider than those of the monogenic ones: every monogenic is also holomorphic Cliffordian, but the reciprocate is false. Indeed, if $Df = 0$, then $D\Delta^m f = \Delta^m Df = 0$ because Δ^m is a scalar operator. The simplest example of a holomorphic Cliffordian function which is not monogenic is the identity $x \mapsto x$. Actually, one can prove that all entire powers of x are holomorphic Cliffordian, while they are not monogenic.

Note that f is holomorphic Cliffordian if and only if $\Delta^m f$ is monogenic.

There is a simple way to construct holomorphic Cliffordian functions which is based on the Fueter principle [4].

Lemma 2.1. *If $u : \mathbf{R}^2 \to \mathbf{R}$ is harmonic, then $u(x_0, |\overrightarrow{x}|)$, where $x = x_0 + \overrightarrow{x}$ and $|\overrightarrow{x}|^2 = \sum_{i=1}^{2m+1} x_i^2$ is $(m+1)$-harmonic, i. e.*

$$\Delta^{m+1} u(x_0, |\overrightarrow{x}|) = 0.$$

Lemma 2.2. *If $f : (\xi, \eta) \mapsto f = u + iv$ is a holomorphic function, then*

$$F(x) = u(x_0, |\overrightarrow{x}|) + \frac{\overrightarrow{x}}{|\overrightarrow{x}|} v(x_0, |\overrightarrow{x}|)$$

is a holomorphic Cliffordian function.

If we summarize: from an usual holomorphic function f, with real part u, we construct the associated $(m+1)$-harmonic $u(x_0, |\overrightarrow{x}|)$ and then it sufficies to take $D^* u(x_0, |\overrightarrow{x}|)$ in order to get a holomorphic Cliffordian one. This receipt is very well adapted for the construction of trigonometric or exponential functions [10] in $\mathbf{R}_{0,2m+1}$. Note also the set of holomorphic Cliffordian functions is stable under any directional derivation.

It is natural to ask for an integral representation formula but in this case, the operator $D\Delta^m$ being of order $2m+1$, such a formula would be much more complicated. Anyway, the first step is to exhibit an analog to the Cauchy kernel. Remember the fundamental solution of the iterated

Laplacian $\Delta^{m+1}h(x) = 0$ for $x \in S \oplus V - \{0\}$ is known: that is $h(x) = \ln|x|$. Hence $D^*(1/2\ \ln(xx^*))$ must be holomorphic Cliffordian. But:

$$D^*\left(\frac{1}{2}\ln(xx^*)\right) = \frac{1}{2}\frac{D^*(|x|^2)}{|x|^2} = \frac{x^*}{|x|^2} = x^{-1}.$$

By the way, we found the first holomorphic Cliffordian function with an isolated punctual singularity at the origin.

It is remarkable that, after computations, one get: $\Delta^m(x^{-1}) = (-1)^m 2^{2^m}(m!)^2\omega_m E(x)$. It becomes natural to introduce a new kernel:

$$N(x) = \varepsilon_m x^{-1},$$

with the suitable choice for the constant $\varepsilon_m = (-1)^m[2^{2m+1}m!\pi^{m+1}]^{-1}$.

So, N is the natural Cauchy kernel for holomorphic Cliffordian functions and we have:

$$D\Delta^m N(x) = DE(x) = \delta, \quad x \in S \oplus V.$$

Integral representation formula. Let B be the unit ball in $\mathbf{R}^{2m+2}$, x an interior point of B, $\frac{\partial}{\partial n}$ means the derivation in the direction of the outward normal. Thus, we have:

$$\begin{aligned} f(x) = &\int_{\partial B} (\Delta^m N(y-x)\gamma(y)f(y) \\ &- \sum_{k=1}^{m}\int_{\partial B}\left(\frac{\partial}{\partial n}\Delta^{m-k}N(y-x)\right)D\Delta^{k-1}f(y)d\sigma_y \\ &+ \sum_{k=1}^{m}\int_{\partial B}(\Delta^{m-k}N(y-x))\frac{\partial}{\partial n}D\Delta^{k-1}f(y)d\sigma_y. \end{aligned}$$

The above formula involves $2m+1$ integrals on ∂B, which means one can deduce the values of f inside B knowing the values on ∂B of $f, D\Delta^{k-1}f$ and $\frac{\partial}{\partial n}D\Delta^{k-1}f, \quad k = 1, 2, \ldots, m$.

Recall that all integer powers of a paravector x are solutions of $D\Delta^m = 0$, and we saw also that $D\Delta^m(x^{-1}) = \delta$. Those facts can be proved directly following straightforward computations. Let $x = x_0 + \vec{x}$ be a paravector in a general Clifford algebra of antieuclidean type $\mathbf{R}_{0,d}$, $d \in \mathbf{N}$. Very fastidious calculations give:

$$D\Delta^m(x^{-1}) = (-1)2^m m! \prod_{j=0}^{m}(2j+1-d)(|x|)^{-(2m+2)}.$$

The right hand side is a scalar vanishing for $d = 1, 3, 5, \ldots, 2m+1$. Moreover, we have:

$$D\Delta^m(x^{2n+1}) = \prod_{j=0}^{m}(2j+1-d)\sum_{q=m}^{n}\prod_{k=1}^{m}(2q-2k+2)C_{2n+1}^{2q+1}x_0^{2n-2q}\vec{x}^{2q-2m},$$

and a similar formula for an even power x^{2n} of x. In both cases, the right hand sides are again scalars vanishing for $d = 1, 3, 5, \ldots, 2m+1$.

Polynomial solutions of $D\Delta^m = 0$. Set

$$\alpha = (\alpha_0, \alpha_1, \ldots, \alpha_{2m+1}), \qquad \alpha_j \in \mathbf{N}, \qquad |\alpha| = \sum_{j=0}^{2m+1} \alpha_j.$$

Consider the set

$$\{e_\nu\} = \{e_0, \ldots, e_0, e_1, \ldots, e_1, \ldots, e_{2m+1}, \ldots, e_{2m+1}\},$$

where e_0 is written α_0 times, $e_i : \alpha_i$ times. Then set:

$$P_\alpha(x) = \sum_{\Theta}\prod_{\nu=1}^{|\alpha|-1}(e_{\sigma(\nu)}x)e_{\sigma(|\alpha|)},$$

the sum being expanded over all distinguishable elements σ of the permutation group Θ of the set $\{e_\nu\}$. The P_α are polynomials of degree $|\alpha| - 1$. A straightforward calculation carried on them shows that P_α is equal up to a rational constant to $\partial^{|\alpha|}(x^{2|\alpha|-1})$. Thus, it follows that the P_α are holomorphic Cliffordian functions, which are left and right, thanks to the symmetrization process.

The classical way for getting Taylor's series of a holomorphic function is to expand the Cauchy kernel in the integral representation formula. The same procedure is available here:

$$\begin{aligned}(y-x)^{-1} &= (y(1-y^{-1}x))^{-1} = (1-y^{-1}x)y^{-1}\\ &= y^{-1} + y^{-1}xy^{-1} + y^{-1}xy^{-1}xy^{-1} + \cdots + (y^{-1}x)^n y^{-1} + \cdots .\end{aligned}$$

Obviously, we have $y^{-1} = y^*(|y|)^{-2}$, and thus:

$$(y-x)^{-1} = \sum_{k=0}^{\infty}\frac{(y^*x)^k y^*}{|y|^{2k+2}}.$$

It is not difficult to observe the polynomials P_α appear again. Finally, as in the classical case, we can deduce the expansion in a "power series" of any holomorphic Cliffordian function f under the form:

$$f(x) = \sum_{|\alpha|=1}^{\infty} P_\alpha(x)C_\alpha,$$

where $C_\alpha \in \mathbf{R}_{0,2m+1}$.

Further, let us mention that a function f which is holomorphic Cliffordian in a punctured neighborhood of the origin possesses a Laurent expansion [7,8]. So, the set of meromorphic Cliffordian functions with isolated singularities is well defined. For this reason, it was possible to generalize the classical theory of elliptic functions to the Cliffordian case [8,9]. Note that only the additive properties of the Weierstrass ζ and $\wp$ functions, also for the Jacobi sn, cn and dn, are used for this purpose. The multiplication of two holomorphic Cliffordian functions is not yet so clear.

What about the case $\mathbf{R}_{0,2m}$? More precisely, could we pass from holomorphic Cliffordian functions in $\mathbf{R}_{0,2m+1}$ to their restrictions in $\mathbf{R}_{0,2m}$? Roughly speaking, it is the same as between $\mathbf{C}$ and $\mathbf{R}$. In the general case of a Clifford algebra of type $\mathbf{R}_{0,d}$, we can observe that the set of homogeneous polynomials of degree n which are holomorphic Cliffordian is a right $\mathbf{R}_{0,d}$-module generated by the monomials $(ax)^n a$, where a is a paravector.

Now, let us consider $f : \mathbf{R} \oplus \mathbf{R}^d \to \mathbf{R}_{0,d}$.

Case 1: d is odd. The function f will be holomorphic Cliffordian if $D\Delta^{(d-1)/2} f = 0$.

Case 2: d is even. We say that f is holomorphic (analytic) Cliffordian [6] if there is a holomorphic Cliffordian function of one more variable $F : \mathbf{R} \oplus \mathbf{R}^{d+1} \to \mathbf{R}_{0,d+1}$, which is even with respect to e_{d+1} and such that $F\mid_{x_{d+1}=0} = f$.

Hypermonogenic functions. In this part we will briefly discuss another class of functions, named *hypermonogenic*, which were introduced by H. Leutwiler [11,12] and studied by himself and Sirkka-Liisa Eriksson-Bique [3].

Recall that any element $a \in \mathbf{R}_{0,n}$ may be uniquely decomposed as $a = b + ce_n$ for $b, c \in \mathbf{R}_{0,n-1}$. This should be compared with the classical decomposition of a complex number $a = b + ic$. Using the above decomposition, one introduces the projections $P : \mathbf{R}_{0,n} \to \mathbf{R}_{0,n-1}$ and $Q : \mathbf{R}_{0,n} \to \mathbf{R}_{0,n-1}$ given by $Pa = b, Qa = c$.

Now define the following modification of the Dirac operator D as follows:

$$Mf = Df + \frac{n-1}{x_n}(Qf)^*,$$

where * denotes the main involution introduced above.

Definition 2.3. An infinitely differentiable function $f : \Omega \to \mathbf{R}_{0,n}, \Omega$ being an open subset of $\mathbf{R}^{n+1}$, such that $Mf = 0$ on $\Omega - \{x : x_n = 0\}$ is called a (left) *hypermonogenic* function.

Decomposing f into $f = Pf + (Qf)e_n$, its P-part satisfies the Laplace-Beltrami equation:

$$x_n \Delta(Pf) - (n-1)\frac{\partial(Pf)}{\partial x_n} = 0,$$

associated to the hyperbolic metric, defined on the upper half space $\mathbf{R}_+^{n+1}$ by $ds^2 = x_n^{-2}(dx_0^2 + dx_1^2 + \cdots + dx_n^2)$.

Its Q-part solves the eigenvalue equation:

$$x_n^2 \Delta(Qf) - (n-1)x_n \frac{\partial(Qf)}{\partial x_n} + (n-1)Qf = 0.$$

It turns out that hypermonogenic functions are stable by derivations in all possible directions excepted this one on e_n and that, for any $m \in \mathbf{N}$, the maps $x \mapsto x^m$ and $x \mapsto x^{-m}$ are hypermonogenic in $\mathbf{R}^{n+1}$, resp. $\mathbf{R}^{n+1} - \{0\}$.

Clearly, hypermonogenic functions generalize usual holomorphic functions of a complex variable. But what about the relations of this class with the class of holomorphic Cliffordian?

Let us study this problem in the case $n = 3$.

We can prove that hypermonogenic are holomorphic Cliffordian. Assuming $f : \Omega \to R_{0,3}$ is hypermonogenic, $Df = -\frac{2}{x_3}(Qf)^*$ and hence $D(\Delta f) = -2\Delta\left[\frac{(Qf)^*}{x_3}\right]$. On account of the above eigenvalue equation for the Q-part of f, we conclude that

$$\Delta\left[\frac{(Qf)^*}{x_3}\right] = \frac{1}{x_3}\left[\Delta(Qf) - \frac{2}{x_3}\frac{\partial(Qf)}{\partial x_3} + 2\frac{Qf}{x_3^2}\right]^* = 0$$

forcing $D(\Delta f) = 0$.

In contrast to left hypermonogenic functions, the left holomorphic Cliffordian ones may also be multiplied from the right by e_3 and not just by e_1 and e_2. In addition, partial differentiation with respect to x_3 does not lead out of the class of left holomorphic Cliffordian functions.

On the other hand, it can be shown [3] that if $f : \Omega \to \mathbf{R}_{0,3}$ is a holomorphic Cliffordian function in $\Omega \subset \mathbf{R}^4$, then there exist, locally, hypermonogenic functions g_1, g_2, g_3 and g_4 such that

$$f = g_1 + g_2 e_3 + \frac{\partial}{\partial x_3}(g_3 + g_4 e_3).$$

The conclusion is clear: both theories agree.

Recently, the last two statements were improved and they give now a better interconnection between hypermonogenic and holomorphic Cliffordian functions in $\mathbf{R}_{0,n}$. Let say also the theory of hypermonogenic functions is provided with an integral representation formula and that the expansion in power series is generated by polynomials which are deeply related to the P_α above.

Acknowledgements

The largest part of the mentioned results concerning the holomorphic Cliffordian functions was obtained thanks to joint works with Guy Laville in the *Laboratoire de Mathématiques Nicolas Oresme, CNRS UMR 6139, Département de Mathématiques, Université de Caen Basse-Normandie.*

References

1. F. Brackx, R. Delanghe, F. Sommen, *Clifford Analysis*, Pitman (1982).
2. R. Delanghe, *Clifford Analysis: History and Perspective*, Comput. Methods and Function theory 1 (2001), 107–153.
3. S.-L. Eriksson-Bique, H. Leutwiler, *Hypermonogenic functions*, Clifford algebras and their applications in Math. Physics 2, Birkhausen, Boston (2000), 287–302.
4. R. Fueter, *Die Funktionnentheorie der Differentialgleichgungen* $\Delta u = 0$ *und* $\Delta\Delta u = 0$ *mit vier reellen Variablen*, Comment. Math. Helv. 7 (1935), 307–330.
5. G. Laville, *Some topics in Clifford Analysis*, 7-th International Conf. on Clifford Algebras (2005), Toulouse.
6. G. Laville, E. Lehman, *Analytic Cliffordian functions*, Annales Acad. Sci. Fennicae Math. 20 (2004), 251–268.
7. G. Laville, I. Ramadanoff, *Holomorphic Cliffordian functions*, Advances in Applied Clifford Analysis 8 (2) (1998), 323–340.
8. G. Laville, I. Ramadanoff, *Elliptic Cliffordian functions*, Complex variables 45 (4), (2001), 297–318.
9. G. Laville, I. Ramadanoff, *Jacobi elliptic Cliffordian functions*, Complex Variables 47 (9), (2002), 787–802.
10. G. Laville, I. Ramadanoff, *An integral transform generating elementary functions in Clifford analysis*, Math. Meth. Appli. Sci. 29 (2006), 637–654.
11. H. Leutwiler, *Modified Clifford analysis*, Complex Variables Theory Appl. 20 (1992), 19–51.
12. H. Leutwiler, *Generalized Function Theory*, Proceedings Conference in Tampere (2004).
13. P. Lounesto, *Clifford Algebras and Spinors*, Cambridge (1996).
14. I. Porteous, *Clifford Algebras and the Classical Groups*, Cambridge (1995).

ON SOME CLASSES OF EXACT SOLUTIONS OF EIKONAL EQUATION

Ł. T. STĘPIEŃ

Department of Computer Sciences and Computer Methods,
Pedagogical University,
Krakow, 30-084, Poland
E-mail: lstepien@inf.ap.krakow.pl, lstepien@ap.krakow.pl, stepien50@poczta.onet.pl

Some classes of exact solutions of eikonal equation has been found by using some decomposition method. There is also prsented its application to quantum mechanics.

Keywords: Nonlinear partial differential equations; Exact solutions; Eikonal equation; WKB method

1. Introduction

Eikonal equation

$$u_{,\mu}u^{,\mu} = n \tag{1}$$

appears in many branches of physics, as:

- describing a propagation of wave front or so called characteristic surface [1,2],
- in classical electromagnetic field [3], and in algebrodynamics [4,5],
- in semiclassical approximation [6],
- a version of relativistic Hamilton-Jacobi equation [7].

In this paper we obtain some new classes of exact solutions of the eikonal equation, by using a method, called decomposition method or decomposition scheme. We also discuss the properties of some of these solutions. The paper is organized as follows. In section 2 we briefly describe the main idea of the mentioned method. We present its extension and its applications for some selected cases of eikonal equation in the section 3. Section 4 includes an applicaion of extended decomposition method in quantum mechanics.

2. The main idea of decomposition method

The essential version of this method has been presented in [8]. If we want to use it for solving eikonal equation (1), we must extend this method.

We want to solve analytically a partial differential equation:

$$F(x^{\mu}, u_1, \dots, u_m, u_{1,x^{\mu}}, u_{1,x^{\mu}x^{\nu}}, \dots, u_{m,x^{\mu}}, u_{m,x^{\mu}x^{\nu}}, \dots) = 0, \qquad (2)$$

where $m \in N$, $\mu, \nu = 0, 1, 2, 3$ and u is in general an unknown function of class C^k, $k \in N$.

Let us limit in this introduction to the case: $m = 1$, $k = 2$.

(1) we examine, whether there is possible to decompose this equation into such fragments, which are characterized by a homogenity of the derivatives of the unknown function u. A meaning of "homogenity of the derivatives": the mentioned fragments should be products of at least two factors:
A. arbitrary expression, which may depend on the unknown function u, its derivatives (this expression may be obviously a constant) and the independent variables (in general: $x^{\mu}, \mu = 0, 1, 2, 3$);
B. a sum of: at least two derivatives and (or) at least two products of the derivatives, so that there we can find for any derivative and (or) any product some other derivative and (or) other product, which has the same degree and order.

(2) For example, the investigated equation may be as follows:

$$F_1 \cdot [(u_{,x})^2 + (u_{,y})^2] + F_2 \cdot [u_{,xx} + u_{,xy}] = 0, \qquad (3)$$

where F_1 and F_2 may depend on $x^{\mu}, u, u_{x^{\mu}}, \dots$.

(3) we insert some ansatzes, presented below.

- The ansatz of first kind, being a nonlinear superposition of travelling waves:

$$u(x, y, z, t) = \beta_1 + f(a_{\mu}x^{\mu} + \beta_2, b_{\nu}x^{\nu} + \beta_3, c_{\rho}x^{\rho} + \beta_4), \qquad (4)$$

where: $v_{\alpha}x^{\alpha} = -v_0x^0 + v_kx^k$, $a_{\mu}, b_{\nu}, c_{\rho}, \mu, \nu, \rho = 0, 1, 2, 3$, $k = 1, 2, 3$, are some constants to be determined later, $\beta_i, i = 1, 2, 3, 4$, are arbitrary constants and $x^{\mu} \in R$.

A talk delivered during the conference "9th International Workshop on Complex Structures, Integrability and Vector Fields", in 25-29 August 2008, in Sofia, Bulgaria.

- The ansatz of second kind:

$$u(x,y,t) = \beta_1 + f(A_0x^0x^1 + A_1x^0x^2 + A_2x^0x^3 + A_3x^1x^2 + A_4x^1x^3 + A_5x^2x^3 + \beta_2) \tag{5}$$

  where $A_\lambda, \lambda = 0, 1, \ldots, 5$ may be in general complex constants and they are to be determined later, $x^\mu \in R$.
- The combined ansatz

$$u(x,y,z,t) = \beta_1 + f(a_\mu x^\mu + \beta_2, A_0x^0x^1 + A_1x^0x^2 + A_2x^0x^3 + A_3x^1x^2 + A_4x^1x^3 + A_5x^2x^3 + \beta_2), \tag{6}$$

  where $a^\mu, \ldots, A^\lambda, \mu = 0, 1, 2, 3,\ \lambda = 0, \ldots, 5$ and $x^0 = t$, $x^1 = x$, $x^2 = y$, $x^3 = z$.

As far as the function f is concerned, we assume only that it is *arbitrary* function of class $\mathcal{C}^2$.
After inserting these ansatzes, we try to collect all algebraic terms, appearing by the variables x_μ and the derivatives of the function f with respect to the arguments like: $a_\mu x^\mu + \beta_2$ and (or) $A_0x^0x^1 + A_1x^0x^2 + A_2x^0x^3 + A_3x^1x^2 + A_4x^1x^3 + A_5x^2x^3 + \beta_2$. We require vanishing of all such algebraic terms. Therefore we obtain a system of algebraic equations, which we call *determining system.* The coefficients $a_\mu, \ldots, A_\lambda$ are to be determined as the solutions of such obtained system of algebraic equations.
For example, after inserting a two-dimensional version of ansatz of first kind (4) into (3), we get:

$$\begin{aligned} F_1 \cdot [&(a_1^2 + a_2^2)(D_1f)^2 + (b_1^2 + b_2^2)(D_2f)^2 + (c_1^2 + c_2^2)(D_3f)^2 \\ &+ 2(a_1b_1 + a_2b_2)D_1fD_2f + 2(a_1c_1 + a_2c_2)D_1fD_3f && (7) \\ &+ 2(b_1c_1 + b_2c_2)D_2fD_3f] + F_2 \cdot [(a_1^2 + a_1a_2)D_{1,1}f && (8) \\ &+ (2a_1b_1 + a_1b_2 + a_2b_1)D_{1,2}f + (2a_1c_1 + a_1c_2 + a_2c_1)D_{1,3}f \\ &+ (b_1^2 + b_1b_2)D_{2,2}f + (2b_1c_1 + b_1c_2 + b_2c_1)D_{2,3}f \\ &+ (c_1^2 + c_1c_2)D_{3,3}f] = 0, \end{aligned}$$

where D_if denotes the derivative of function f with respect to *i-nary* argument and $D_{i,j}f$ denotes the mixed derivative of this function with respect to: *i-nary* and *j-nary* argument.

To sum up:

- this method (or scheme) does not use any other known analytic method (i.e. Lie group method) of solving of PDE's,

- the main goal of the method is to obtain the wide classes of solutions of the NPDE, as general as possible,
- it assumes applying some ansatzes,
- the necessary conditions for using this method are: the homogenity of the derivatives in the investigated NPDE and possibility of obtaining some system of algebraic equations.

The above procedure constitutes the main version of decomposition method. We have found the classes of the solutions, given by the all above ansatzes for 4-dimensional real eikonal equation and complex eikonal equation, in the case, when $n = 0$. The appropriate values of the coefficients $a_\mu, \ldots, A_\lambda$, $\mu = 0, \ldots, 3$, $\lambda = 0, \ldots, 5$ are (in the case of complex eikonal equation in Minkowski spacetime, when n = 0): for the solution of first kind:

$$a_0 = 0, \quad a_1 = 0, \quad a_3 = ia_2, \quad b_0 = b_1, b_3 = ib_2, \tag{9}$$

$$c_1 = c_0, \quad c_3 = ic_2, \tag{10}$$

for the solutions of second kind:

$$A_1 = iA_0, \quad A_3 = A_0, \quad A_5 = iA_0, \quad A_2 = 0, A_4 = A_0, \tag{11}$$

and for the combined solutions:

$$A_0 = A_3, \quad A_1 = 0, \quad A_2 = iA_3, \quad A_5 = iA_3, \tag{12}$$

$$A_4 = 0, \quad a_2 = -a_0, \quad a_3 = ia_1. \tag{13}$$

However, if we want to apply this method for solving the more general case of eikonal equation, i.e. when $n \neq 0$, we must extend this method.

3. An extension of decomposition method

Now, let us look at the eikonal equation (1) in flat space-time:

$$-(u_{,0})^2 + (u_{,1})^2 + (u_{,2})^2 + (u_{,3})^2 = n(x^i), \quad i \in \{1, 2, 3\} \tag{14}$$

We will be engaged now with more general cases $n = const$ and $n = n(x^i)$, $i \in \{1, 2, 3\}$ and we will want to find the class of solutions of first kind (4). Then, let us come back to the idea of an extension of the decomposition method.

This idea is such that for $n \neq 0$ we split the unknown function u : $u(x, y, z, t) = f_1(x^\alpha) + f_2(x^\beta)$, and next balance the term n by second power of the derivative of f_1. It has turned out that in the case, when $n = const \neq 0$, there does not exist the classes of solutions of II- and combined kind. When $n = n(x, y, z)$, the solutions of all kinds exist [9]. However, owing to limited amount of the space of this paper we limit our considerations to the class of solutions of I kind in three cases:

(1) If $n = const$, then f_1 is a linear function of at least, two independent variables, for example y and z: $f_1 = -\gamma_0 t + \gamma_1 z$, $f_2 = \beta_1 + f(a_\mu x^\mu + \beta_2, b_\nu x^\nu + \beta_3, c_\rho x^\rho + \beta_4)$, γ_ξ are to be determined later as the coefficients in the function f_1, $\xi = 1, 2$, $\mu, \nu, \rho = 0, 1, 2, 3$. Then after inserting $u = f_1 + f_2$ into (1) and collecting appropriate algebraic terms, we obtain determining system:

$$\gamma_1^2 - \gamma_0^2 = n, \quad \gamma_1 c_3 - \gamma_0 c_0 = 0, \tag{15}$$

$$\gamma_1 b_3 - \gamma_0 b_0 = 0, \quad \gamma_1 a_3 - \gamma_0 a_0 = 0, \tag{16}$$

$$a_\mu b^\mu = 0, \; a_\mu c^\mu = 0, \quad b_\mu c^\mu = 0, \tag{17}$$

$$a_\mu a^\mu = 0, \; b_\mu b^\mu = 0, \quad c_\mu c^\mu = 0. \tag{18}$$

The solutions of this above system are:

$$\gamma_1 = a_0 \sqrt{\frac{n}{a_0^2 - a_3^2}}, \quad a_1 = \frac{a_3 c_1}{c_3}, \tag{19}$$

$$a_2 = \frac{\sqrt{(a_0^2 - a_3^2)c_3^2 - a_3^2 c_1^2}}{c_3}, \tag{20}$$

$$b_0 = \frac{a_0 b_3}{a_3}, \; b_1 = \frac{b_3 c_1}{c_3}, \tag{21}$$

$$b_2 = \frac{b_3 \sqrt{(a_0^2 - a_3^2)c_3^2 - a_3^2 c_1^2}}{a_3 c_3}, \tag{22}$$

$$c_0 = \frac{a_0 c_3}{a_3}, \; c_2 = \frac{\sqrt{(a_0^2 - a_3^2)c_3^2 - a_3^2 c_1^2}}{a_3}, \tag{23}$$

$$\gamma_0 = \frac{\gamma_1 a_3}{a_0} \tag{24}$$

where the values of the coefficients have be such that the square roots are real and of course $a_0, a_3, c_3 \neq 0$.
If we put $n = 0$ in (1) and in formulas of corresponding coefficients above, we obtain the solution of (1) for special case $n = 0$.

(2) In the case of $n = n(x, y)$, we apply separation of variables and we put $u(x, y, z, t) = \beta_1 + f_1(x^\alpha) + f_2(a_\rho x^\rho + \beta_2, b_\sigma x^\sigma + \beta_3, c_\varsigma x^\varsigma + \beta_4)$, where $\alpha = 1, 2$ and $\rho, \sigma, \varsigma = 0, 3$. The values of the coefficients are determined accordingly with the main decomposition method. These values are:

$$a_1 = -a_0, \quad b_1 = -b_0, \quad c_1 = -c_0. \tag{25}$$

The function $f_1(x, y)$ must satisfy the equation:

$$(f_{1,1})^2 + (f_{1,2})^2 = n(x, y), \tag{26}$$

where $f_{1,i} = \frac{\partial f_1}{\partial x^i}$, $i = 1, 2$.

(3) Separately in this paper we treat 3-dimensional complex eikonal equation, when $n = n(x, y, z)$:

$$(u_{,x})^2 + (u_{,y})^2 + (u_{,z})^2 = n(x, y, z), \tag{27}$$

for which we apply also extended decomposition method, but without separation of variables (section 4).

4. Some applications to quantum mechanics

Now, we apply the extended decomposition method for the case $n = n(x, y, z)$, considered previously, to some situation from quantum mechanics.

As it is known, the most simple idea of limit transition from quantum mechanics to classical mechanics is to choose a wave-function of the form [6]:

$$\psi(\vec{r}, t) = exp\left(\frac{i}{\hbar} S(\vec{r}, t)\right), \quad S \in \mathbb{C}. \tag{28}$$

After inserting it into Schroedinger equation [6]:

$$i\hbar \frac{\partial \psi}{\partial t} = -\frac{\hbar^2}{2m} \nabla^2 \psi + V(\vec{x})\psi \tag{29}$$

we get:

$$-\frac{\partial S}{\partial t} = \frac{(\nabla S)^2}{2m} + V(\vec{x}) - i\frac{\hbar^2}{2m} \nabla^2 S. \tag{30}$$

If we pass over the third term on the right hand of (30), we get Hamilton-Jacobi equation:

$$-\frac{\partial S}{\partial t} = \frac{(\nabla S)^2}{2m} + V(\vec{x}). \tag{31}$$

So, as it is well-known, the limit transition to classical mechanics, is realized by transition to the limit $\hbar \to 0$. Let's investigate a case of stationary state:

$$S(\vec{x}, t) = u(\vec{x}) - Et. \tag{32}$$

Then, the equation (30) is [6]:

$$\frac{(\nabla u)^2}{2m} + V(\vec{x}) - E - \frac{i\hbar}{2m} \nabla^2 u = 0. \tag{33}$$

If $|\vec{p}|^2 \gg \hbar |\vec{\nabla}\cdot\vec{p}|$, then [6]:

$$\frac{(\nabla u)^2}{2m} + V(\vec{x}) - E = 0. \tag{34}$$

Usually the equation (34) is to be solved by WKB method.

Now, in order to solve this equation, we apply the results, obtained for the eikonal equation in the previous part of this talk.

Another approach towards the extension of decomposition method will be now applied directly to quantum mechanics.

We will now illustrate, on the example of this equation, some extension of decomposition method. Namely, we assume that the potential $U(x,y,z) = U(c_k x^k + \beta_4)$, $k = 1,2,3$, $\beta_4 = const$. Now we split the function ω:

$$u = \beta_1 + f_1(c_k x^k + \beta_4) + f_2(a_i x^i + \beta_2, b_j x^j + \beta_3), \quad i,j,k = 1,2,3, \tag{35}$$

so it is a special case of the ansatz of first kind (4) and: $X = c_k x^k + \beta_4$.

If we insert this function into (34), we get:

$$\begin{aligned}&\underline{(a_1^2+a_2^2+a_3^2)}f_{2,1}^2 + \underline{(b_1^2+b_2^2+b_3^2)}f_{2,2}^2 + \underline{2(a_1b_1+a_2b_2+a_3b_3)}f_{2,1}f_{2,2}\\ &+ \underline{2(a_1c_1+a_2c_2+a_3c_3)}\frac{df_1}{dX}f_{2,1} + \underline{2(b_1c_1+b_2c_2+b_3c_3)}\frac{df_1}{dX}f_{2,2}\\ &+ (c_1^2+c_2^2+c_3^2)\left(\frac{df_1}{dX}\right)^2 + 2m(V(X)-E) = 0.\end{aligned} \tag{36}$$

Now, similarly to the main version of decomposition method, we require vanishing of all underlined terms in the ebove equation. Then we get some system of equations:

$$a_1^2 + a_2^2 + a_3^2 = 0, \quad b_1^2 + b_2^2 + b_3^2 = 0, \tag{37}$$

$$a_1b_1 + a_2b_2 + a_3b_3 = 0, \quad a_1c_1 + a_2c_2 + a_3c_3 = 0, \tag{38}$$

$$b_1c_1 + b_2c_2 + b_3c_3 = 0, \tag{39}$$

$$(c_1^2+c_2^2+c_3^2)\left(\frac{df_1}{dX}\right)^2 + 2m(V(X)-E) = 0. \tag{40}$$

The roots of the system of algebraic equations are:

$$a_3 = i\sqrt{a_1^2+a_2^2}, \quad b_1 = \frac{a_1b_2}{a_2}, \quad b_3 = i\frac{b_2\sqrt{a_1^2+a_2^2}}{a_2}, \tag{41}$$

$$c_1 = 0, \quad c_3 = i\frac{a_2c_2}{\sqrt{a_1^2+a_2^2}} \tag{42}$$

and the solution of the ordinary differential equation (40) is:

$$f_1(X) = \pm \int \frac{\sqrt{2}\sqrt{m(E - V(X))}}{\sqrt{c_1^2 + c_2^2 + c_3^2}} dX + C. \tag{43}$$

The simplest example: $m = 1$, $V(X) = X \equiv c_1x + c_2y + c_3z$, where $c_1 = 0$, $c_3 = i\frac{a_2c_2}{\sqrt{a_1^2+a_2^2}}$

$$f_1(X) = \pm\frac{2\sqrt{2}}{3}\frac{(E - X)^{\frac{3}{2}}}{\sqrt{c_2^2 - \frac{a_2^2c_2^2}{a_1^2+a_2^2}}} \tag{44}$$

so we have the exact solution of (34) for $V(y,z) = c_2y + i\frac{a_2c_2}{\sqrt{a_1^2+a_2^2}}z$.

$$u(x,y,z) = + \pm\frac{2\sqrt{2}}{3}\frac{(E - (c_2y + i\frac{a_2c_2}{\sqrt{a_1^2+a_2^2}}z))^{\frac{3}{2}}}{\sqrt{c_2^2 - \frac{a_2^2c_2^2}{a_1^2+a_2^2}}} \tag{45}$$

$$+ f_2(a_1x + a_2y + i\sqrt{a_1^2 + a_2^2}z + \beta_2, \frac{a_1b_2}{a_2}x + b_2y + i\frac{b_2\sqrt{a_1^2 + a_2^2}}{a_2}z + \beta_3).$$

However, the potential V should be real function and thus the coefficients c_i must be real. Hence we chose $a_2 = ip$, where p and a_1 must be real numbers.

5. Summary

The presented decomposition method and its extension give some chance for obtaining exact solutions of eikonal equation.

As far as the application of the decomposition method to quantum mechanics is concerned, let us note that when we apply WKB method, there in return points or where $V(X) = E$, the wave-function is singular [6]. Namely, here, for example, when $V(X) = X$, the wave-function is *non-singular*, even when $V(X)$ attains the value E (return points).

More exact discussion of physical context of the results obtained by extended decomposition method, in among others, quantum mechanics, will be included in separate paper [9].

6. Resources

The computations was carried out by using Waterloo MAPLE 12 Software on computers: MARS and SATURN (No. of grants *MNiI/IBM$_B$C$_H$S21/AP/057/2008* and *MNiI/Sun6800/WSP/008/2005*, correspondingly) in ACK-Cyfronet AGH in Krakow.

References

1. T. Taniuti, *Prog. Theor. Phys. Suppl.* No. 9, 69 (1959).
2. G.B. Whitham, *Linear and nonlinear waves*, John Viley & Sons, 1974.
3. R. Magnanini and G. Talenti, *Contemp. Math.* **238**, 203 (1999).
4. V. Kassandrov, *arXiv:math-ph/0311006 v1.*
5. V. Kassandrov, *Grav. & Cosm., Suppl. 2*, **8**, 57 (2002).
6. A.S. Davydov, *Quantum mechanics*, PWN, Warszawa 1969 (in Polish).
7. L.D. Landau, E.M. Lifshitz, *Classical theory of fields*, PWN, Warszawa 1980 (in Polish).
8. Ł.T. Stępień, *Some decomposition method for analytic solving of certain non-linear partial differential equations in physics with applications, A talk delivered during the conference "Special Functions, Information Theory and Mathematical Physics"*, An interdisciplinary conference in honor of Prof. Jesús S. Dehesa's 60th birthday, Sept 17–19th 2007, Granada, Spain. An extended version of this talk will be published in a special issue of *J. Comp. Appl. Math.*
9. Ł.T. Stępień, *A paper in preparation.*

DIRICHLET PROPERTY FOR TESSELLATIONS OF TILING-TYPE 4 ON A PLANE BY CONGRENT PENTAGONS

YUSUKE TAKEO

Division of Mathematics and Mathematical Science,
Graduate School of Engineering, Nagoya Institute of Technology,
Gokiso, Nagoya, 466-8555, JAPAN
E-mail: morgen1029@zelus.ics.nitech.ac.jp

TOSHIAKI ADACHI*

Department of Mathematics, Nagoya Institute of Technology
Gokiso, Nagoya, 466-8555, JAPAN
E-mail: adachi@nitech.ac.jp

We take a convex pentagon $R = \mathrm{ABCDE}$ satisfying $\overline{\mathrm{AB}} = \overline{\mathrm{BC}}$, $\overline{\mathrm{AE}} = \overline{\mathrm{ED}}$ and $\angle\mathrm{B} = \angle\mathrm{E} = \pi/2$. By use of 4 pentagons congruent to R we make a fundamental hexagon and give a tessellation of a Euclidean plane, which is called a tessellation of tiling-type 4 by congruent pentagons. This tessellation has 3-valent and 4-valent vertices. We study Dirichlet property for such tessellations of a plane. We also consider similar tessellations by congruent original and reversed pentagons which have fundamental regions made by 4 pieces and study their Dirichlet property.

Keywords: Dirichlet tessellations; Even-valent vertices; Congruent pentagons; Tiling-type 4; Fundamental region made by 4 pentagons.

1. Introduction

The aim of this paper is to study the Dirichlet property of some tessellations by congruent convex polygons which have even-valent vertices. In the preceding paper [5] we studied the Dirichlet property of periodic tessellations by congruent quadrangles. We are hence interested in getting other Dirichlet tessellations by congruent convex polygons.

Classifying patterns of tiling on a Euclidean plane by congruent convex

*The second author is partially supported by Grant-in-Aid for Scientific Research (C) (No. 20540071) Japan Society of Promotion Science.

polygons itself is an interesting problem. It was shown by Reinhardt [3] that when we consider a convex polygon with $n\,(\geq 7)$ edges then there are no patterns of tiling on a plane all of whose tiles are congruent to this polygon. For about tiling by convex triangles and tiling by convex quadrangles, there are infinitely many patterns. But for tiling by convex pentagons and tiling by convex hexagons their situations are not the same. It is known that there are (at least) 14 types of tiling on a Euclidean plane by congruent convex pentagons (see [4] for example).

Our meaning of "tessellation" which is due to Ash and Bolker [1] is a bit different from the meaning of tiling (for detail see section 2). Being different from a tiling, if two regions of a tessellation have an intersection it should be either their vertices or their edges. But we can make use of reversed polygons. In this sense tiling of types $1, 2, 3, 10, 11, 12$ and 14, are not tessellations in the sense of [1] in general. In this paper, we restrict ourselves on tessellations by congruent convex pentagons whose tiling-type is 4 and some similar tessellations and study their Dirichlet property.

2. Dirichlet tessellations

A convex *tessellation* $\mathcal{R} = \{R_\alpha\}$ of a Euclidean plane $\mathbb{R}^2$ is a family of convex closed polygons which satisfies the following two conditions:

1) $\mathbb{R}^2 = \bigcup_\alpha R_\alpha$,
2) if $R_\alpha \cap R_\beta \neq \emptyset$, then this set $R_\alpha \cap R_\beta$ is either an edge of each R_α and R_β or a vertex of each of them.

We do not arrow that a vertex of some region $R_\alpha \in \mathcal{R}$ lies on an edge of other region R_β where it is not a vertex of R_β. We call a convex tessellation $\mathcal{R} = \{R_\alpha\}$ of $\mathbb{R}^2$ *Dirichlet* if there is a point $\mathrm{P}_\alpha \in R_\alpha$ for each α and each polygon R_α coincides with the set

$$\bigcap_{\beta\neq\alpha} \{\mathrm{P} \in \mathbb{R}^2 \mid d(\mathrm{P},\mathrm{P}_\alpha) \leq d(\mathrm{P},\mathrm{P}_\beta)\}.$$

These points $\{\mathrm{P}_\alpha\}$ are called *cites* of $\mathcal{R}$.

It is an interesting problem to know when a convex tessellation is Dirichlet. A vertex V of a convex tessellation $\mathcal{R}$, which is a vertex of some $R_\alpha \in \mathcal{R}$, is said to be n-valent if n edges of polygons in $\mathcal{R}$ meet at V. Under the assumption that all vertices of a tessellation are odd-valent, Ash and Bolker [1] gave a necessary and sufficient condition for this tessellation to be Dirichlet. But if a tessellation contains an even-valent vertex, the situation is not simple and we still do not have any general conditions. From probabilistic point of view, tessellations containing even-valent vertices are exceptional. But from the viewpoint of applications it seems useful to treat tessellations containing 4-valent vertices, if one imagines cages in a zoo. We hence study tessellations with even-valent vertices whose regions are congruent to each other. As we obtain Dirichlet tessellations by convex quadrangles in [5], we investigate tessellations by congruent pentagons.

3. A tessellation of tiling-type 4 by congruent pentagons

We give tessellations by congruent pentagons whose tiling-type is 4 in this section. Let $R = \mathrm{ABCDE}$ be a convex pentagon whose lengths of edges satisfy $\overline{\mathrm{AB}} = \overline{\mathrm{BC}}$, $\overline{\mathrm{DE}} = \overline{\mathrm{EA}}$ and whose angles at vertices satisfy $\angle\mathrm{B} = \angle\mathrm{E} = \pi/2$. In order to make a tessellation of a Euclidean plane $\mathbb{R}^2$, we construct a "fundamental region" in the following manner. We take a pentagon $R_1 = \mathrm{A_1B_1C_1D_1E_1}$ which is congruent to R. We rotate R_1 centered at the vertex $\mathrm{B_1}$ with angle $-\pi/2$ and obtain a pentagon $R_2 = \mathrm{A_2B_2C_2D_2E_2}$. We also rotate original R_1 centered at the vertex $\mathrm{E_1}$ with angle $\pi/2$ and obtain another pentagon $R_4 = \mathrm{A_4B_4C_4D_4E_4}$. Let $R_3 = \mathrm{A_3B_3C_3D_3E_3}$ be a pentagon which is symmetric to R_1 with respect to the midpoint of the edge $\mathrm{C_1D_1}$ (see Figure 2).

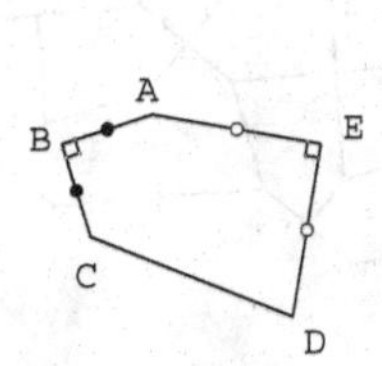

Fig. 1. original pentagon

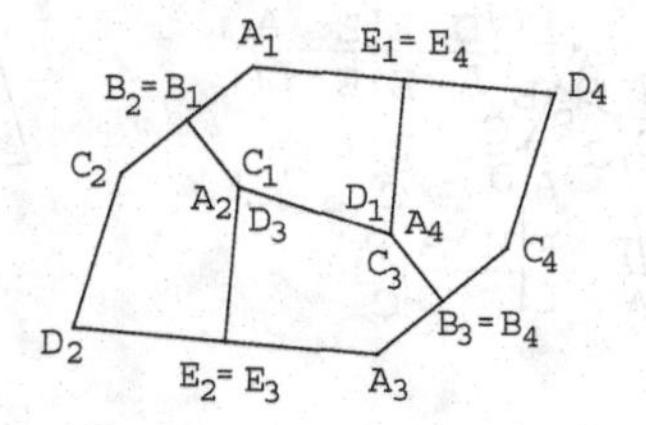

Fig. 2. Fundamental region

Lemma 3.1. *For about the region $R_1 \cup R_2 \cup R_3 \cup R_4$ we can conclude the following:*

(1) *The pentagons R_2, R_3 are adjacent to each other by the edge* $\mathrm{A_2E_2} = \mathrm{D_3E_3}$.

(2) *The pentagons R_3, R_4 are adjacent to each other by the edge* $\mathrm{A_4B_4} = \mathrm{C_3B_3}$.

Proof. Since R is a convex pentagon and $\angle\mathrm{B} = \angle\mathrm{E} = \pi/2$, we find that $\angle\mathrm{A} + \angle\mathrm{C} + \angle\mathrm{D} = 2\pi$. As we have $\overline{\mathrm{AB}} = \overline{\mathrm{CB}}$ and $\overline{\mathrm{AE}} = \overline{\mathrm{DE}}$, we get our conclusions. □

By this lemma we find that $S = R_1 \cup R_2 \cup R_3 \cup R_4$ forms a hexagon. We shall call this S a fundamental region of our tessellation. We denote S by FGHKMN, where vertices are named as $\mathrm{F} = \mathrm{A}_1$, $\mathrm{G} = \mathrm{C}_2$, $\mathrm{H} = \mathrm{D}_2$, $\mathrm{K} = \mathrm{A}_3$, $\mathrm{M} = \mathrm{C}_4$, $\mathrm{N} = \mathrm{D}_4$. We here note that this fundamental region S is symmetric with respect to the mid point of the edge $\mathrm{C}_1\mathrm{D}_1 = \mathrm{D}_3\mathrm{C}_3$. Moreover, the motion which gives this symmetry preserves the partitions by 4 pentagons. In particular, we have $\overline{\mathrm{GH}} = \overline{\mathrm{MN}}\,(= \overline{\mathrm{CD}})$. Thus if we take hexagons $S_i = \mathrm{F}_i\mathrm{G}_i\mathrm{H}_i\mathrm{K}_i\mathrm{M}_i\mathrm{N}_i$ $(i = 0, \pm 1, \pm 2, \ldots)$ which are congruent to the fundamental region S so that $\mathrm{G}_i\mathrm{H}_i = \mathrm{N}_{i-1}\mathrm{M}_{i-1}$, then they form a band. We consider infinitely many such bands. We note that these bunds are congruent to each other by parallel translations. Since the angles of the hexagon satisfy $\angle\mathrm{F} = \angle\mathrm{K} = \angle\mathrm{A}$, $\angle\mathrm{G} = \angle\mathrm{M} = \angle\mathrm{C}$, $\angle\mathrm{H} = \angle\mathrm{N} = \angle\mathrm{D}$, we have

$$\angle\mathrm{F} + \angle\mathrm{H} + \angle\mathrm{M} = \pi, \quad \angle\mathrm{G} + \angle\mathrm{K} + \angle\mathrm{N} = \pi.$$

As we have $\overline{\mathrm{FG}} = \overline{\mathrm{MK}}\,(= 2\overline{\mathrm{AB}})$ and $\overline{\mathrm{FN}} = \overline{\mathrm{HK}}\,(= 2\overline{\mathrm{AE}})$, we obtain a tessellation $\mathcal{R} = \{R_\alpha\}$ of $\mathbb{R}^2$ which consists of congruent pentagons by using bands like Figure 3. This tessellation is said to be of tiling-type 4.

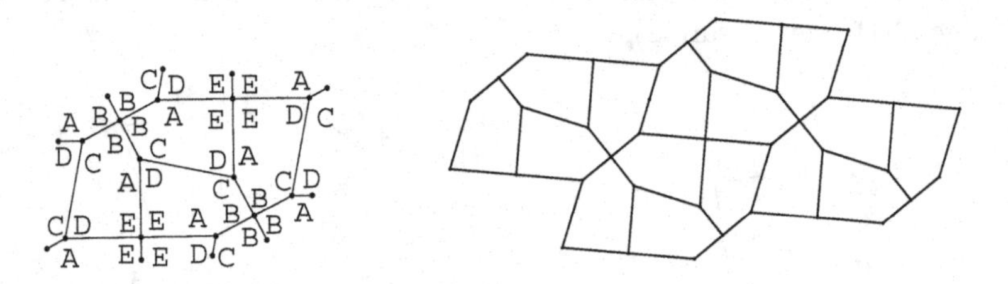

Fig. 3. Tessellation of tiling-type 4

We here note that all 3-valent vertices of this tessellation $\mathcal{R}$ are formed by vertices corresponding to A, C and D. They are the same type in the following sence: If $R_1 \cup R_2 \cup R_3$ and $\widehat{R}_1 \cup \widehat{R}_2 \cup \widehat{R}_3$ form 3-valent vertices, then they are congruent to each other with respect to parallel movements and rotations. In order to see this we are enough to investigate at the vertices in the inside of a fundamental region S and at vertices $\mathrm{F}, \mathrm{N} \in S$. In view of Figure 3, our assertion is clearly true. We also note that in $\mathcal{R}$ meeting pairs of edges are uniquely determined in the following sense:

i) Edges corresponding to AB only meet with edges corresponding to CB,
ii) edges corresponding to CD only meet with edges corresponding to DC,
iii) edges corresponding to DE only meet with edges corresponding to AE.

4. Dirichlet property

In this section we show our tessellation $\mathcal{R}$ of a Euclidean given in section 2 satisfies the Dirichlet condition.

Theorem 4.1. *Let $R = \mathrm{ABCDE}$ be a convex pentagon satisfying $\overline{\mathrm{AB}} = \overline{\mathrm{BC}}$, $\overline{\mathrm{DE}} = \overline{\mathrm{EA}}$ and $\angle\mathrm{B} = \angle\mathrm{E} = \pi/2$. Let $\mathcal{R}$ be a tessellation of $\mathbb{R}^2$ given in section 2 all of whose regions are congruent to R. Then this tessellation $\mathcal{R}$ of tiling-type 4 is Dirichlet. Cite for each $R \in \mathcal{R}$ is uniquely determined.*

For a Dirichlet tessellation with odd-valent vertices, positions of cites are restricted by information on alternative sum of angles formed by edges around odd-valent vertices (see [1]).

Lemma 4.1. *There are three vertices $\mathrm{P}_0, \mathrm{P}_1, \mathrm{P}_2$ and three edges OA_0, OA_1 and OA_2 which meet at the vertex O on $\mathbb{R}^2$. We suppose the line OA_i is the perpendicular bisector of the segment $\mathrm{P}_{i-1}\mathrm{P}_i$, where indices are considered in modulo 3. We then have $\overline{\mathrm{OP}_0} = \overline{\mathrm{OP}_1} = \overline{\mathrm{OP}_2}$ and*

$$\angle\mathrm{P}_i\mathrm{OA}_i = \frac{1}{2}(\theta_i - \theta_{i+1} + \theta_{i+2}),$$

where $\theta_i = \angle\mathrm{A}_i\mathrm{OA}_{i+1}$.

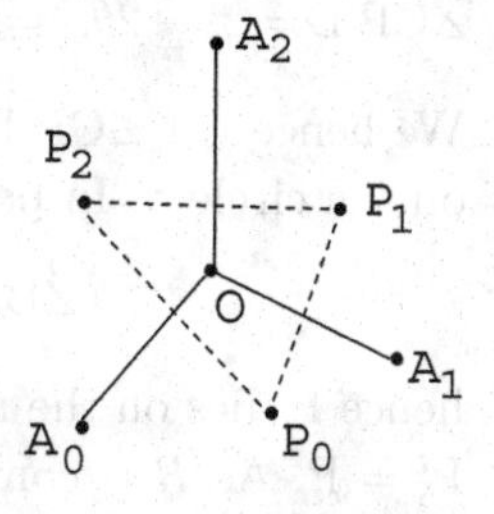

Fig. 4. Positions of cites around 3-valent vertex

Proof. We denote by Q_i the crossing point of edges $\mathrm{P}_{i-1}\mathrm{P}_i$ and OA_i. As triangles $\Delta\mathrm{P}_i\mathrm{OQ}_i$ and $\Delta\mathrm{P}_{i-1}\mathrm{OQ}_i$ are congruent to each other, we see $\overline{\mathrm{OP}_0} = \overline{\mathrm{OP}_1} = \overline{\mathrm{OP}_2}$ and

$$\theta_i = \angle\mathrm{P}_i\mathrm{OA}_i + \angle\mathrm{P}_{i+1}\mathrm{OA}_{i+1},$$

which leads us to the conclusion. □

As was pointed out in section 2, in this tessellation every vertex of order 3 is a point where vertices corresponding to original A, C and D of R meet (recall Figure 2). We hence consider on the original R. For the pentagon R we take a ray $\lambda(\mathrm{A}; R)$ which is emanating from A whose angle between the edge AB is the half of the alternative sum $\theta_\mathrm{A} = (\angle\mathrm{A} - \angle\mathrm{D} + \angle\mathrm{C})/2$ of the angles around the vertex A, where we measure the angle anticlockwisely from the edge AB. We also consider a ray $\lambda(\mathrm{C}; R)$ which is emanating from C whose angle between the edge CD is $\theta_\mathrm{C} = (\angle\mathrm{C} - \angle\mathrm{A} + \angle\mathrm{D})/2$, and a ray $\lambda(\mathrm{D}; R)$ which is emanating from D whose angle between the edge DE is $\theta_\mathrm{D} = (\angle\mathrm{D} - \angle\mathrm{C} + \angle\mathrm{A})/2$. Since R is convex, each angle is less than π, hence these angles are positive because $\angle\mathrm{A} + \angle\mathrm{C} + \angle\mathrm{D} = 2\pi$.

Lemma 4.2. *Three rays $\lambda(\mathrm{A}; R), \lambda(\mathrm{C}; R)$ and $\lambda(\mathrm{D}; R)$ meet at one point in R. If we denote this point by P, then it satisfies $\overline{\mathrm{PA}} = \overline{\mathrm{PC}} = \overline{\mathrm{PD}}$ and it lies in the interior of R.*

Proof. The angle between the ray $\lambda(\mathrm{C}; R)$ and the edge CD is θ_{C}, and the angle between the ray $\lambda(\mathrm{D}; R)$ and the edge CD is $\angle\mathrm{D} - \theta_{\mathrm{D}} = \theta_{\mathrm{C}}$. As $\overline{\mathrm{AB}} = \overline{\mathrm{BC}}$, $\overline{\mathrm{DE}} = \overline{\mathrm{EA}}$ and $\angle\mathrm{B} = \angle\mathrm{E} = \pi/2$, we find $\angle\mathrm{A} > \angle\mathrm{BAC} + \angle\mathrm{DAE} = \pi/2$. Hence we have

$$\theta_{\mathrm{C}} + (\angle\mathrm{D} - \theta_{\mathrm{D}}) = \angle\mathrm{C} - \angle\mathrm{A} + \angle\mathrm{D} = 2\pi - 2\angle\mathrm{A} < \pi.$$

We therefore see that two rays $\lambda(\mathrm{C}; R)$, $\lambda(\mathrm{D}; R)$ meet at some point. We denote this point by P′. Since $\angle\mathrm{P'CD} = \angle\mathrm{P'DC} = \theta_{\mathrm{C}}$, we have $\overline{\mathrm{P'C}} = \overline{\mathrm{P'D}}$.

Draw a circle γ of radius $\overline{\mathrm{P'C}}$ centered at P′. As $\overline{\mathrm{AB}} = \overline{\mathrm{BC}}$, $\overline{\mathrm{DE}} = \overline{\mathrm{EA}}$ and $\angle\mathrm{B} = \angle\mathrm{E} = \pi/2$, we see that

$$\angle\mathrm{CAD} = \angle\mathrm{A} - \angle\mathrm{BAC} - \angle\mathrm{DAE} = \angle\mathrm{A} - \frac{\pi}{2}.$$

On the other hand, we have

$$\angle\mathrm{CP'D} = \pi - 2\theta_{\mathrm{C}} = \pi - (2\pi - 2\angle\mathrm{A}) = 2\angle\mathrm{A} - \pi.$$

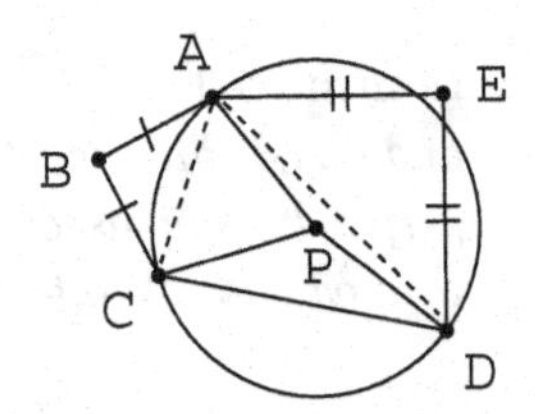

Fig. 5.

We hence get $\angle\mathrm{CP'D} = 2\angle\mathrm{CAD}$ and therefore find that the vertex A lies on the circle γ. In particular, we have $\overline{\mathrm{P'A}} = \overline{\mathrm{P'C}}$. Moreover, we have

$$\angle\mathrm{BAP'} = \angle\mathrm{P'CA} + \frac{\pi}{4} = \angle\mathrm{C} - \theta_{\mathrm{C}} = \theta_{\mathrm{A}},$$

hence P′ lies on the ray $\lambda(\mathrm{A}; R)$. Thus we find three rays meet at the point $\mathrm{P'} = \mathrm{P}$. As R is convex, by the way of taking three rays we find it lies in the interior of R. □

We are now in the position to prove Theorem 4.1. For each region $R_\alpha \in \mathcal{R}$ we take the point $\mathrm{P}_\alpha \in R_\alpha$ corresponding to the point $\mathrm{P} \in R$. This means that $(R_\alpha, \mathrm{P}_\alpha)$ is pointed congruent to (R, P). More precisely, there is a congruent transformation $\varphi : R_\alpha \to R$ satisfying $\varphi(\mathrm{P}_\alpha) = \mathrm{P}$.

For the sake or readers' convenience, we shall start by studying inside the fundamental region $S = R_1 \cup R_2 \cup R_3 \cup R_4$ (recall Figure 2). We consider R_1 and R_2. By Lemma 4.2 we have $\overline{\mathrm{P_1C_1}} = \overline{\mathrm{P_2A_2}}$. As $\mathrm{A_2} = \mathrm{C_1}$ and $\angle\mathrm{P_1C_1B_1} = \angle\mathrm{C} - \theta_{\mathrm{C}} = \theta_{\mathrm{A}} = \angle\mathrm{P_2A_2B_2}$, we find the line $\mathrm{B_1C_1}$ is the perpendicular bisector of the segment $\mathrm{P_1P_2}$. Next we consider R_1 and R_4 by the same way. By Lemma 4.2 we have $\overline{\mathrm{P_1D_1}} = \overline{\mathrm{P_4A_4}}$. As $\mathrm{A_4} = \mathrm{D_1}$ and $\angle\mathrm{P_4A_4E_4} = \angle\mathrm{A} - \theta_{\mathrm{A}} = \theta_{\mathrm{D}} = \angle\mathrm{P_1D_1E_1}$, we find the line $\mathrm{D_1E_1}$ is the perpendicular bisector of the segment $\mathrm{P_1P_4}$. Hence we also obtain the following;

- the line A_2E_2 is the perpendicular bisector of the segment P_2P_3,
- the line B_3C_3 is the perpendicular bisector of the segment P_3P_4.

We consider R_1 and R_3. As we have $\overline{PC} = \overline{PD}$ and $\angle PCD = \angle PDC$, we find the line C_1D_1 is the perpendicular bisector of the segment P_1P_3.

Next we study P_α's beyond the boundary of a fundamental region. Suppose that $R_5 \in \mathcal{R}$ is adjacent to R_1 by the edge $A_1B_1 = C_5B_5$, and that $R_6 \in \mathcal{R}$ is adjacent to R_1 by the edge $A_1E_1 = D_6E_6$. Since the situation of R_1, R_5 and R_6 is completely the same as the situation of R_2, R_1 and R_3, we find the following;

- the line A_1B_1 is the perpendicular bisector of the segment P_1P_5,
- the line A_1E_1 is the perpendicular bisector of the segment P_1P_6,
- the line C_5D_5 is the perpendicular bisector of the segment P_5P_6.

These guarantee that $\mathcal{R}$ is Dirichlet with cites $\{P_\alpha\}$. The uniqueness of choosing cites is guaranteed by Lemma 4.1 because each $R \in \mathcal{R}$ has three 3-valent vertices. We hence get our conclusion.

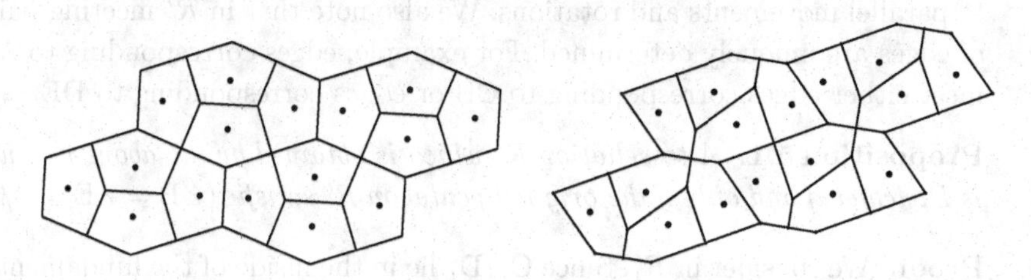

Fig. 6. Cites in tessellations of tiling-type 4

5. A tessellation of semi-tiling type 4

We consider another tessellation of a plane by congruent pentagons whose pattern of tiling is closely related to that of a tessellation of tiling-type 4. Let $R' = ABCDE$ be a convex pentagon satisfying $\overline{AB} = \overline{BC} = \overline{DE} = \overline{EA}$ and $\angle B + \angle E = \pi$. Since $\angle A + \angle C + \angle D = 2\pi$, we can construct a fundamental region $S = R'_1 \cup R'_2 \cup R'_3 \cup R'_4$ by just the same way as in section 2. When $\angle B \neq \pi/2$, this fundamental region is a decagon, but we can obtain a bund of fundamental regions in the same manner. We take two bunds and

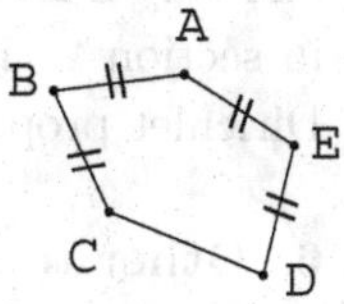

Fig. 7.

reverse one of them. We then obtain a thick bund like Figure 7, hence get a tessellation $\mathcal{R}' = \{R'_\alpha\}$ of $\mathbb{R}^2$.

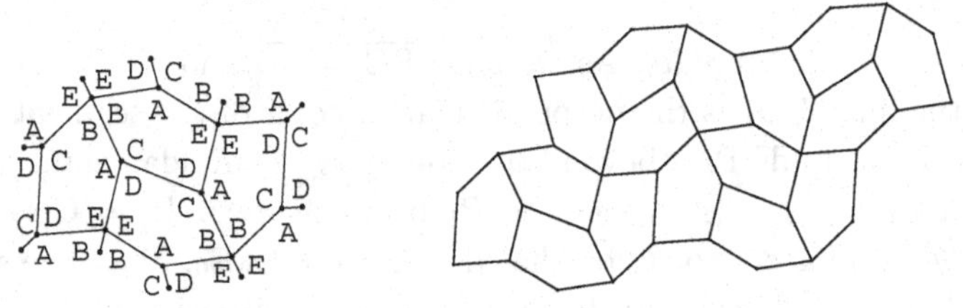

Fig. 8. Tessellation made by original and reversed fundamental regions

Unfortunately, this tessellation $\mathcal{R}'$ is essentially not Dirichlet. We should note that we have two types of 3-valent vertices in general. One is the type of vertices in the inside of a fundamental region, and the other is the type of some vertices on the boundary of a fundamental region. More precisely, there are $R'_1 \cup R'_2 \cup R'_3$ and $\widehat{R}'_1 \cup \widehat{R}'_2 \cup \widehat{R}'_3$ which form 3-valent vertices and which are congruent to each other with respect to parallel movements, rotations and axial symmetries but not congruent to each other with respect to parallel movements and rotations. We also note that in $\mathcal{R}'$ meeting pairs of edges are uniquely determined. For example, edges corresponding to AB meet either edges corresponding to CB or edges corresponding to DE.

Proposition 5.1. *A tessellation $\mathcal{R}'$ which is obtained by the above manner is Dirichlet if and only if the original pentagon R' satisfies $\angle \mathrm{B} = \angle \mathrm{E} = \pi/2$.*

Proof. We consider in R'_1. Since $\mathrm{C}_1, \mathrm{D}_1$ lie in the inside of the fundamental region $\mathcal{S}$, we take two rays $\lambda(\mathrm{C}_1; R'_1)$ and $\lambda(\mathrm{D}_1; R'_1)$ which are the same as in Lemma 4.2. As we can see in Figure 8, because A_1 lies on the boundary of $\mathcal{S}$, we have to consider a ray $\lambda(\mathrm{A}_1; R_1)$ whose angle between the edge $\mathrm{A}_1\mathrm{B}_1$ is $(\angle \mathrm{A} - \angle \mathrm{C} + \angle \mathrm{D})/2$. In view of the proof of Lemma 4.2, these three rays meet at one point if and only if $\angle \mathrm{C} = \angle \mathrm{D}$. By the assumption that $\overline{\mathrm{AB}} = \overline{\mathrm{BC}} = \overline{\mathrm{DE}} = \overline{\mathrm{EA}}$ we find that $\angle \mathrm{B} = \angle \mathrm{E}$ and get the conclusion. □

This proposition shows that a tessellation $\mathcal{R}'$ in this section is Dirichlet if and only if it can be regard as the type in section 2. Though the tessellation in section 2 and for the above tessellation contain 4-valent vertices, their Dirichlet property and their cites are determined by 3-valent vertices.

6. Other tessellations by congruent pentagons

Since we can reverse tiles we can consider many such patterns of tiling. We here give some examples of patterns of tiling on a plane by congruent pentagons which have a fundamental region made by 4 original pentagons, and consider their Dirichlet property.

As for the first example, we take a convex pentagon $R'' = \mathrm{ABCDE}$ whose lengths of edges satisfy $\overline{\mathrm{AB}} = \overline{\mathrm{CD}} = \overline{\mathrm{DE}} = \overline{\mathrm{EA}}$ and whose angles satisfy $2\angle\mathrm{A} + \angle\mathrm{E} = 2\pi$, $2\angle\mathrm{B} + \angle\mathrm{D} + \angle\mathrm{E} = 2\pi$, $2\angle\mathrm{C} + \angle\mathrm{D} = 2\pi$. We construct a fundamental region in the following manner. We take a pentagon $R_1'' = \mathrm{A_1B_1C_1D_1E_1}$ which is congruent to R''. Let $R_2'' = \mathrm{A_2B_2C_2D_2E_2}$ be a pentagon which is symmetric to R_1'' with respect to the line $\mathrm{B_1C_1} = \mathrm{B_2C_2}$ and $R_3'' = \mathrm{A_3B_3C_3D_3E_3}$ be a pentagon which is symmetric to R_1'' with respect to the mid point of the edge $\mathrm{C_1D_1} = \mathrm{D_3C_3}$. Let $R_4'' = \mathrm{A_4B_4C_4D_4E_4}$ be a pentagon which is symmetric to R_3'' with respect to the line $\mathrm{B_3C_3} = \mathrm{B_4C_4}$. We note that the faces of R_2'' and R_4'' are the reverse of the face of R_1''. By the assumption that $\overline{\mathrm{CD}} = \overline{\mathrm{DE}}$ and $2\angle\mathrm{C} + \angle\mathrm{D} = 2\pi$ we find these 4 pentagons form a region like Figure 10. We can make a bund of fundamental regions by just the same way as in section 2. We take two bunds and reverse one of them. As $\overline{\mathrm{AB}} = \overline{\mathrm{DE}} = \overline{\mathrm{EA}}$ and $2\angle\mathrm{B}+\angle\mathrm{D}+\angle\mathrm{E} = 2\pi$, $2\angle\mathrm{A}+\angle\mathrm{E} = 2\pi$, we can obtain a thick bund like Figure 11, hence we obtain a tessellation $\mathcal{R}'' = \{R_\alpha''\}$ of $\mathbb{R}^2$.

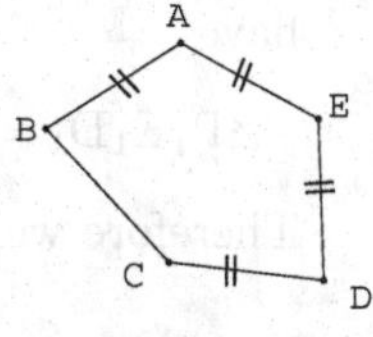

Fig. 9.

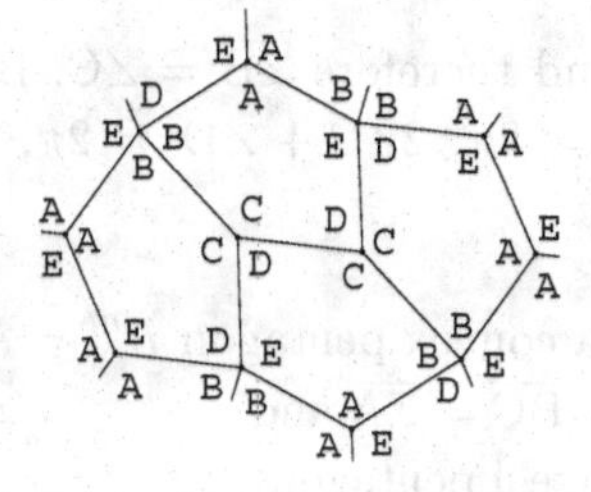

Fig. 10. Fudamental region

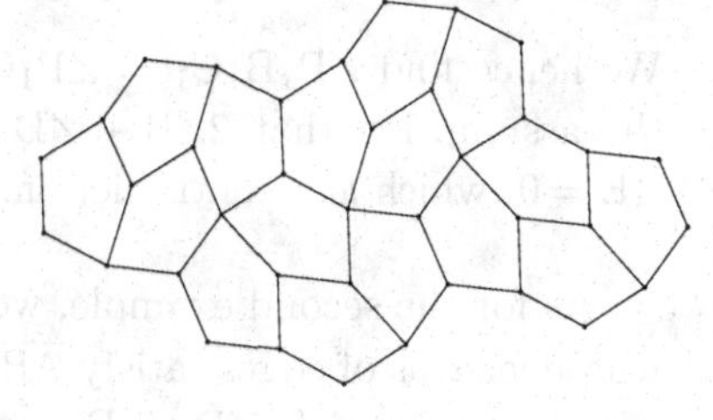
Fig. 11. Tessellation made by original and reversed fundamental regions

Proposition 6.1. *A tessellation $\mathcal{R}''$ which is obtained by the above manner is not Dirichlet.*

Proof. We suppose this tessellation $\mathcal{R}''$ to be Dirichlet. We take pentagons R_1'', R_2'', R_3'' and R_4'' as we mentioned above. Let $R_5'' = \mathrm{A_5B_5C_5D_5E_5}$ be a pentagon which is adjacent to R_1'' by the edge $\mathrm{A_1B_1} = \mathrm{E_5D_5}$. Let $R_6'' = \mathrm{A_6B_6C_6D_6E_6}$ be a pentagon which is adjacent to R_2'' by the edge $\mathrm{A_2B_2} = \mathrm{A_6E_6}$ and is adjacent to R_5'' by the edge $\mathrm{C_5D_5} = \mathrm{D_6E_6}$. We denote by P_i $(i = 1, 2, \ldots, 6)$ the cites of these pentagons.

Since A_1, C_1, D_1 are 3-valent in this tessellation, calculating alternative sums in Lemma 4.1 at these vertices we find that

$$\angle P_1C_1D_1 = \angle P_1D_1C_1 = \angle P_1D_1E_1 = \angle D/2,\ \angle P_1A_1B_1 = \angle E/2.$$

On the other hand, it is clear that two pairs (R_1'', P_1), (R_2'', P_2) and (R_3'', P_3), (R_4'', P_4) are pointed congruent to each other. In view of their 3-valent vertices, we see by Lemma 4.1 that (R_1'', P_1), (R_3'', P_3) are pointed congruent to each other. Since a fundamental region $R_1'' \cup R_2'' \cup R_3'' \cup R_4''$ of $\mathcal{R}''$ contains two 3-valent vertices in its interior, cites are uniquely determined. Recalling the way of constructing the tessellation we hence find (R_i'', P_i) $(i = 1, \ldots, 6)$ are pointed congruent to each other. As we suppose $\mathcal{R}''$ to be Dirichlet, we have

$$\angle P_1E_1D_1 = \angle P_4D_4C_4,\ \angle P_1B_1A_1 = \angle P_5D_5E_5,\ \angle P_2B_2A_2 = \angle P_6E_6A_6.$$

Therefore we see

$$\angle P_1B_1A_1 = \angle P_1E_1A_1 = \angle P_1E_1D_1 = \angle D/2,$$

which leads us to $\angle D = \angle E$ and hence to $\angle P_1A_1B_1 = \angle D/2$. As we have $\overline{A_1B_1} = \overline{C_1D_1} = \overline{D_1E_1}$, these equalities on angles show that three triangles $\Delta P_1A_1B_1$, $\Delta P_1C_1D_1$, $\Delta P_1D_1E_1$ are congruent to each other. Thus we have

$$\overline{P_1A_1} = \overline{P_1B_1} = \overline{P_1C_1} = \overline{P_1D_1} = \overline{P_1E_1}.$$

We hence find $\angle P_1B_1C_1 = \angle P_1C_1B_1$ and therefore $\angle B = \angle C$. By use of the assumption that $2\angle B + \angle D + \angle E = 2\pi$, $2\angle C + \angle D = 2\pi$, we have $\angle E = 0$, which is a contradiction. □

As for the second example, we take a convex pentagon $R''' = ABCDE$ whose lengths of edges satisfy $\overline{AB} = \overline{DE}$, $\overline{BC} = \overline{EA}$ and whose angles satisfy $\angle B + \angle E = \pi$. We take 4 pentagons $R_i''' = A_iB_iC_iD_iE_i$ $(i = 1, 2, 3, 4)$ which are congruent to R'''. Two pentagons R_1''', R_3''' are adjacent to each other by the edge $C_1D_1 = D_3C_3$ and are symmetric to each other with respect to the mid point of this edge.

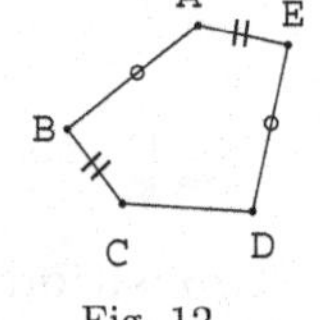

Fig. 12.

We reverse the tiles R_2''' and R_4'''. By the condition on angles we have $\angle A + \angle D + \angle E = 2\pi$. Therefore with the conditions on lengths of edges, we can construct a fundamental region by 4 pentagons like Figure 13: R_2''' is adjacent to R_1''' by the edge $B_1C_1 = E_2A_2$ and R_4''' is adjacent to R_1''' by the edge $D_1E_1 = A_4B_4$. We can make a bund of fundamental regions by just the same way as in section 2. By the conditions on lengths of edges and on angles we can obtain a tessellation $\mathcal{R}'''$ of $\mathbb{R}^2$ by use of such bunds without reversing them (see Figure 14). This tessellation $\mathcal{R}'''$ has a similar property as of tessellations of tiling-type 4. In $\mathcal{R}'''$ meeting pairs of edges are uniquely determined:

i) Edges corresponding to AB only meet with edges corresponding to DE,
ii) edges corresponding to BC only meet with edges corresponding to EA,
iii) edges corresponding to CD only meet with edges corresponding to DC.

But being different from tessellation of tiling-type 4, it has two types of 3-valent vertices; vertices in the interior and those on the boundary of a fundamental region.

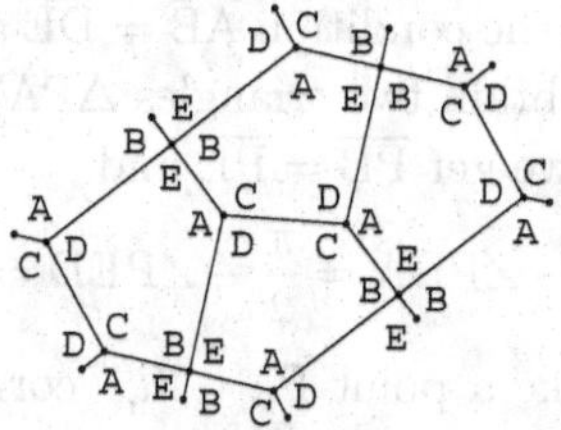

Fig. 13. Fudamental region

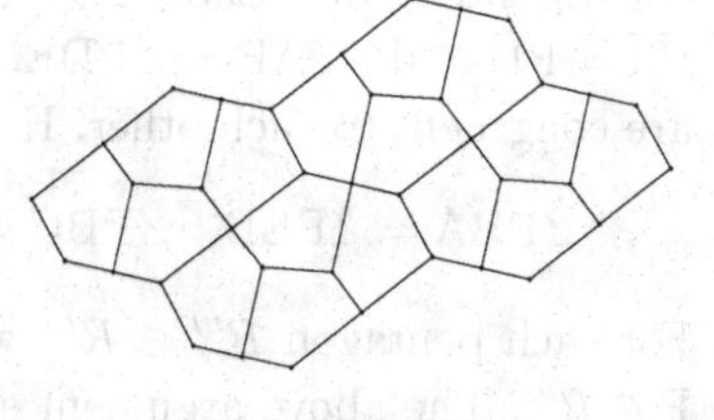

Fig. 14.

Theorem 6.1. *A tessellation $\mathcal{R}'''$ which is obtained by the above manner is Dirichlet if and only if $\angle B = \angle E = \pi/2$. When $\mathcal{R}'''$ is Dirichlet, its cites are uniquely determined.*

Proof. We suppose $\mathcal{R}'''$ is Dirichlet. We take pentagons R_1''', R_2''' and R_4''' as we mentioned above. In view of 3-valent vertices in these pentagons, Lemma 4.1 shows that (R_1''', P_1), (R_2''', P_2), (R_4''', P_4) are pointed congruent to each other. Since $\mathcal{R}'''$ is Dirichlet, we have

$$\angle P_1B_1C_1 = \angle P_2E_2A_2, \ \angle P_1E_1D_1 = \angle P_4B_4A_4,$$

which gurantee that $\angle P_1B_1C_1 = \angle P_1E_1A_1$, $\angle P_1B_1A_1 = \angle P_1E_1D_1$. Thus we obtain $\angle B = \angle E$. As $\angle B + \angle E = \pi$, we get $\angle B = \angle E = \pi/2$.

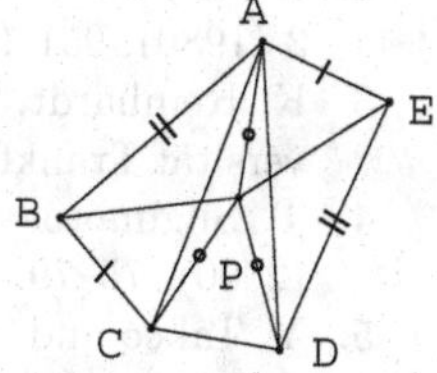

Fig. 15.

We now show the converse. We suppose $\angle B = \angle E = \pi/2$. With the conditions $\overline{AB} = \overline{DE}$, $\overline{BC} = \overline{EA}$ this condition on angles shows two triangles ΔABC and ΔDEA are congruent to each other. In particular, we see $\overline{AC} = \overline{AD}$ and $\angle BAC = \angle EDA$. Let $P \in R'''$ be the center of circumcircle of the triangle ΔACD.

We shall show that points corresponding to P are cites of our tessellation $\mathcal{R}'''$. We study Dirichlet property of $\mathcal{R}'''$ at 3-valent vertices each of which is formed by vertices corresponding to A, C and D. Since P is the center of circumcircle of ΔACD, we have $\overline{\text{PA}} = \overline{\text{PC}} = \overline{\text{PD}}$. This gurantees that

$$\angle\text{PAC} = \angle\text{PCA}, \ \angle\text{PAD} = \angle\text{PDA}, \ \angle\text{PCD} = \angle\text{PDC}.$$

As ΔACD is an isosceles triangle, we have $\angle$PAC = $\angle$PAD, hence we obtain

$$\angle\text{PAB} = \angle\text{PAC} + \angle\text{BAC} = \angle\text{PDA} + \angle\text{EDA} = \angle\text{PDE},$$
$$\angle\text{PAE} = \angle\text{PAD} + \angle\text{DAE} = \angle\text{PCA} + \angle\text{ACB} = \angle\text{PCB}.$$

Next we study at 4-valent vertices each of which is formed by vertices corresponding to B and D. By use of the condition $\overline{\text{AB}} = \overline{\text{DE}}$ and equalities $\overline{\text{PA}} = \overline{\text{PD}}$ and $\angle$PAB = $\angle$PDE, we obtain two triangles ΔPAB and ΔPDE are congruent to each other. Hence we get $\overline{\text{PB}} = \overline{\text{PE}}$ and

$$\angle\text{PBA} = \angle\text{PED}, \ \angle\text{PBC} = \frac{\pi}{2} - \angle\text{PBA} = \frac{\pi}{2} - \angle\text{PED} = \angle\text{PEA}.$$

For each pentagon $R'''_\alpha \in \mathcal{R}'''$ we take a point $\text{P}_\alpha \in R'''_\alpha$ corresponding to $\text{P} \in R'''$. The above argument shows that these $\{\text{P}_\alpha\}$ are cites of $\mathcal{R}'''$. □

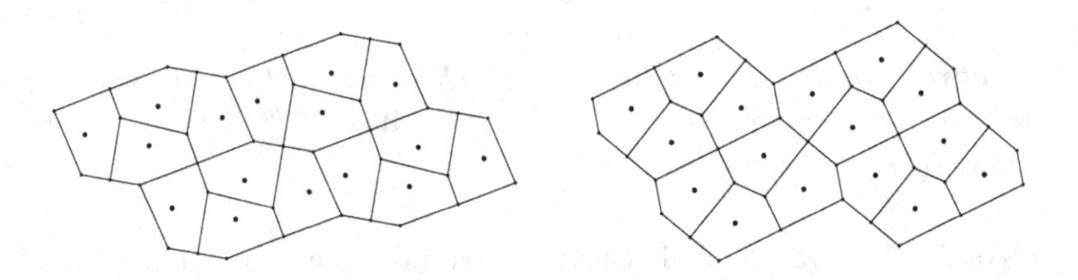

Fig. 16. Cites in a tessellation of the type in Theorem 6.1

References

1. P. Ash and E.D. Bolker, *Recognizing Dirichlet tessellations*, Geom. Dedicata 19 (1985), 175–206.
2. B. Grunbaum and G.C. Shephard, *Tilings with congruent tiles*, Bull. A. M. S. 3 (1980), 951–973.
3. K. Reinhardt, *Über die Zerlegung der Ebene in Polygone*, Dissertation, Universität Frankfult, 1918.
4. T. Sugimoto and T. Ogawa, *Tiling problem of convex pentagon*, Forma 15 (2000), 75–79.
5. Y. Takeo and T. Adachi, *Dirichlet tessellations of a plane by congruent quadrangles*, to appear in New Zealand J. Math.

ALMOST COMPLEX CONNECTIONS ON ALMOST COMPLEX MANIFOLDS WITH NORDEN METRIC

MARTA TEOFILOVA

Department of Geometry, Faculty of Mathematics and Informatics,
University of Plovdiv,
236 Bulgaria Blvd., Plovdiv 4003, Bulgaria
E-mail: marta.teofilova@gmail.com

A four-parametric family of linear connections preserving the almost complex structure is defined on an almost complex manifold with Norden metric. Necessary and sufficient conditions for these connections to be natural are obtained. A two-parametric family of complex connections is studied on a conformal Kähler manifold with Norden metric. The curvature tensors of these connections are proved to coincide.

Keywords: Almost complex manifold; Norden metric; Complex connection.

Introduction

Almost complex manifolds with Norden metric were first studied by A. P. Norden [9]. These manifolds are introduced in [6] as generalized B-manifolds. A classification of the considered manifolds with respect to the covariant derivative of the almost complex structure is obtained in [2] and two equivalent classifications are given in [3,5].

An important problem in the geometry of almost complex manifolds with Norden metric is the study of linear connections preserving the almost complex structure or preserving both, the structure and the Norden metric. The first ones are called *almost complex* connections, and the second ones are called *natural* connections. A special type of a natural connection is the canonical one. In [3] it is proved that on an almost complex manifold with Norden metric there exists a unique canonical connection. The canonical connection and its conformal group on a conformal Kähler manifold with Norden metric are studied in [5].

In [10] we have studied the Yano connection on a complex manifold with Norden metric and in [11] we have proved that the curvature tensors of the canonical connection and the Yano connection coincide on a conformal Kähler manifold with Norden metric.

In the present paper we define a four-parametric family of almost complex connections on an almost complex manifold with Norden metric. We find necessary and sufficient conditions for these connections to be natural. By this way we obtain a two-parametric family of natural connections on an almost complex manifold with Norden metric. We study a two-parametric family of complex connections on a conformal Kähler manifold with Norden metric, obtain the form of the Kähler curvature tensor corresponding to each of these connections and prove that these tensors coincide.

1. Preliminaries

Let (M, J, g) be a $2n$-dimensional almost complex manifold with Norden metric, i.e. J is an almost complex structure and g is a metric on M such that

$$J^2X = -X, \qquad g(JX, JY) = -g(X, Y) \tag{1}$$

for all differentiable vector fields X, Y on M, i.e. $X, Y \in \mathfrak{X}(M)$.

The associated metric $\widetilde{g}$ of g, given by $\widetilde{g}(X,Y) = g(X, JY)$, is a Norden metric, too. Both metrics are necessarily neutral, i.e. of signature (n, n).

Further, X, Y, Z, W (x, y, z, w, respectively) will stand for arbitrary differentiable vector fields on M (vectors in T_pM, $p \in M$, respectively).

If ∇ is the Levi-Civita connection of the metric g, the tensor field F of type $(0, 3)$ on M is defined by $F(X, Y, Z) = g((\nabla_X J)Y, Z)$ and has the following symmetries

$$F(X, Y, Z) = F(X, Z, Y) = F(X, JY, JZ). \tag{2}$$

Let $\{e_i\}$ $(i = 1, 2, \ldots, 2n)$ be an arbitrary basis of T_pM at a point p of M. The components of the inverse matrix of g are denoted by g^{ij} with respect to the basis $\{e_i\}$. The Lie 1-forms θ and θ^* associated with F, and the Lie vector Ω, corresponding to θ, are defined by, respectively

$$\theta(z) = g^{ij}F(e_i, e_j, z), \qquad \theta^* = \theta \circ J, \qquad \theta(z) = g(z, \Omega). \tag{3}$$

The Nijenhuis tensor field N for J is given by $N(X, Y) = [JX, JY] - [X, Y] - J[JX, Y] - J[X, JY]$. The corresponding tensor of type (0,3) is given by $N(X, Y, Z) = g(N(X, Y), Z)$. In terms of ∇J this tensor is expressed in the following way

$$N(X, Y) = (\nabla_X J)JY - (\nabla_Y J)JX + (\nabla_{JX} J)Y - (\nabla_{JY} J)X. \tag{4}$$

It is known [8] that the almost complex structure is complex if and only if it is integrable, i.e. $N = 0$. The associated tensor $\widetilde{N}$ of N is defined by [2]

$$\widetilde{N}(X,Y) = (\nabla_X J)JY + (\nabla_Y J)JX + (\nabla_{JX} J)Y + (\nabla_{JY} J)X, \qquad (5)$$

and the corresponding tensor of type (0,3) is given by $\widetilde{N}(X,Y,Z) = g(\widetilde{N}(X,Y),Z)$.

A classification of the almost complex manifolds with Norden metric is introduced in [2], where eight classes of these manifolds are characterized according to the properties of F. The three basic classes $\mathcal{W}_i$ $(i = 1,2,3)$ and the class $\mathcal{W}_1 \oplus \mathcal{W}_2$ are given by

- the class $\mathcal{W}_1$:

$$\begin{aligned} F(X,Y,Z) = \tfrac{1}{2n}\,[g(X,Y)\theta(Z) + g(X,JY)\theta(JZ) \\ +g(X,Z)\theta(Y) + g(X,JZ)\theta(JY)]; \end{aligned} \qquad (6)$$

- the class $\mathcal{W}_2$ of *the special complex manifolds with Norden metric*:

$$F(X,Y,JZ) + F(Y,Z,JX) + F(Z,X,JY) = 0, \quad \theta = 0; \qquad (7)$$

- the class $\mathcal{W}_3$ of *the quasi-Kähler manifolds with Norden metric*:

$$F(X,Y,Z) + F(Y,Z,X) + F(Z,X,Y) = 0 \Leftrightarrow \widetilde{N} = 0; \qquad (8)$$

- the class $\mathcal{W}_1 \oplus \mathcal{W}_2$ of *the complex manifolds with Norden metric*:

$$F(X,Y,JZ) + F(Y,Z,JX) + F(Z,X,JY) = 0 \Leftrightarrow N = 0. \qquad (9)$$

The special class $\mathcal{W}_0$ of *the Kähler manifolds with Norden metric* is characterized by $F = 0$.

A $\mathcal{W}_1$-manifold with closed Lie 1-forms θ and θ^* is called *a conformal Kähler manifold with Norden metric.*

Let R be the curvature tensor of ∇, i.e. $R(X,Y)Z = \nabla_X\nabla_Y Z - \nabla_Y\nabla_X Z - \nabla_{[X,Y]}Z$ and $R(X,Y,Z,W) = g\,(R(X,Y)Z,W)$.

A tensor L of type (0,4) is said to be *curvature-like* if it has the properties of R, i.e. $L(X,Y,Z,W) = -L(Y,X,Z,W) = -L(X,Y,W,Z)$, $L(X,Y,Z,W) + L(Y,Z,X,W) + L(Z,X,Y,W) = 0$. Then, the Ricci tensor $\rho(L)$ and the scalar curvatures $\tau(L)$ and $\tau^*(L)$ of L are defined by:

$$\begin{aligned} \rho(L)(y,z) = g^{ij}L(e_i,y,z,e_j), \\ \tau(L) = g^{ij}\rho(L)(e_i,e_j), \quad \tau^*(L) = g^{ij}\rho(L)(e_i,Je_j). \end{aligned} \qquad (10)$$

A curvature-like tensor L is called *a Kähler tensor* if $L(X,Y,JZ,JW) = -L(X,Y,Z,W)$.

Let S be a tensor of type (0,2). We consider the following tensors [5]:

$$\begin{aligned}\psi_1(S)(X,Y,Z,W) &= g(Y,Z)S(X,W) - g(X,Z)S(Y,W)\\ &\quad + g(X,W)S(Y,Z) - g(Y,W)S(X,Z),\\ \psi_2(S)(X,Y,Z,W) &= g(Y,JZ)S(X,JW) - g(X,JZ)S(Y,JW)\\ &\quad + g(X,JW)S(Y,JZ) - g(Y,JW)S(X,JZ),\\ \pi_1 = \frac{1}{2}\psi_1(g), \qquad \pi_2 &= \frac{1}{2}\psi_2(g), \qquad \pi_3 = -\psi_1(\widetilde{g}) = \psi_2(\widetilde{g}).\end{aligned} \tag{11}$$

The tensor $\psi_1(S)$ is curvature-like if S is symmetric, and the tensor $\psi_2(S)$ is curvature-like is S is symmetric and hybrid with respect to J, i.e. $S(X,JY) = S(Y,JX)$. The tensors $\pi_1 - \pi_2$ and π_3 are Kählerian.

2. Almost complex connections on almost complex manifolds with Norden metric

In this section we study almost complex connections and natural connections on almost complex manifolds with Norden metric. First, let us recall the following

Definition 2.1 ([7]). *A linear connection ∇' on an almost complex manifold (M,J) is said to be* almost complex *if $\nabla' J = 0$.*

Theorem 2.1. *On an almost complex manifold with Norden metric there exists a 4-parametric family of almost complex connections ∇' with torsion tensor T defined by, respectively:*

$$\begin{aligned}&g(\nabla'_X Y - \nabla_X Y, Z)\\ &= \frac{1}{2}F(X,JY,Z) + t_1\{F(Y,X,Z) + F(JY,JX,Z)\}\\ &\quad + t_2\{F(Y,JX,Z) - F(JY,X,Z)\} + t_3\{F(Z,X,Y)\\ &\qquad + F(JZ,JX,Y)\} + t_4\{F(Z,JX,Y) - F(JZ,X,Y)\},\end{aligned} \tag{12}$$

$$\begin{aligned}&T(X,Y,Z)\\ &= t_1\{F(Y,X,Z) - F(X,Y,Z) + F(JY,JX,Z) - F(JX,JY,Z)\}\\ &\quad + (\frac{1}{2} - t_2)\{F(X,JY,Z) - F(Y,JX,Z)\} + t_2\{F(JX,Y,Z)\\ &\qquad - F(JY,X,Z)\} + 2t_3F(JZ,JX,Y) + 2t_4F(Z,JX,Y),\end{aligned} \tag{13}$$

where $t_i \in \mathbb{R}$, $i = 1,2,3,4$.

Proof. By (12), (2) and direct computation, we prove that $\nabla' J = 0$, i.e. the connections ∇' are almost complex. □

By (8) and (9) we obtain the form of the almost complex connections ∇' on the manifolds belonging to the classes $\mathcal{W}_1 \oplus \mathcal{W}_2$ and $\mathcal{W}_3$ as follows, respectively

Corollary 2.1. *On a complex manifold with Norden metric there exists a 2-parametric family of complex connections ∇' defined by*

$$\begin{aligned} \nabla'_X Y &= \nabla_X Y + \frac{1}{2}(\nabla_X J)JY \\ &\quad + p\{(\nabla_Y J)X + (\nabla_{JY} J)JX\} + q\{(\nabla_Y J)JX - (\nabla_{JY} J)X\}, \end{aligned} \tag{14}$$

where $p = t_1 + t_3$, $q = t_2 + t_4$.

Corollary 2.2. *On a quasi-Kähler manifold with Norden metric there exists a 2-parametric family of almost complex connections ∇' defined by*

$$\begin{aligned} \nabla'_X Y &= \nabla_X Y + \frac{1}{2}(\nabla_X J)JY \\ &\quad + s\{(\nabla_Y J)X + (\nabla_{JY} J)JX\} + t\{(\nabla_Y J)JX - (\nabla_{JY} J)X\}, \end{aligned} \tag{15}$$

where $s = t_1 - t_3$, $t = t_2 - t_4$.

Definition 2.2 ([3]). *A linear connection ∇' on an almost complex manifold with Norden metric (M, J, g) is said to be* natural *if*

$$\nabla' J = \nabla' g = 0 \quad (\Leftrightarrow \nabla' g = \nabla' \widetilde{g} = 0). \tag{16}$$

Lemma 2.1. *Let (M, J, g) be an almost complex manifold with Norden metric and let ∇' be an arbitrary almost complex connection defined by (12). Then*

$$\begin{aligned} (\nabla'_X g)(Y, Z) &= (t_2 + t_4)\widetilde{N}(Y, Z, X) - (t_1 + t_3)\widetilde{N}(Y, Z, JX), \\ (\nabla'_X \widetilde{g})(Y, Z) &= -(t_1 + t_3)\widetilde{N}(Y, Z, X) - (t_2 + t_4)\widetilde{N}(Y, Z, JX). \end{aligned} \tag{17}$$

Then, by help of Theorem 2.1 and Lemma 2.1 we prove

Theorem 2.2. *An almost complex connection ∇' defined by (12) is natural on an almost complex manifold with Norden metric if and only if $t_1 = -t_3$ and $t_2 = -t_4$, i.e.*

$$\begin{aligned} g(\nabla'_X Y - \nabla_X Y, Z) &= \frac{1}{2}F(X, JY, Z) \\ &\quad + t_1 N(Y, Z, JX) - t_2 N(Y, Z, X). \end{aligned} \tag{18}$$

The equation (18) defines a 2-parametric family of natural connections on an almost complex manifold with Norden metric and non-integrable almost complex structure. In particular, by (8) and (17) for the manifolds in the class $\mathcal{W}_3$ we obtain

Corollary 2.3. *Let (M, J, g) be a quasi-Kähler manifold with Norden metric. Then, the connection ∇' defined by (15) is natural for all $s, t \in \mathbb{R}$.*

If (M, J, g) is a complex manifold with Norden metric, then from (9) and (18) it follows that there exists a unique natural connection ∇' in the family (14) which has the form

$$\nabla'_X Y = \nabla_X Y + \frac{1}{2}(\nabla_X J)JY. \tag{19}$$

Definition 2.3 ([3]). *A natural connection ∇' with torsion tensor T on an almost complex manifold with Norden metric is said to be* canonical *if*

$$T(X, Y, Z) + T(Y, Z, X) - T(JX, Y, JZ) - T(Y, JZ, JX) = 0. \tag{20}$$

Then, by applying the last condition to the torsion tensors of the natural connections (18), we obtain

Proposition 2.1. *Let (M, J, g) be an almost complex manifold with Norden metric. A natural connection ∇' defined by (18) is canonical if and only if $t_1 = 0$, $t_2 = \frac{1}{8}$. In this case (18) takes the form*

$$2g\big(\nabla'_X Y - \nabla_X Y, Z\big) = F(X, JY, Z) - \frac{1}{4}N(Y, Z, X).$$

Let us remark that G. Ganchev and V. Mihova [3] have proven that on an almost complex manifold with Norden metric there exists a unique canonical connection. The canonical connection of a complex manifold with Norden metric has the form (19).

Next, we study the properties of the torsion tensors (13) of the almost complex connections ∇'.

The torsion tensor T of an arbitrary linear connection is said to be *totally antisymmetric* if $T(X, Y, Z) = g(T(X, Y), Z)$ is a 3-form. The last condition is equivalent to

$$T(X, Y, Z) = -T(X, Z, Y). \tag{21}$$

Then, having in mind (13) we obtain that the torsion tensors of the almost complex connections ∇' defined by (12) satisfy the condition (21) if and only if $t_1 = t_2 = t_3 = 0$, $t_4 = \frac{1}{4}$. Hence, we prove the following

Theorem 2.3. *Let (M, J, g) be an almost complex manifold with Norden metric and non-integrable almost complex structure. Then, on M there exists a unique almost complex connection ∇' in the family (12) whose torsion*

tensor is a 3-form. This connection is defined by

$$g(\nabla'_X Y - \nabla_X Y, Z) \\ = \frac{1}{4}\{2F(X, JY, Z) + F(Z, JX, Y) - F(JZ, X, Y)\}. \tag{22}$$

By Corollary 2.2 and the last theorem we obtain

Corollary 2.4. *On a quasi-Kähler manifold with Norden metric there exists a unique natural connection ∇' in the family (15) whose torsion tensor is a 3-form. This connection is given by*

$$\nabla'_X Y = \nabla_X Y + \frac{1}{4}\{2(\nabla_X J)JY - (\nabla_Y J)JX + (\nabla_{JY} J)X\}. \tag{23}$$

Let us remark that the connection (23) can be considered as an analogue of the Bismut connection [1,4] in the geometry of the almost complex manifolds with Norden metric.

Let us consider symmetric almost complex connections in the family (12). By (13) and (4) we obtain

$$T(X, Y) - T(JX, JY) = \frac{1}{2}N(X, Y). \tag{24}$$

From (24), (13) and (4) it follows that $T = 0$ if and only if $N = 0$ and $t_1 = t_3 = t_4 = 0$, $t_2 = \frac{1}{4}$. Then, it is valid the following

Theorem 2.4. *Let (M, J, g) be a complex manifold with Norden metric. Then, on M there exists a unique complex symmetric connection ∇' belonging to the family (14) which is given by*

$$\nabla'_X Y = \nabla_X Y + \frac{1}{4}\{(\nabla_X J)JY + 2(\nabla_Y J)JX - (\nabla_{JX} J)Y\}. \tag{25}$$

The connection (25) is known as the Yano connection [12,13].

We give a summery of the obtained results for the 4-parametric family of almost complex connections ∇' in the following Table.

3. Complex connections on conformal Kähler manifolds with Norden metric

Let (M, J, g) be a $\mathcal{W}_1$-manifold with Norden metric and consider the 2-parametric family of complex connections ∇' defined by (14). By (6) we

	Class manifolds		
Connection type	$\mathcal{W}_1 \oplus \mathcal{W}_2 \oplus \mathcal{W}_3$	$\mathcal{W}_1 \oplus \mathcal{W}_2$	$\mathcal{W}_3$
almost complex	$t_1, t_2, t_3, t_4 \in \mathbb{R}$	$p, q \in \mathbb{R}$	$s, t \in \mathbb{R}$
natural	$t_1 = -t_3,$ $t_2 = -t_4$	$p = q = 0$	$s, t \in \mathbb{R}$
canonical	$t_1 = t_3 = 0,$ $t_2 = -t_4 = 1/8$	$p = q = 0$	$s = 0,\ t = 1/4$
T is a 3-form	$t_1 = t_2 = t_3 = 0,$ $t_4 = 1/4$	$\nexists$	$s = 0,\ t = -1/4$
symmetric	$\nexists$	$p = 0,\ q = 1/4$	$\nexists$

obtain the form of ∇' on a $\mathcal{W}_1$-manifold as follows

$$\begin{aligned}\nabla'_X Y = \nabla_X Y + \frac{1}{4n}\{g(X,JY)\Omega - g(X,Y)J\Omega + \theta(JY)X \\ - \theta(Y)JX\} + \frac{p}{n}\{\theta(X)Y + \theta(JX)JY\} + \frac{q}{n}\{\theta(JX)Y - \theta(X)JY\}.\end{aligned} \tag{26}$$

Then, by (26) and straightforward computation we prove

Theorem 3.1. *Let (M, J, g) be a conformal Kähler manifold with Norden metric and ∇' be an arbitrary complex connection in the family (14). Then, the Kähler curvature tensor R' of ∇' has the form*

$$R' = R - \frac{1}{4n}\{\psi_1 + \psi_2\}(S) - \frac{1}{8n^2}\psi_1(P) - \frac{\theta(\Omega)}{16n^2}\{3\pi_1 + \pi_2\} + \frac{\theta(J\Omega)}{16n^2}\pi_3,$$

where S and P are defined by, respectively:

$$\begin{aligned}S(X,Y) &= (\nabla_X\theta)JY + \frac{1}{4n}\{\theta(X)\theta(Y) - \theta(JX)\theta(JY)\}, \\ P(X,Y) &= \theta(X)\theta(Y) + \theta(JX)\theta(JY).\end{aligned} \tag{27}$$

By (26) we prove the following

Lemma 3.1. *Let (M, J, g) be a $\mathcal{W}_1$-manifold and ∇' be an arbitrary complex connection in the family (14). Then, the covariant derivatives of g and $\widetilde{g}$ are given by*

$$\begin{aligned}(\nabla'_X g)(Y,Z) = &- \frac{2}{n}\{[p\theta(X) + q\theta(JX)]\, g(Y,Z) \\ &+ [p\theta(JX) - q\theta(X)]\, g(Y,JZ)\}, \\ (\nabla'_X \widetilde{g})(Y,Z) = &\frac{2}{n}\{[p\theta(JX) - q\theta(X)]\, g(Y,Z) \\ &- [p\theta(X) + q\theta(JX)]\, g(Y,JZ)\}.\end{aligned} \tag{28}$$

It is well-known [7] that the curvature tensor R' and the torsion tensor T of an arbitrary linear connection ∇' satisfy the second Bianchi identity, i.e.

$$\underset{X,Y,Z}{\mathfrak{S}}\{(\nabla'_X R')(Y,Z,W) + R'(T(X,Y),Z,W)\} = 0, \tag{29}$$

where $\mathfrak{S}$ is the cyclic sum over X, Y, Z.

From (26) it follows that the torsion tensor of an arbitrary connection ∇' in the family (14) has the following form on a $\mathcal{W}_1$-manifold

$$\begin{aligned} T(X,Y) = \frac{1-4q}{4n}&\{\theta(X)JY - \theta(Y)JX - \theta(JX)Y + \theta(JY)X\} \\ +\frac{p}{n}&\{\theta(X)Y - \theta(Y)X + \theta(JX)JY - \theta(JY)JX\}. \end{aligned} \tag{30}$$

Let us denote $\tau' = \tau(R')$ and $\tau'^* = \tau^*(R')$. We establish the following

Theorem 3.2. *Let (M, J, g) be a conformal Kähler manifold with Norden metric, and τ' and τ'^* be the scalar curvatures of the Kähler curvature tensor R' corresponding to the complex connection ∇' defined by (14). Then, the function $\tau' + i\tau'^*$ is holomorphic on M and the Lie 1-forms θ and θ^* are defined in a unique way by τ' and τ'^* as follows:*

$$\theta = 2n\,d\left(\arctan\frac{\tau'}{\tau'^*}\right), \qquad \theta^* = -2n\,d\left(\ln\sqrt{\tau'^2 + \tau'^*2}\right). \tag{31}$$

Proof. By (29) and (30) we obtain

$$\begin{aligned} &(\nabla'_X R')(Y,Z,W) + (\nabla'_Y R')(Z,X,W) + (\nabla'_Z R')(X,Y,W) \\ &= \frac{4q-1}{2n}\{\theta(X)R'(JY,Z,W) - \theta(JX)R'(Y,Z,W) - \theta(Y)R'(JX,Z,W) \\ &\quad + \theta(JY)R'(X,Z,W) + \theta(Z)R'(JX,Y,W) - \theta(JZ)R'(X,Y,W)\} \\ &\quad - \frac{2p}{n}\{\theta(X)R'(Y,Z,W) + \theta(JX)R'(JY,Z,W) - \theta(Y)R'(X,Z,W) \\ &\quad - \theta(JY)R'(JX,Z,W) + \theta(Z)R'(X,Y,W) + \theta(JZ)R'(JX,Y,W)\}. \end{aligned} \tag{32}$$

Then, having in mind the equalities (28) and their analogous equalities for g^{ij}, we find the total traces of the both sides of (32) and get

$$d\tau' = \frac{1}{2n}\{\tau'^*\theta - \tau'\theta^*\}, \qquad d\tau'^* = -\frac{1}{2n}\{\tau'\theta + \tau'^*\theta^*\}. \tag{33}$$

From (33) it follows immediately that $d\tau'^* \circ J = -d\tau'$, i.e. the function $\tau' + i\tau'^*$ is holomorphic on M and the equalities (31) hold. □

Acknowledgement

The author is supported by The Fund for Scientific Research of the University of Plovdiv "Paisii Hilendarski", Project for young researchers MU-5, contract No. 494/08.

References

1. J.-M. Bismut, *A local index theorem for non-Kähler manifolds*, Math. Ann. 284 (1989), 681–699.
2. G. Ganchev, A. Borisov, *Note on the almost complex manifolds with a Norden metric*, Compt. Rend. Acad. Bulg. Sci. 39 (5) (1986), 31–34.
3. G. Ganchev, V. Mihova, *Canonical connection and the canonical conformal group on an almost complex manifold with B-metric*, Ann. Univ. Sofia Fac. Math. Inform. 81 (1) (1987), 195–206.
4. P. Gauduchon, *Hermitian connections and Dirac operators*, Bollettino U. M. I. 11 (1997), 257–288.
5. G. Ganchev, K. Gribachev, V. Mihova, *B-connections and their conformal invariants on conformally Kähler manifolds with B-metric*, Publ. Inst. Math. (Beograd) (N.S.) 42 (56) (1987), 107–121.
6. K. Gribachev, D. Mekerov, G. Djelepov, *Generalized B-manifolds*, Compt. Rend. Acad. Bulg. Sci. 38 (3) (1985), 299–302.
7. S. Kobayshi, K. Nomizu, *Foundations of differential geometry* vol. 1, 2, Intersc. Publ., New York (1963), (1969).
8. A. Newlander, L. Niremberg, *Complex analytic coordinates in almost complex manifolds*, Ann. Math. 65 (1957), 391–404.
9. A.P. Norden, *On a class of four-dimensional A-spaces*, Russian Math. (Izv VUZ) 17 (4) (1960), 145–157.
10. M. Teofilova, *Complex connections on complex manifolds with Norden metric*, Contemporary Aspects of Complex Analysis, Differential Geometry and Mathematical Physics, eds. S. Dimiev and K. Sekigawa, World Sci. Publ., Singapore (2005), 326–335.
11. M. Teofilova, *Curvature properties of conformal Kähler manifolds with Norden metric*, Math. Educ. Math., Proc. of 35th Spring Conference of UBM, Borovec (2006), 214–219.
12. K. Yano, *Affine connections in an almost product space*, Kodai Math. Semin. Rep. 11 (1) (1959), 1–24.
13. K. Yano, *Differential geometry on complex and almost complex spaces*, Pure and Applied Math. vol. 49, Pergamon Press Book, New York (1965).

PSEUDO-BOSON COHERENT AND FOCK STATES

D. A. TRIFONOV

Institute for Nuclear Research and Nuclear Energy
Bulgarian Academy of Sciences
Tzarigradsko chausse Blv, 1784 Sofia, Bulgaria
E-mail: dtrif@inrne.bas.bg

Coherent states (CS) for non-Hermitian systems are introduced as eigenstates of pseudo-Hermitian boson annihilation operators. The set of these CS includes two subsets which form bi-normalized and bi-overcomplete system of states. The subsets consist of eigenstates of two complementary lowering pseudo-Hermitian boson operators. Explicit constructions are provided on the example of one-parameter family of pseudo-boson ladder operators. The wave functions of the eigenstates of the two complementary number operators, which form a bi-orthonormal system of Fock states, are found to be proportional to new polynomials, that are bi-orthogonal and can be regarded as a generalization of standard Hermite polynomials.

Keywords: Non-Hermitian quantum mechanics; Coherent states; Fock states; Bi-orthogonal polynomials.

1. Introduction

In the last decade a growing interest is shown in the non-Hermitian PT-symmetric (or pseudo-Hermitian) quantum mechanics. For a review with an enlarged list of references see the recent papers [1,2]. This trend of interest was triggered by the papers of Bender and coauthors [3], where the Bessis conjecture about the reality and positivity of the spectrum of Hamiltonian $H = p^2 + x^2 + ix^3$ was proven ('using extensive numerical and asymptotic studies') and argued that the reality of the spectrum is due to the unbroken PT-symmetry. The Bessis-Zinn Justin conjecture about the reality of the spectrum of the PT-symmetric Hamiltonian $p^2 - (ix)^{2\nu}$ for $\nu \geq 1$ has been proven in Ref. 4. A criterion for the reality of the spectrum of non-Hermitian PT-symmetric Hamiltonians is provided in Ref. 5. Mustafazadeh [6] has soon noted that all the PT-symmetric non-Hermitian Hamiltonians studied in the literature belong to the class of pseudo-Hermitian Hamiltonians. A

Hamiltonian H was said to be pseudo-Hermitian if it obeys the relation [6]

$$H^{\#} := \eta^{-1} H^{\dagger} \eta = H, \tag{1}$$

where η is an invertible Hermitian operator. $H^{\#}$ was called *η-pseudo-Hermitian conjugate* to H, shortly *η-pseudo-adjoint* to H. The PT-symmetric Hamiltonian $H = p^2 - (ix)^{2\nu}$, examined in Refs. 3,4, obeys (1) with η equal to the parity operator P. The spectrum of a diagonalizable pseudo-Hermitian H is either real or comes in complex conjugate pairs. A diagonalizable (non-Herimitian) Hamiltonian H has a real spectrum iff it is pseudo-Hermitian with respect to positive definite η [7]. In terms of P and T operations the reality of the spectrum of H occurs if the PT symmetry is exact (not spontaneously broken) (see e.g. Refs. 1,2 and references therein). In fact many of the later developments in the field are anticipated in the paper by Scholtz et al [8] (see comments in Ref. 1).

In the present paper we address the problem of construction of pseudo-Hermitian boson (shortly pseudo-boson) creation and annihilation operators and related Fock states and coherent states (CS). For pseudo-fermion system ladder operator CS have been introduced by Cherbal et al [11] on the example of two-level atom interacting with a monochromatic em field in the presence of level decays. For non-Hermitian PT-symmetric system CS of Gazeau-Klauder type have been constructed, on the example of Scarf potential, by Roy et al [9]. Annihilation and creation operators in non-Hermitian (supersymmetric) quantum mechanics were considered by Znojil [13]. For the bosonic PT symmetric singular oscillator (which depicts a double series of real energy eigenvalues) [14] ladder operators and eigenstates of the annihilation operators have been built up by Bagchi and Quesne [15]. Our main aim here is the construction of overcomplete families (in fact bi-overcomplete) of ladder operator CS for pseudo-bosons. The problem of pseudo-boson ladder operators is considered in section 2. In the third section we consider the construction of eigenstates of pseudo-boson number operators. The pseudo-boson CS are introduced and discussed in section 4. Explicit example of pseudo-boson ladder operators and related eigenstates is provided and briefly commented in section 5. Outlook over the main results is given in the Conclusion.

2. Pseudo-boson ladder operators

With the aim to construct pseudo-Hermitian boson (shortly *pseudo-boson*) coherent states (CS) we address the problem of ladder operators that are *pseudo-adjoint* to each other. In analogy with the pseudo-Hermitian fermion (phermion) annihilation and creation operators [10] the η-pseudo-boson ladder operators b, $b^{\#} := \eta^{-1}b^{\dagger}\eta$ can be defined by means of the commutation relation

$$[b, b^{\#}] \equiv bb^{\#} - b^{\#}b = 1. \tag{2}$$

If $\eta = 1$ the standard boson operators a, $a^{\dagger}$ are recovered.

From (2) it follows that the pseudo-Hermitian (pseudo-selfadjoint) operator $b^{\#}b \equiv N$ commutes with b and $b^{\#}$ according to

$$[b, N] = b, \qquad [b^{\#}, N] = -b^{\#}, \tag{3}$$

and could be regarded as *pseudo-boson number operator*. For a pair of non-Hermitian operators b, $\tilde{b}$ with commutator $[b, \tilde{b}] = 1$, the existence of η such that $\tilde{b} = \eta^{-1}b^{\dagger}\eta \equiv b^{\#}$, stems from the existence of b-vacuum. We have the following

Proposition 2.1. *If the operators b and $\tilde{b}$ and a state $|0\rangle$ satisfy*

$$[b, \tilde{b}] = 1, \qquad b|0\rangle = 0, \tag{4}$$

then $\tilde{b}$ is η-adjoint to b with

$$\eta = \sum_{n=0} |\varphi_n\rangle\langle\varphi_n|, \tag{5}$$

where $|\varphi_n\rangle$ are eigenstates of $N' = b^{\dagger}\tilde{b}^{\dagger}$.

Proof. Note first that the non-Hermitian operator $N = \tilde{b}b$ is diagonalizable and with real and discrete spectrums. It's eigenstates can be constructed acting on the b-vacuum by the operators $\tilde{b}$ correspondingly (see the next section). The spectrum of a diagonalizable non-Hermitian operator H is real iff H is η-pseudo-Hermitian (theorem of Ref. 6), and this η may be chosen as a sum of projectors onto the eigenstates of $H^{\dagger}$ [6]. In our cases $H = N$ and $H^{\dagger} = N^{\dagger} = b^{\dagger}\tilde{b}^{\dagger}$. This ends the proof of the Proposition. □

For the sake of completeness however we provide in the next section the construction (and brief discussion) of the eigenstates of N, N'.

Remark 2.1. A similar Proposition can be formulated and proved for a pair of non-Hermitian nilpotent operators g, $\tilde{g}$, $g^2 = 0$ with anticommutator $\{g, \tilde{g}\} = 1$: just replace b, $\tilde{b}$, $[b, \tilde{b}]$ with g, $\tilde{g}$, $\{g, \tilde{g}\}$.

3. Pseudo-boson Fock states

The eigenstates of pseudo-boson number operators $N = \tilde{b}b$ can be constructed by acting on the ground states $|0\rangle$ by the raising operator $\tilde{b}$:

$$|\psi_n\rangle = \frac{1}{\sqrt{n!}}\tilde{b}^n|0\rangle, \tag{6}$$

$$N|\psi_n\rangle = \tilde{b}b|\psi_n\rangle = n|\psi_n\rangle. \tag{7}$$

However in view of $\tilde{b} \neq b^\dagger$ these number states are not orthogonal. It turned out that a *complementary pair* of pseudo-boson ladder operators and number operator exist, such that the system of the two complementary sets of number states form the so-called *bi-orthogonal and bi-complete sets*. Indeed, if $\tilde{b}$ is creating operator related to b, then, on the symmetry ground, we could look for new operators b' for which $b^\dagger$ is the creating operator,

$$[b', b^\dagger] = 1. \tag{8}$$

The pairs of "prime"-ladder operators b', $b'^\dagger$ is just $\tilde{b}^\dagger$, $b^\dagger$, and the "prime" number operator is

$$N' = b'^\dagger b' = b^\dagger \tilde{b}^\dagger. \tag{9}$$

The eigenstates of N' are constructed in a similar way acting with $b^\dagger$ on the b'-vacuums $|0\rangle'$:

$$|\varphi_n\rangle = \frac{1}{\sqrt{n!}}(b^\dagger)^n|0\rangle'. \tag{10}$$

The existence of the b'-vacuum $|0\rangle' = |\varphi_0\rangle$ follows from the properties of the pseudo-Hermitian operators H with real spectra [6,7]: the spectrum of H and $H^\dagger$ coincide since they are related via a similarity transformation. In our case $H = \tilde{b}b$, $H^\dagger = b^\dagger\tilde{b}^\dagger \equiv b^\dagger b'$.

Using the commutation relations of the above described ladder operators one can easily check that if $\langle 0|0\rangle' = 1$ then the "prime" number-states $|\varphi_n\rangle$ are *bi-orthonormalized* to $|\psi_n\rangle$ (that is $\langle\psi_n|\varphi_m\rangle = \delta_{nm}$), and form together the *bi-complete* system of states $\{|\psi_n\rangle, |\varphi_n\rangle\}$:

$$\sum_n |\psi_n\rangle\langle\varphi_n| = 1 = \sum_n |\varphi_n\rangle\langle\psi_n|. \tag{11}$$

The set $\{|\psi_n\rangle, |\varphi_n\rangle\}$ can be called the set of Fock states for pseudo-Hermitian boson system (shortly *pseudo Fock states*). In terms of the projectors on these states the pseudo-boson ladder operators b, $\tilde{b}$ can be represented as follows

$$\begin{aligned} b &= \sum_n \sqrt{n}|\psi_{n-1}\rangle\langle\varphi_n|, & \tilde{b} &= \sum_n \sqrt{n}|\psi_n\rangle\langle\varphi_{n-1}|, \\ b' &= \sum_n \sqrt{n}|\varphi_{n-1}\rangle\langle\psi_n|, & b^\dagger &= \sum_n \sqrt{n}|\varphi_n\rangle\langle\psi_{n-1}|. \end{aligned} \tag{12}$$

Now consider the operator [6]

$$\eta = \sum_n |\varphi_n\rangle\langle\varphi_n|. \tag{13}$$

This is Hermitian, positive and invertible operator, $\eta^{-1} = \sum_n |\psi_n\rangle\langle\psi_n|$. From the above expressions of η and η^{-1} one can see that $|\varphi_n\rangle = \eta|\psi_n\rangle$.

Finally one can easily check (using (12)) and (13) that $\tilde{b}$ is η-pseudo-adjoint to b, b' is η^{-1}-pseudo-adjoint to $b^\dagger$,

$$\tilde{b} = \eta^{-1}b^\dagger\eta, \quad b' = \eta b\eta^{-1}, \tag{14}$$

and N and N' are η- and η^{-1} pseudo-Hermitian: $N^\# := \eta^{-1}N^\dagger\eta = N$, $(N')^\# := (\eta')^{-1}(N')^\dagger\eta' = N'$, $\eta' = \eta^{-1}$.

4. Pseudo-boson coherent states

We define coherent states (CS) for the pseudo-Hermitian boson systems as eigenstates of the corresponding pseudo-boson annihilation operators. In this aim we introduce the pseudo-unitary displacement operator $D(\alpha) = \exp(\alpha b^\# - \alpha^* b)$ and construct eigenstates of b as displaced ground state $|0\rangle$,

$$b|\alpha\rangle = \alpha|\alpha\rangle, \quad \rightarrow \quad |\alpha\rangle = D(\alpha)|0\rangle, \tag{15}$$

where $\alpha \in C$. Using BCS formula one gets the expansion

$$|\alpha\rangle = e^{-\frac{1}{2}|\alpha|^2}\sum_n \frac{\alpha^n}{\sqrt{n!}}|\psi_n\rangle. \tag{16}$$

The structures of the above two formulas are the same as those for the case Hermitian boson CS (the Glauber canonical CS [16]), but the properties of our $D(\alpha)$ and $|\alpha\rangle$ are different. First note that $D(\alpha)$ is not unitary. Therefore $|\alpha\rangle$ are not normalized. Second, the set $\{|\alpha\rangle; \alpha \in C\}$ is not overcomplete (since $|\psi_n\rangle$ are not orthogonal).

The way out of this impasse is to consider the eigenstates of the dual ladder operator $b' = (b^\#)^\dagger$, which take the analogous to (15) form,

$$b'|\alpha\rangle' = \alpha|\alpha\rangle', \quad \rightarrow \quad |\alpha\rangle' = D'(\alpha)|0\rangle', \tag{17}$$

where $D'(\alpha) = \exp(\alpha b^\dagger - \alpha^* b')$ is the complementary pseudo-unitary displacement operator. Therefore the eigenstates $|\alpha\rangle'$ are again non-normalized and do not form overcomplete set. However they are *bi-normalized* to $|\alpha\rangle$ (that is $\langle\alpha|\alpha\rangle' = 1$) and the system $\{|\alpha\rangle, |\alpha\rangle'; \alpha \in C\}$ is {bi-overcomplete} in the following sense

$$\frac{1}{\pi}\int d^2\alpha\, |\alpha\rangle'\, \langle\alpha| = 1, \qquad \frac{1}{\pi}\int d^2\alpha\, |\alpha\rangle\, '\langle\alpha| = 1. \tag{18}$$

It is this bi-overcomplete set that we call *pseudo-boson CS*, or CS of pseudo-boson systems. More precisely they are η- pseudo-boson CS. When $\eta = 1$ these states recover the famous Glauber CS $|\alpha\rangle = \exp(\alpha a^\dagger - \alpha^* a)|0\rangle$, where a, $a^\dagger$ are canonical boson annihilation and creation operators.

5. Example

In this section we illustrate the above described scheme of construction of pseudo-boson Fock states and CS on the example of the following one-parameter family of non-Hermitian operators,

$$\begin{aligned} b(s) &= a + s a^\dagger, \\ \tilde{b}(s) &= s a + (1 + s^2) a^\dagger, \end{aligned} \tag{19}$$

where $s \in (-1, 1)$ and a, $a^\dagger$ are Bose annihilation and creation operators: $[a, a^\dagger] = 1$. It is clear that $b(0) = a$, $\tilde{b}(0) = a^\dagger$ and $\tilde{b}(s) \neq b^\dagger(s)$. In this way the parameter s could be viewed as a measure of deviation of $b(s)$ and $b^{\#}(s)$ from the canonical boson operators a and $a^\dagger$.

5.1. *Pseudo-boson Fock state wave functions*

The $b(s)$- and b'-vacuums $|0\rangle$ and $|0\rangle'$ do exist. Using the coordinate representation of $b(s)$ and $\tilde{b}(s)$,

$$b(s) = \frac{1}{\sqrt{2}}\left((1+s)x - (1-s)\frac{d}{dx}\right), \tag{20}$$

$$\tilde{b}(s) = \frac{1}{\sqrt{2}}\left((s+1+s^2)x - (s-1-s^2)\frac{d}{dx}\right), \tag{21}$$

we find the wave functions of $|0\rangle$ and $|0\rangle'$:

$$\psi_0(x, s) = \mathcal{N}(s)\exp\left(-\frac{1+s}{2(1-s)}x^2\right), \tag{22}$$

$$\varphi_0(x, s) = \mathcal{N}(s)\exp\left(-\frac{1+s+s^2}{2(1+s^2-s)}x^2\right), \tag{23}$$

$$\mathcal{N}(s) = \left(\pi(1-s)(1-s+s^2)\right)^{-1/4}.$$

The above two wave functions are bi-normalized, that is $\langle\varphi_0|\psi_0\rangle = 1$, if the parameter s is restricted in the interval $(-1, 1)$, that is $-1 < s < 1$.

Therefore, according to Proposition 1 (and the related development in the previous section) $\tilde{b}(s)$ is η-pseudo-adjoint to $b(s)$ and (for $s^2 < 1$) the bi-orthonormalized Fock states and bi-overcomplete CS can be explicitly realized. The wave functions of the pseudo-boson Fock states (6) and (10) are obtained in the following form:

$$\psi_n(x,s) = \frac{1}{\sqrt{2^n n!}} P_n(x,s)\psi_0(x,s), \qquad \varphi_n(x,s) = \frac{1}{\sqrt{2^n n!}} Q_n(x,s)\varphi_0(x,s), \tag{24}$$

where $P_n(x,s)$, $Q_n(x,s)$ are polynomials of degree n in x, defined by means of the following recurrence relations

$$P_n = \frac{2x}{1-s}P_{n-1} + (n-1)\frac{2(s-s^2-1)}{1-s}P_{n-2}, \qquad Q_n = \frac{2x}{1+s^2-s}Q_{n-1} + (n-1)\frac{2(s-1)}{1+s^2-s}Q_{n-2}. \tag{25}$$

For the first three values of n, $n = 0, 1, 2$, the polynomials $P_n(x,s)$ and $Q_n(x,s)$ read:

$$P_0 = 1, \quad P_1 = \frac{2}{1-s}x, \quad P_2 = \frac{4x^2}{(1-s)^2} + \frac{2(s-s^2-1)}{1-s},$$

$$Q_0 = 1, \quad Q_1 = \frac{2}{1-s+s^2}x, \quad Q_2 = \frac{4x^2}{(1-s+s^2)^2} + \frac{2(s-1)}{1-s+s^2}.$$

At $s = 0$ these two polynomials $P_n(x,0)$ and $Q_n(x,0)$ coincide and recover the known Hermite polynomials $H_n(x)$. Therefore $P_n(x,s)$ and $Q_n(x,s)$ can be viewed as two different generalizations of $H_n(x)$. They are not orthogonal. Instead of the orthogonality they satisfy the *bi-orthonality* relations, the weight function being $w(x,s) = \psi_0(x,s)\varphi_0(x,s)$,

$$\int P_n(x,s)Q_m(x,s)\,\psi_0(x)\varphi_0(x) = n!2^n\delta_{nm}. \tag{26}$$

Therefore $P_n(x,s)$ and $Q_n(x,s)$ realize *bi-orthogonal generalization* of $H_n(x)$. The relations between these three types of polynomials may be illustrated by the following diagram (where $\mu(s) = 1/(1-s)(1-s+s^2)$):

$$H_n(x) \longrightarrow \int H_n H_m e^{-x^2}dx = \sqrt{\pi}\,2^n n!\,\delta_{nm}$$

$$\swarrow \quad \searrow$$

$$P_n(x) \quad Q_n(x) \longrightarrow \int P_n Q_m e^{-\mu(s)x^2}dx = \mathcal{N}^{-2}(s)2^n n!\,\delta_{nm}$$

It is worth emphasizing that the above bi-orthogonal generalization of $H_n(x)$ is not unique. If instead of (19), we take another pair of b, $\tilde{b}$ (satisfying the Proposition 2.1), say

$$\begin{aligned} b_2(s) &= a + sa^\dagger, \\ \tilde{b}_2(s) &= -sa + (1-s^2)a^\dagger, \end{aligned} \tag{27}$$

and apply the above described scheme, we would get another similar pair of bi-orthogonal polynomials.

5.2. *Pseudo-boson CS wave functions*

Equations (15), (17) for the eigenstates of $b(s)$ and $b'(s)$, given in (20), (21) in the coordinate representation, lead to the following wave functions for any $\alpha \in C$,

$$\psi_\alpha(x,s) = \mathcal{N} \exp\left[-\frac{1+s}{2(1-s)}x^2 + \frac{\sqrt{2}\alpha}{1-s}x\right], \tag{28}$$

$$\varphi_\alpha(x,s) = \mathcal{N}' \exp\left[-\frac{1+s+s^2}{2(1-s+s^2)}x^2 + \frac{\sqrt{2}\alpha}{1-s+s^2}x\right], \tag{29}$$

where $\mathcal{N}$ and $\mathcal{N}'$ are bi-normalization constants. Up to constant phase factors they are determined by the bi-normalization condition $\langle\psi_\alpha|\varphi_\alpha\rangle = 1$. We put $\alpha = \alpha_1 + i\alpha_2$ and find

$$\begin{aligned} &(N^*N')^{-1} \\ &= \int \exp\left[-\frac{x^2}{(1-s)(1-s+s^2)} + \sqrt{2}x\frac{(2-2s+s^2)\alpha_1 + i\alpha_2 s^2}{(1-s)(1-s+s^2)}\right] dx \\ &= \sqrt{\pi(1-s)(1-s+s^2)} \exp\left[\frac{\left(2\alpha_1 - 2\alpha_1 s + \alpha_1 s^2 + i\alpha_2 s^2\right)^2}{2(1-s)(1-s+s^2)}\right]. \end{aligned} \tag{30}$$

At $s = 0$ we get, up to a constant phase factor, $\mathcal{N} = \mathcal{N}' = \pi^{-1/4}\exp(-\alpha_1^2)$. At $s = 0$ both $\psi_\alpha(x,0)$ and $\varphi_\alpha(x,0)$ recover the wave functions of Glauber canonical CS.

The constructed Fock states and CS are time independent, and can be used as initial states (initial conditions) of pseudo-boson systems. Important question arises of the *temporal stability* of these states. In analogy with the case of Hermitian mechanics we define temporal stability of a given set of states by the requirement the time evolved states to belong to the same set. This means that, up to a time-dependent phase factor, the time-dependence of any state from the initial set should be included in the time-dependent parameters only. Clearly the time dependent parameters should remain in the same domain as defined initially.

For our CS the temporal stability means that the time evolved wave functions $\psi_\alpha(x,s,t)$, $\varphi_\alpha(x,s,t)$ should keep the form

$$\begin{aligned} \psi_\alpha(x,s,t) &= e^{i\chi(t)}\psi_{\alpha(t)}(x,s), \\ \varphi_\alpha(x,s,t) &= e^{i\chi'(t)}\varphi_{\alpha(t)}(x,s)), \end{aligned} \tag{31}$$

where $\chi(t), \chi'(t) \in R$, $\alpha(t) \in C$ and $s(t)^2 < 1$. It is clear, that if the time evolved CS obey (31) they remain eigenstates of the same ladder operators $b(s)$ and $b'(s)$. (Let us recall at this point that we are in the Schrödinger picture, where operators are time-independent). As an illustration consider now the time evolution of CS governed by the simple pseudo-Hermitian Hamiltonian

$$H_{\mathrm{po}} = \omega\left(b^{\#}(s)b(s) + \frac{1}{2}\right), \tag{32}$$

where $b^{\#}(s) = \eta^{-1}b^{\dagger}(s)\eta$, $\omega \in R_+$. System with Hamiltonian of the type (32) should be called *pseudo-Hermitian oscillator*. At $s = 0$ it coincides with the Hermitian harmonic oscillator of frequency ω. In pseudo-Hermitian mechanics the time evolution of initial $\psi_\alpha(x,s)$ and $\varphi_\alpha(x,s)$, by definition, is given by

$$\begin{aligned} \psi_\alpha(x,s,t) &= e^{-iHt}\psi_\alpha(x,s), \\ \varphi_\alpha(x,s,t) &= e^{-iH^{\dagger}t}\varphi_\alpha(x,s). \end{aligned} \tag{33}$$

For $H = H_{\mathrm{po}}$ equations (33) produce

$$\begin{aligned} \psi_\alpha(x,s,t) &= e^{-i\omega t/2}\psi_{\alpha(t)}(x,s), \\ \varphi_\alpha(x,s,t) &= e^{-i\omega t/2}\varphi_{\alpha(t)}(x,s), \qquad \alpha(t) = \alpha e^{-i\omega t}, \end{aligned} \tag{34}$$

which shows that the evolution of CS, governed by the pseudo-Hermitian oscillator Hamiltonian H_{po} is temporally stable. Comparing (34) and (31) we see that for the system (32) the time-evolved CS remain stable with constant s, $\alpha(t) = e^{-i\omega t}$ and $\chi = -i\omega t/2$.

Finally it is worth noting that the pseudo-Hermitian Hamiltonian H_{po} has *unbroken PT*-symmetry [1]. Indeed, from (20) and (20) it follows that $PTH_{\mathrm{po}}PT = H_{\mathrm{po}}$, and from (24) and (25) it follows that all eigenstates of H_{po} (and of $H^{\dagger}_{\mathrm{po}}$ as well) are eigenvectors of PT with eigenvalue $+1$ or -1.

Conclusion

We have shown that if the commutator of two non-Hermitian operators b and $\tilde{b}$ equals 1 and b annihilates a state ψ_0 then $\tilde{b}$ is η-pseudo-Hermitian agjoint $b^{\#}$ of b and $b^\dagger$ is η^{-1}-pseudo-adjoint of $(\tilde{b}^{\#})^\dagger$. Eigenstates of the pseudo-boson number operator $b^{\#}b$ and its adjoint $b^\dagger(b^{\#})^\dagger$ form a bi-orthonormal system of pseudo-boson Fock states, while eigenstates of b and its complementary lowering operator $b' = (b^{\#})^\dagger$ are shown to form *bi-normalized and bi-overcomplete* system.This system of states is regarded as system of coherent states (CS) for pseudo-Hermitian bosons. We have provided a simple one-parameter family of ladder operators $b(s)$ and $\tilde{b}(s)$ that possess the above described properties and constructed the wave functions of the related Fock states and CS. Fock state wave functions are obtained as product of an exponential of a quadratic form of x and one of the two new polynomials $P_n(x)$, $Q_n(x)$ that are bi-orthogonal and at $s = 0$ recover the standard Hermite orthogonal polynomials. The pseudo-boson CS are shown to be temporally stable for the pseudo-boson oscillator Hamiltonian $\omega\left(b^{\#}(s)b(s) + 1/2\right)$.

References

1. C.M. Bender, *Rep. Prog. Phys.* **70**, 947 (2007); C.M. Bender and P.D. Mannheim, *Phys. Rev.* **D78**, 025022 (2008).
2. P. Dorey, C. Dunning, and R. Tateo, *J. Phys.* **A40**, R205 (2007).
3. C.M. Bender and S. Boettcher, *Phys. Rev. Lett.* **80**, 5243 (1998); C.M. Bender, S. Boettcher, and P.N. Meisinger, *J. Math. Phys.* **40**, 2201 (1999).
4. P. Dorey, C. Dunning and R. Tateo, *J. Phys.* **A 34**, L391 (2001); P. Dorey, C. Dunning and R. Tateo, *J. Phys.* **A34**, 5679 (2001).
5. E. Caliceti, S. Graffi and J. Sjostrand, *J. Phys.* **A38** 185 (2005).
6. A. Mostafazadeh, *J. Math. Phys.* **43**, 205 (2002).
7. A. Mostafazadeh, *J. Math. Phys.* **43**, 3944 (2002).
8. F.G. Scholtz, H.B. Geyer and F.J.W. Hahne, *Ann. Phys. (NY)* **213**, 74 (1992).
9. B. Roy and P. Roy, *Phys. Lett.* **A359**, 110 (2006).
10. A. Mostafazadeh, *Nucl. Phys.* **B640**, 419 (2002); *J. Phys.* **A37**, 10193 (2004).
11. O. Cherbal, M. Drir, M. Maamache, and D.A. Trifonov, *J. Phys.* **A40**, 1835 (2007);
12. Y. Ben-Aryeh, A. Mann and I. Yaakov, *J. Phys.* **A37**, 12059 (2004).
13. M. Znojil, *Annihilation and creation operators in non-Hermitian supersymmetric quantum mechanics*, hep-th/0012002.
14. M. Znojil, *Phys. Lett.* **A259**, 220 (1999).
15. B. Bagchi and C. Quesne, *Mod. Phys. Lett.* **A16**, 2449 (2001).
16. R.J. Glauber, *Phys. Rev.* **131**, 2766 (1963).

NEW INTEGRABLE EQUATIONS OF MKDV TYPE

T. I. VALCHEV
Institute for Nuclear Research and Nuclear Energy,
Bulgarian Academy of Sciences,
Sofia, Bulgaria
E-mail: valtchev@inrne.bas.bg

We present new examples of multicomponent mKdV type equations associated with symmetric spaces of **BD**.**I** series. These are integrable equations obtained by using the method of reduction group. Their 1-soliton solutions are derived as well by applying the dressing procedure with an appropriately chosen dressing factor.

Keywords: MKdV; Symmetric spaces; Method of reductions; Dressing method.

1. Introduction

The modified Korteweg-de Vries equation (mKdV)

$$v_t + v_{xxx} + 6v^2 v_x = 0 \tag{1}$$

is one of the classical soliton equations which describes the propagation of ion-acoustic waves in warm plasma. It was Wadati [1] who solved mKdV for the first time by means of the inverse scattering problem.

One of the purposes of this report is to present multicomponent S-integrable generalizations of the equation (1) which are to be expected to find physical applications as well. In order to achieve this we apply the reduction group method [2] on the generic Lax representation of mKdV associated with symmetric spaces of the series **BD**.**I** [3]. Generic multicomponent mKdV equations associated with different types of symmetric spaces are considered for the first time by Athorne and Fordy [4]. Another purpose of the report is to demonstrate how one can generate soliton solutions to these reduced mKdV equations by applying the well-known Zakharov-Shabat's dressing method [5] with an appropriate dressing factor. This dressing factor has to be compatible with the structure of the symmetric space and the action of the reduction group.

2. Some facts from the theory of solitons

In the present paragraph we are going to expose some basic facts from the theory of solitons, for more details see Ref. 6. We consider nonlinear evolution equations[a] which admit a zero curvature representation (Lax representation)

$$[L, M] = 0 \tag{2}$$

where L and M are linear differential operators of the type

$$L = i\partial_x + U(x,t,\lambda) = i\partial_x + q(x,t) - \lambda J, \tag{3}$$

$$M = i\partial_t + V(x,t,\lambda) = i\partial_t + \sum_k \lambda^k V_k(x,t). \tag{4}$$

The matrix-valued functions q and V_k take values in a simple Lie algebra $\mathfrak{g}$ and the real constant matrix J belongs to its Cartan subalgebra $\mathfrak{h} \subset \mathfrak{g}$. We shall resrict ourselves with considering only zero boundary conditions for q, i.e. q is inifinitely smooth and satisfies

$$\lim_{|x|\to\infty} |x|^n q(x,t) = 0$$

for any n. Since L and M commute there exist functions $\psi(x,t,\lambda)$ taking values in the Lie group G corresponding to the Lie algebra $\mathfrak{g}$ which satisfy

$$i\partial_x\psi + (q - \lambda J)\psi = 0, \tag{5}$$

$$i\partial_t\psi + \sum_k \lambda^k V_k\psi = \psi C(\lambda). \tag{6}$$

The linear problem (5) is called generalized Zakharov-Shabat problem (GZS).

An important role in the scattering theory is played by fundamental solutions of GZS to obey the property

$$\lim_{x\to\pm\infty} \psi_\pm(x,t,\lambda)e^{i\lambda Jx} = \mathbb{1}. \tag{7}$$

These are the so-called Jost solutions. By choosing

$$C(\lambda) = \lim_{x\to\pm\infty} V(x,t,\lambda)$$

one can ensure the relevance of definition (7). The transition matrix

$$\psi_-(x,t,\lambda) = \psi_+(x,t,\lambda)T(t,\lambda)$$

---

[a] These equations are sometimes called S-integrable, see Ref. 7.

between the two Jost solutions is the scattering matrix. Its time evolution is determined by the equality

$$T(t,\lambda) = e^{if(\lambda)t}T(0,\lambda)e^{-if(\lambda)t}$$

where $f(\lambda) := \lim_{x\to\pm\infty} V(x,t,\lambda)$ is the dispersion law of the nonlinear equation.

Another important notion in the theory of solitons is provided by the so-called fundamental analytic solutions introduced by Shabat [8]. These are fundamental solutions which have analytic properties in the upper and lower half plane of the λ-plane. They can be constructed from the Jost solutions in the following manner

$$\chi^{\pm}(x,t,\lambda) = \psi_{-}(x,t,\lambda)S^{\pm}(t,\lambda) = \psi_{+}(x,t,\lambda)T^{\mp}(t,\lambda)D^{\pm}(\lambda) \qquad (8)$$

where $S^{\pm}$, $T^{\mp}$ and $D^{\pm}$ are factors in the Gauss decomposition of the scattering matrix

$$T(t,\lambda) = T^{\mp}(t,\lambda)D^{\pm}(\lambda)(S^{\pm}(t,\lambda))^{-1}.$$

As a direct consequence of (8) it follows that $\chi^{\pm}$ can be viewed as solutions of a local Riemann-Hilbert problem

$$\chi^{+}(x,t,\lambda) = \chi^{-}(x,t,\lambda)G(t,\lambda), \qquad G(t,\lambda) = (S^{-}(t,\lambda))^{-1}S^{+}(t,\lambda).$$

This is a manifestation of the deep connection between the inverse scattering method and the Riemann-Hilbert problem.

One of the basic applications of the fundamental analytic solutions is in the dressing method. The idea that underlies the dressing method is deriving a new solution q_1 from a known one q_0 taking into account the existence of auxiliary linear problems. Given an auxiliary linear problem

$$L_0\psi_0 = i\partial_x\psi_0 + (q_0 - \lambda J)\psi_0 = 0 \qquad (9)$$

for some known potential q_0. In principle, one can obtain a fundamental solution, say ψ_0, and then construct a function $\psi_1 = g\psi_0$ which is assumed to be a fundamental solution of a linear problem

$$L_1\psi_1 = i\partial_x\psi_1 + (q_1 - \lambda J)\psi_1 = 0 \qquad (10)$$

with a potential q_1 to be found. Combining (9) and (10) we convince ourselves that the dressing factor g satisfies

$$i\partial_x g + q_1 g - g q_0 - \lambda[J,g] = 0. \qquad (11)$$

Due to reasons concerning the relation between the inverse scattering method and the Riemann-Hilbert problem we choose the dressing factor in the form

$$g = \mathbb{1} + \frac{A}{\lambda - \lambda^+} + \frac{B}{\lambda - \lambda^-}, \qquad \lambda^\pm \in \mathbb{C}_\pm. \tag{12}$$

After taking the limit $\lambda \to \infty$ we obtain the following interrelation

$$q_1 = q_0 + [J, A + B].$$

If we know the residues A and B we are able to generate another solution q_1 from q_0. A more detailed analysis of the definition $gg^{-1} = \mathbb{1}$ shows that the matrices A and B admit a decomposition of the type

$$A = XF^T, \qquad B = YG^T,$$

where X and Y are given by

$$X = (\lambda^+ - \lambda^-)[Z - W(G^T W)^{-1}\tilde{\beta}][F^T Z - \tilde{\alpha}(G^T W)^{-1}\tilde{\beta}]^{-1}, \tag{13}$$

$$Y = (\lambda^- - \lambda^+)[W - Z(F^T Z)^{-1}\tilde{\alpha}][G^T W - \tilde{\beta}(F^T Z)^{-1}\tilde{\alpha}]^{-1}. \tag{14}$$

As a consequence of (11) and the corresponding equation for g^{-1} all quantities above can be expressed via the fundamental analytic solutions $\chi_0^\pm$ of the initial linear problem

$$\begin{aligned} F^T(x) &= F_0^T[\chi_0^+(x,\lambda^+)]^{-1}, \qquad & G^T(x) &= G_0^T[\chi_0^-(x,\lambda^-)]^{-1}, \\ W(x) &= \chi_0^+(x,\lambda^+)W_0, & Z(x) &= \chi_0^-(x,\lambda^-)Z_0, \end{aligned}$$

$$\tilde{\alpha}(x) = -(\lambda^+ - \lambda^-)F_0^T[\chi_0^+(x,\lambda^+)]^{-1}\partial_\lambda\chi_0^+(x,\lambda^+)W_0 + \tilde{\alpha}_0,$$

$$\tilde{\beta}(x) = -(\lambda^- - \lambda^+)G_0^T[\chi_0^-(x,\lambda^-)]^{-1}\partial_\lambda\chi_0^-(x,\lambda^-)Z_0 + \tilde{\beta}_0,$$

where all matrices with a subscript 0 are constant. The dressing factor take values in $SL(n)$. This factor generalizes those obtained by Zakharov and Mikhailov [9] and Ivanov [10] which are suited for the case of symplectic and orthogonal algebras. We can apply the dressing procedure on q_1 and thus obtain another solution q_2 and so on. In particular, when $q_0 \equiv 0$ the dressed solution q_1 is called 1-soliton solution, q_2 — 2-soliton solution etc.

Consider a discrete group G_R (reduction group) acting on the set of fundamental solutions as follows

$$\mathcal{K} : \psi(x,\lambda) \mapsto \tilde{\psi}(x,\lambda) = \mathbf{K}\left[\psi\left(x, \kappa^{-1}(\lambda)\right)\right],$$

where $\mathbf{K} \in \mathrm{Aut}(G)$ and $\kappa : \mathbb{C} \mapsto \mathbb{C}$ is some conformal mapping. This action yields another action on the Lax operators

$$L \mapsto \tilde{L} = \mathcal{K}L\mathcal{K}^{-1}, \qquad M \to \tilde{M} = \mathcal{K}M\mathcal{K}^{-1}.$$

From the G_R-invariance condition of the set of fundamental solutions one derives certain restrictions on the potential U. We require that the dressing factor is G_R-invariant

$$(\mathcal{K}g)(x,\lambda) = \mathbf{K}\left[g\left(x,\kappa^{-1}(\lambda)\right)\right] = g(x,\lambda). \tag{15}$$

For example, let us consider the following action of $\mathbb{Z}_2$

$$\mathcal{K} : \psi(x,\lambda) \to \tilde{\psi}(x,\lambda) = K\left(\psi^{\dagger}\left(x,\lambda^{*}\right)\right)^{-1} K^{-1}, \tag{16}$$

provided $K \in G$ and $K^2 = \pm\mathbb{1}$. Therefore the potential fulfills the symmetry condition

$$KU^{\dagger}(x,\lambda^{*})K^{-1} = U(x,\lambda) \;\Rightarrow\; Kq^{\dagger}(x)K^{-1} = q(x), \quad KJK^{-1} = J \tag{17}$$

and the dressing factor satisfies the equality

$$K[g^{\dagger}(x,\lambda^{*})]^{-1}K^{-1} = g(x,\lambda). \tag{18}$$

3. Multicomponent mKdV associated with symmetric spaces of BD.I series

The generic system of mKdV type equations associated with symmetric spaces $SO(2r+1)/SO(2)\times SO(2r-1)$ have the following form

$$\begin{aligned}
&\vec{q}_t + \vec{q}_{xxx} + 3(\vec{p},\vec{q})\vec{q}_x + 3\left(\vec{q}_x,\vec{p}\right)\vec{q} - 3\left(\vec{q}_x s_0 \vec{q}\right) s_0\vec{p} = 0,\\
&\vec{p}_t + \vec{p}_{xxx} + 3(\vec{p},\vec{q})\vec{p}_x + 3\left(\vec{p}_x,\vec{q}\right)\vec{p} - 3\left(\vec{p}_x s_0 \vec{p}\right) s_0\vec{q} = 0,
\end{aligned}$$

where

$$\vec{q} = (q_2, q_3, \ldots, q_{2r})^T, \quad \vec{p} = (p_2, p_3, \ldots, p_{2r})^T, \quad (s_0)_{ij} = (-1)^{i-1}\delta_{i,2r-j}.$$

The connection with the symmetric spaces of the mentioned kind is via its Lax pair

$$\begin{aligned}
L &= i\partial_x + q(x,t) - \lambda J,\\
M &= i\partial_t + V_0(x,t) + \lambda V_1(x,t) + \lambda^2 V_2(x,t) - \lambda^3 J.
\end{aligned}$$

It is known [3,4,11] that with any symmetric space G/K it can be associated a splitting $\mathfrak{g} = \mathfrak{k} + m$ where $\mathfrak{k} \subset \mathfrak{g}$ is the Lie subalgebra corresponding to K and m is its complement to $\mathfrak{g}$. J is the element whose centralizer is $\mathfrak{k}$ while $q \in m$. In the case under consideration $J = H_{e_1} = \mathrm{diag}\,(1,0,\ldots 0,-1)$ and

$$q = \begin{pmatrix} 0 & \vec{q}^T & 0 \\ \vec{p} & 0 & s_0\vec{q} \\ 0 & \vec{p}^T s_0 & 0 \end{pmatrix}, \qquad V_2 = q, \qquad V_1 = i\,\mathrm{ad}_{J}\partial_x q + \frac{1}{2}\left[\mathrm{ad}_{J}q, q\right],$$

$$V_0 = -\partial^2_{xx}q + \frac{1}{2}\left[\mathrm{ad}_{J}q, \left[\mathrm{ad}_{J}q, q\right]\right] + i\left[\partial_x q, q\right].$$

From now on we shall focus on the case when $r = 2$. Then the potential q and the Cartan element J associted with $SO(5)/SO(2) \times SO(3)$ are presented by

$$q = \begin{pmatrix} 0 & q_2 & q_3 & q_4 & 0 \\ p_2 & 0 & 0 & 0 & q_4 \\ p_3 & 0 & 0 & 0 & -q_3 \\ p_4 & 0 & 0 & 0 & q_2 \\ 0 & p_4 & -p_3 & p_2 & 0 \end{pmatrix}, \qquad J = \mathrm{diag}\,(1, 0, 0, 0, -1).$$

Let us consider several examples.

Example 3.1. Let the following $\mathbb{Z}_2$ reduction be given

$$KU^\dagger(x, \lambda^*)K^{-1} = U(x, \lambda) \quad \Rightarrow \quad Kq^\dagger(x)K^{-1} = q(x), \quad KJK^{-1} = J \quad (19)$$

with $K = \mathrm{diag}\,(\epsilon_1, \epsilon_2, 1, \epsilon_2, \epsilon_1)$, $\epsilon_{1,2} = \pm 1$. Hence we have the following relations

$$p_2 = \epsilon_1\epsilon_2 q_2^*, \qquad p_3 = \epsilon_1 q_3^*, \qquad p_4 = \epsilon_1\epsilon_2 q_4^*.$$

As a result one derives the following 3-component mKdV type system

$$\begin{aligned} &q_{2,t} + q_{2,xxx} + 3\epsilon_1(q_2q_3)_x q_3^* + 3\epsilon_1\epsilon_2 q_3 q_4^* q_{3,x} + 6\epsilon_1\epsilon_2|q_2|^2 q_{2,x} = 0, \\ &q_{3,t} + q_{3,xxx} + 3\epsilon_1(q_2q_4)_x q_3^* + 3\epsilon_1\epsilon_2(q_2q_3)_x q_2^* + 3\epsilon_1\epsilon_2(q_3q_4)_x q_4^* \\ &\quad + 3\epsilon_1|q_3|^2 q_{3,x} = 0, \\ &q_{4,t} + q_{4,xxx} + 3\epsilon_1(q_3q_4)_x q_3^* + 3\epsilon_1\epsilon_2 q_2^* q_3 q_{3,x} + 6\epsilon_1\epsilon_2|q_4|^2 q_{4,x} = 0. \end{aligned}$$

In order to find its soliton solution we apply the dressing procedure with a dressing factor to take values in $SO(2r+1)$, i.e. $g^T S g = S$ is satisfied. More specifically, we choose $S_{ij} = (-1)^{i-1}\delta_{i\,6-j}$. This implies that

$$Z = S^{-1}G, \qquad W = S^{-1}F, \qquad \tilde{\alpha}^T = -\tilde{\alpha}, \qquad \tilde{\beta}^T = -\tilde{\beta}.$$

Taking into account the $\mathbb{Z}_2$ symmetry condition (18) we obtain that the dressing factor (12) in the simplest case when rank(X) = rank(F) = 1 reads

$$g = \mathbb{1} + \frac{A}{\lambda - \lambda^+} + \frac{KSA^*S^{-1}K^{-1}}{\lambda - (\lambda^+)^*}.$$

The resudue A admit the following decomposition

$$A = XF^T, \qquad X = \frac{2i\nu KF^*(x,t)}{F^\dagger(x,t)KF(x,t)}, \qquad F^T(x) = F_0^T[\chi^+(x, \lambda^+)]^{-1}.$$

In the soliton case, i.e. $q = 0$ we have

$$\chi_0^+(x, \lambda) = e^{-i\lambda Jx} \quad \Rightarrow \quad F(x) = e^{i\lambda^+ Jx}F_0.$$

Thus we obtain a reflectionless potential. To derive the soliton solution it remains to recover the time dependence using the formula

$$F_0 \to e^{i(\lambda^+)^3 Jt} F_0.$$

Taking into account all these facts we can derive the following expressions for the 1-soliton solution

$$q_2(x,t) = i\nu e^{-i\mu(x-vt-\delta_0)} \frac{\epsilon_1 e^{-\nu(x-ut-\xi_0)}\mathcal{F}_2 + \epsilon_2 e^{\nu(x-ut-\xi_0)}\mathcal{F}_4^*}{\epsilon_1 \cosh 2\nu(x-ut-\xi_0) + \mathcal{C}},$$

$$q_3(x,t) = i\nu e^{-i\mu(x-vt-\delta_0)} \frac{\epsilon_1 e^{-\nu(x-ut-\xi_0)}\mathcal{F}_3 - e^{\nu(x-ut-\xi_0)}\mathcal{F}_3^*}{\epsilon_1 \cosh 2\nu(x-ut-\xi_0) + \mathcal{C}},$$

$$q_4(x,t) = i\nu e^{-i\mu(x-vt-\delta_0)} \frac{\epsilon_1 e^{-\nu(x-ut-\xi_0)}\mathcal{F}_4 + \epsilon_2 e^{\nu(x-ut-\xi_0)}\mathcal{F}_2^*}{\epsilon_1 \cosh 2\nu(x-ut-\xi_0) + \mathcal{C}},$$

where $u = \nu^2 - 3\mu^2$, $v = 3\nu^2 - \mu^2$ and

$$\mathcal{C} = \frac{1}{2}[\epsilon_2(|\mathcal{F}_2|^2 + |\mathcal{F}_4|^2) + |\mathcal{F}_3|^2], \qquad \mathcal{F}_k = \frac{F_{0,k}}{\sqrt{|F_{0,1}||F_{0,5}|}},$$

$$\xi_0 = \frac{1}{2\nu}\ln\frac{|F_{0,1}|}{|F_{0,5}|}, \qquad \mu\delta_0 = -\arg F_{0,1} = \arg F_{0,5}.$$

Example 3.2. Let us consider another example of a $\mathbb{Z}_2$ reduction

$$KU^\dagger(-\lambda^*)K^{-1} = U(\lambda) \quad \Rightarrow \quad Kq^\dagger K^{-1} = q, \quad KJK^{-1} = -J. \tag{20}$$

In order to fulfill the condition for J we choose K in the form

$$W_{e_1} = \begin{pmatrix} 0 & 0 & 0 & 0 & -1 \\ 0 & 1 & 0 & 0 & 0 \\ 0 & 0 & -1 & 0 & 0 \\ 0 & 0 & 0 & 1 & 0 \\ -1 & 0 & 0 & 0 & 0 \end{pmatrix} \quad \Rightarrow \quad K = \begin{pmatrix} 0 & 0 & 0 & 0 & -\epsilon_1 \\ 0 & \epsilon_2 & 0 & 0 & 0 \\ 0 & 0 & -1 & 0 & 0 \\ 0 & 0 & 0 & \epsilon_2 & 0 \\ -\epsilon_1 & 0 & 0 & 0 & 0 \end{pmatrix}.$$

As a result the following relations hold true

$$q_4 = -\epsilon_1\epsilon_2 q_2^*, \qquad q_3 = -\epsilon_1 q_3^*, \qquad p_4 = -\epsilon_1\epsilon_2 p_2^*, \qquad p_3 = -\epsilon_1 p_3^*$$

and we have a 4-component mKdV system

$$\begin{aligned} q_{2,t} + q_{2,xxx} + 3(q_2q_3)_x p_3 - 3\epsilon_1\epsilon_2 q_3 p_2^* q_{3,x} + 6q_2p_2q_{2,x} &= 0, \\ q_{3,t} + q_{3,xxx} - 3\epsilon_1\epsilon_2 |q_2|_x^2 p_3 + 3(q_2q_3)_x p_2 + 3(q_2^*q_3)_x p_2^* + 3q_3p_3q_{3,x} &= 0, \\ p_{2,t} + p_{2,xxx} + 3(p_2p_3)_x q_3 - 3\epsilon_1\epsilon_2 q_2^* p_3 p_{3,x} + 6q_2p_2p_{2,x} &= 0, \\ p_{3,t} + p_{3,xxx} - 3\epsilon_1\epsilon_2 |p_2|_x^2 q_3 + 3(p_2p_3)_x q_2 + 3(p_2^*p_3)_x q_2^* + 3q_3p_3p_{3,x} &= 0. \end{aligned}$$

In this case the dressing factor satisfies

$$K[g^{\dagger}(-\lambda^{*})]^{-1}K^{-1} = g(\lambda).$$

After taking into account the explicit form of g

$$g = \mathbb{1} + \frac{A}{\lambda - \lambda^{+}} + \frac{B}{\lambda - \lambda^{-}}$$

we conclude that

$$\lambda^{\pm} = \pm i\nu^{\pm}, \qquad A = -KSA^{*}S^{-1}K^{-1}, \qquad B = -KSB^{*}S^{-1}K^{-1}.$$

Like before the residues admit the decomposition $A = XF^{T}$, $B = YG^{T}$ where

$$X = i(\nu^{+} + \nu^{-})\frac{SG}{F^{T}SG}, \qquad Y = -i(\nu^{+} + \nu^{-})\frac{SF}{F^{T}SG},$$
$$F(x,t) = e^{-\nu^{+}Jx}F_0, \qquad G(x,t) = e^{\nu^{-}Jx}G_0.$$

After substituting all expressions in

$$q = [J, A + B]$$

we reach to the following expressions for 1-soliton solution

$$q_2 = \frac{i(\nu^{+} + \nu^{-})}{\Delta}\left(e^{-\nu^{-}(x-u^{-}t)}F_{0,2}G_{0,5} - e^{\nu^{+}(x-u^{+}t)}G_{0,2}F_{0,5}\right),$$
$$q_3 = \frac{i(\nu^{+} + \nu^{-})}{\Delta}\left(e^{-\nu^{-}(x-u^{-}t)}F_{0,3}G_{0,5} - e^{\nu^{+}(x-u^{+}t)}G_{0,3}F_{0,5}\right),$$
$$p_2 = \frac{i(\nu^{+} + \nu^{-})}{\Delta}\left(e^{-\nu^{+}(x-u^{+}t)}F_{0,1}G_{0,4} - e^{\nu^{-}(x-u^{-}t)}G_{0,1}F_{0,4}\right),$$
$$p_3 = \frac{i(\nu^{+} + \nu^{-})}{\Delta}\left(e^{\nu^{-}(x-u^{-}t)}G_{0,1}F_{0,3} - e^{-\nu^{+}(x-u^{+}t)}F_{0,1}G_{0,3}\right),$$
$$\Delta = e^{-(\nu^{+}+\nu^{-})(x-ut)}F_{0,1}G_{0,5} + e^{(\nu^{+}+\nu^{-})(x-ut)}F_{0,5}G_{0,1} + \mathcal{C},$$
$$\mathcal{C} = F_{0,3}G_{0,3} - F_{0,2}G_{0,4} - F_{0,4}G_{0,2},$$
$$u^{\pm} = (\nu^{\pm})^{2}, \qquad u = (\nu^{+})^{2} + (\nu^{-})^{2} - \nu^{+}\nu^{-}.$$

Example 3.3. Let us consider now two $\mathbb{Z}_2$ reductions acting simultaneously as follows

$$U^{\dagger}(x,\lambda^{*}) = U(x,\lambda) \qquad \Rightarrow \qquad q^{\dagger}(x) = q(x), \tag{21}$$
$$U^{T}(x,-\lambda) = -U(x,\lambda) \qquad \Rightarrow \qquad q^{T}(x) = -q(x). \tag{22}$$

As a consequence of both reductions we find that $q_k^{*} = -q_k$, $k = 2,3,4$, i.e. $q_k = i\mathbf{q}_k$. These 3 independent fields obey the following system of equations for three real functions:

$$\mathbf{q}_{2,t} + \mathbf{q}_{2,xxx} + 3(\mathbf{q}_2\mathbf{q}_3)_x\mathbf{q}_3 + 3\mathbf{q}_3\mathbf{q}_4\mathbf{q}_{3,x} + 6\mathbf{q}_2^2\mathbf{q}_{2,x} = 0,$$
$$\mathbf{q}_{3,t} + \mathbf{q}_{3,xxx} + 3(\mathbf{q}_2\mathbf{q}_4)_x\mathbf{q}_3 + 3(\mathbf{q}_2\mathbf{q}_3)_x\mathbf{q}_2 + 3(\mathbf{q}_3\mathbf{q}_4)_x\mathbf{q}_4 + 3\mathbf{q}_3^2\mathbf{q}_{3,x} = 0,$$
$$\mathbf{q}_{4,t} + \mathbf{q}_{4,xxx} + 3(\mathbf{q}_3\mathbf{q}_4)_x\mathbf{q}_3 + 3\mathbf{q}_2\mathbf{q}_3\mathbf{q}_{3,x} + 6\mathbf{q}_4^2\mathbf{q}_{4,x} = 0.$$

In the present case there exist two essentially different types of soliton solutions: doublet solitons (2 eigenvalues $\pm i\nu$) and quadruplet solitons (4 eigenvalues $\pm\lambda^+, \pm(\lambda^+)^*$). The situation resembles that of sin-Gordon equation or $\mathbb{Z}_2 \times \mathbb{Z}_2$ -reduced N-wave equations where there also exist two types of solitons [12]. The doublet soliton can be obtained by using a 2-poles dressing factor of the form

$$g = \mathbb{1} + \frac{A}{\lambda - i\nu} + \frac{SA^*S^{-1}}{\lambda + i\nu}, \qquad A^* = -A.$$

The result reads

$$\mathbf{q}_2(x,t) = \frac{\nu}{\cosh 2\nu(x - ut - \xi_0) + \mathcal{C}}\left(e^{-\nu(x-ut-\xi_0)}\mathcal{F}_2 + e^{\nu(x-ut-\xi_0)}\mathcal{F}_4\right),$$
$$\mathbf{q}_3(x,t) = -\frac{2\nu\sinh\nu(x - ut - \xi_0)\mathcal{F}_3}{\cosh 2\nu(x - ut - \xi_0) + \mathcal{C}},$$
$$\mathbf{q}_4(x,t) = \frac{\nu}{\cosh 2\nu(x - ut - \xi_0) + \mathcal{C}}\left(e^{-\nu(x-ut-\xi_0)}\mathcal{F}_4 + e^{\nu(x-ut-\xi_0)}\mathcal{F}_2\right),$$
$$\mathcal{C} = \mathcal{F}_2^2 + \mathcal{F}_3^2 + \mathcal{F}_4^2, \qquad u = \nu^2.$$

The quadruplet solution requires a dressing factor with 4-poles as follows

$$g = \mathbb{1} + \frac{A}{\lambda - \lambda^+} + \frac{SA^*S}{\lambda - (\lambda^+)^*} - \frac{SAS}{\lambda + \lambda^+} - \frac{A^*}{\lambda + (\lambda^+)^*}.$$

Then the quadruplet soliton can be derived using next formula

$$\mathbf{q} = 2\mathrm{Im}\,[J, A - SAS].$$

Like before $A = XF^T$ is fulfilled and it can be proven that the factors now are given by

$$X = \frac{1}{\Delta}\left(a^*F + bF^* - cSF^*\right), \qquad F = e^{[i\mu(x-vt)-\nu(x-ut)]J}F_0,$$
$$\Delta = |a|^2 + b^2 - c^2, \qquad a = \frac{F^TF}{2\lambda^+}, \qquad b = \frac{F^TF^*}{2i\nu}, \qquad c = \frac{F^TSF^*}{2\mu}.$$

Finally we obtain

$$\begin{aligned}\mathbf{q}_k = 2\mathrm{Im}\,[(a^*e^{i\mu(x-vt)-\nu(x-ut)}F_{0,1} + be^{-i\mu(x-vt)-\nu(x-ut)}F_{0,1}^*\\ - ce^{i\mu(x-vt)+\nu(x-ut)}F_{0,5}^*)F_{0,k} + (-1)^k(a^*e^{-i\mu(x-vt)+\nu(x-ut)}F_{0,5}\\ + be^{i\mu(x-vt)+\nu(x-ut)}F_{0,5}^*e^{-i\mu(x-vt)-\nu(x-ut)}F_{0,1}^*)F_{0,\bar{k}}].\end{aligned}$$

There exists another possible combination of two $\mathbb{Z}_2$ reductions: first of the type (20) and the second one in the form (22). In this case one obtains a coupled system of just two mKdV equations, see Ref. 13.

4. Conclusion

New two, three and four component generalizations of the classical mKdV equation have been obtained by imposing additional $\mathbb{Z}_2$ and $\mathbb{Z}_2 \times \mathbb{Z}_2$ reductions on the generic Lax representation. The Lax representation of all equations is associated with symmetric spaces of the type $SO(5)/SO(2) \times SO(3)$. This fact explains the variety of possible reductions which mKdV admits. Following the same procedures one is able to derive mKdV equations associated with symmetric spaces $SO(2r)/SO(2) \times SO(2r-2)$ as well. The only difference consists in the choice of the matrix s_0. By using the dressing technique with a dressing factor which is compatible with the reductions we have derived their soliton solutions.

Acknowledgments

The author would like to thank prof. V. Gerdjikov for posing him the problem and for the fruitful discussions.

References

1. M. Wadati, *J. Phys. Soc. Japan* **34**, 1289–1296 (1973) .
2. A. Mikhailov, *Physica D* **3**, 73–117 (1981).
3. S. Helgason, *Differential Geometry, Lie Groups and Symmetric Spaces*, Academic Press, Toronto, 1978.
4. C. Athorne, A. Fordy, *J. Phys. A: Math. Gen.* **20**, 1377–1386 (1987).
5. V. Zakharov, A. Shabat, *Funkts. Anal. Prilozhen.* **13**, 13–22 (1979).
6. V. Gerdjikov, *Contemporary Math.* **301**, 35–68 (2002).
7. F. Calogero, *Why are Certain Nonlinear PDE's Both Widely Applicable and Integrable?*, ed. V. Zakharov, What is Integrability?, Springer-Verlag, 1990, pp. 1–62.
8. A. Shabat, *Diff. Equations* **15**, 1824–1834 (1979) (In Russian).
9. V. Zakharov, A. Mikhailov, *Commun. Math. Phys.* **74**, 21–40 (1980).
10. R. Ivanov, *Nuclear Physics* B **694**, 509–524 (2004).
11. A. Fordy, P. Kulish, *Commun. Math. Phys.* **89**, 427–443 (1983).
12. V. Gerdjikov, N. Kostov, T. Valchev, *Sigma* **3**, 039, 19 pages (2007).
13. V. Gerdjikov, D. Kaup, N. Kostov, T. Valchev, *J. Phys. A: Math. Theor.* **41**, 36 pages (2008); doi:10.1016/j.physd.2008.06.007.

INTEGRABLE DYNAMICAL SYSTEMS OF THE FRENET-SERRET TYPE

V. M. VASSILEV* and P. A. DJONDJOROV

Institute of Mechanics,
Bulgarian Academy of Sciences,
Acad. G. Bonchev St., Block 4, 1113 Sofia, Bulgaria
**E-mail: vasilvas@imbm.bas.bg*

I. M. MLADENOV

Institute of Biophysics,
Bulgarian Academy of Sciences,
Acad. G. Bonchev St., Block 21, 1113 Sofia, Bulgaria
E-mail: mladenov@obzor.bio21.bas.bg

Let the curvature of a plane curve parametrized by arclength s be given as a function of the Cartesian co-ordinates of the points the curve is passing through in the Euclidean plane. Then, the co-ordinates of its position vector are determined by a system of equations arising from the Frenet-Serret relations that can be regarded as a dynamical system of two degrees of freedom determining the motion (trajectories) of a particle of unit mass, s playing the role of time. Here, two classes of integrable systems of the foregoing type are identified. For that purpose, we explore the variational symmetries of a generic system of this kind with respect to Lie groups of point transformations of the involved variables. As a result, a set of sufficient conditions are found which ensure that such a system possesses two functionally independent integrals of motion and, consecutively, is integrable by quadratures. In each such case, we achieve either an explicit parameterization of the corresponding trajectory curves in terms of their curvatures or, at least, a separation of the dependent variables.

Keywords: Integrability; Dynamical systems; Plane curves; Parameterization.

1. Introduction

Suppose that the curvature κ of a plane curve Γ parametrized by arclength s is given explicitly as a function of the arclength, i.e., the intrinsic equation of the curve Γ is known. Then, it is possible to recover the position vector $\mathbf{x}(s) = (x(s), z(s)) \in \mathbb{R}^2$ of the curve in the plane $\mathbb{R}^2$ (up to a rigid motion) by quadratures in the standard manner.

First, recall that the unit tangent $\mathbf{t}(s)$ and normal $\mathbf{n}(s)$ vectors to the curve Γ

$$\mathbf{t}(s) = (\dot{x}(s), \dot{z}(s)), \qquad \mathbf{n}(s) = (-\dot{z}(s), \dot{x}(s)) \tag{1}$$

are related to the curvature $\kappa(s)$ through the Frenet-Serret formulae

$$\dot{\mathbf{t}}(s) = \kappa(s)\mathbf{n}(s), \qquad \dot{\mathbf{n}}(s) = -\kappa(s)\mathbf{t}(s). \tag{2}$$

Here and throughout this paper, the dots denote derivatives with respect to s. Then, expressions (1) and the Frenet-Serret relations (2) provide the following system of two second-order ordinary differential equations for the components of the position vector

$$\ddot{x} + \kappa(s)\dot{z} = 0, \qquad \ddot{z} - \kappa(s)\dot{x} = 0 \tag{3}$$

which is readily integrable by quadratures to give the parametric equations of the curve Γ. Indeed, in terms of the slope angle $\varphi(s)$ of the curve Γ one has

$$\kappa(s) = \dot{\varphi}(s), \qquad \dot{x}(s) = \cos\varphi(s), \qquad \dot{z}(s) = \sin\varphi(s)$$

and hence, the parametric equations of the curve Γ can be expressed in the form

$$x(s) = \int \cos\varphi(s)\,\mathrm{d}s, \qquad z(s) = \int \sin\varphi(s)\,\mathrm{d}s$$

where

$$\varphi(s) = \int \kappa(s)\,\mathrm{d}s.$$

Suppose now that the intrinsic equation of the curve Γ is not known but its curvature κ is given as a function of the Cartesian co-ordinates (x, z) of the points the curve is passing through in the Euclidean plane $\mathbb{R}^2$, i.e., $\kappa = \mathcal{K}(x, z)$ is a known function. In this case, system (3) reads

$$\ddot{x} + \mathcal{K}(x, z)\dot{z} = 0, \qquad \ddot{z} - \mathcal{K}(x, z)\dot{x} = 0 \tag{4}$$

and here, each system of this form is referred to as a planar Frenet-Serret system as it arises from the Frenet-Serret relations (2) for planar curves. In this setting, unlike the case when the intrinsic equation of the curve Γ is known, the integrability of a system of form (4) by quadratures is not clear in advance. Below, we identify two types of functions $\mathcal{K}(x, z)$ which are such that the solutions of the corresponding systems of form (4) can be expressed explicitly by quadratures. Before proceeding with the details, however, let us note the following.

Although the considered systems (4) have appeared here in a purely geometric context, it is convenient to regard them as dynamical systems of two degrees of freedom determining the motion (trajectories) of particles of unit mass, s playing the role of time. Actually, a generic system of this type coincides with the system of Newton-Lorentz equations [1] describing the planar non-relativistic motion of a charged particle of unit mass in the transverse magnetic field which essential component is $\mathcal{K}(x, z)$. Due to this analogy, it is clear that the system (4) is a Lagrangian system consisting of the Euler-Lagrange equations associated with the action functional

$$A = \int L(x, z, \dot{x}, \dot{z})\,\mathrm{d}s \tag{5}$$

in which the Lagrangian L can be taken in the form

$$L = \frac{1}{2}\left(\dot{x}^2 + \dot{z}^2\right) + F(x, z)\,\dot{x} + G(x, z)\,\dot{z} \tag{6}$$

where the functions $F(x, z)$ and $G(x, z)$ are the non-zero components of the vector potential and

$$\mathcal{K}(x, z) = \frac{\partial F(x, z)}{\partial z} - \frac{\partial G(x, z)}{\partial x}. \tag{7}$$

At the same time, Eqs. (4) are the equations of motion corresponding to the Hamiltonian function

$$H = \frac{1}{2}\left(p_x^2 + p_z^2\right) - F(x, z)\,p_x - G(x, z)\,p_z + \frac{1}{2}\left(F(x, z)^2 + G(x, z)^2\right)$$

which can be established directly by eliminating the momenta p_x and p_z from the associated canonical Hamilton's equations and taking into account relation (7).

This latter property of system (4) gives a hint of what one has to look for in order to find integrable systems of this form. Indeed, according to the famous Liouville theorem [2] if an n-degree-of-freedom Hamiltonian system possesses n functionally independent integrals of motion in involution, then it is integrable by quadratures, i.e., it is possible to express its general solution in terms of integrals.

As far as $v = \dot{x}^2 + \dot{z}^2$ is an obvious integral of motion for any dynamical system of form (4), the problem here is to find conditions under which a system of that form has at least one more appropriate integral of motion. To settle this matter, we explore the variational symmetry of a generic system of form (4) with respect to Lie groups of point transformations of the involved variables. As a result, a set of sufficient conditions are found below which ensure that such a system admits a suitable second integral of motion and, consequently, is integrable by quadratures.

2. Variational Symmetries

By a variational symmetry of a system of form (4) we assume each local one-parameter Lie group $G_{\mathbf{v}}$ of local point transformations of the involved independent s and dependent variables x, z whose generator, which is a vector field $\mathbf{v}$ of the form

$$\mathbf{v} = \xi\,(s,x,z)\,\frac{\partial}{\partial s} + \eta\,(s,x,z)\,\frac{\partial}{\partial x} + \zeta\,(s,x,z)\,\frac{\partial}{\partial z} \tag{8}$$

(here ξ, η and ζ are certain smooth functions of the indicated variables), is an infinitesimal divergence symmetry (see Definition 4.33 in Ref. 3) of any action functional (5) with Eqs. (4) as the associated Euler-Lagrange equations. It should be noted that if two action functionals lead to the same system of Euler-Lagrange equations, then they have the same collection of infinitesimal divergence symmetries.

According to the aforementioned definition and Theorem 4.7 in Ref. 3, a system of form (4) admits a group $G_{\mathbf{v}}$ as a variational symmetry group if and only if its generator $\mathbf{v}$ is such that the relation

$$\mathbf{E}\left[\mathbf{pr}^{(1)}\,\mathbf{v}\,(L) + (\mathbf{D}\,\xi)\,L\right] = 0 \tag{9}$$

holds, where L is the Lagrangian (6). Here, $\mathbf{E}$ is the Euler operator, $\mathbf{pr}^{(1)}\,\mathbf{v}$ is the first prolongation of the vector field $\mathbf{v}$ and $\mathbf{D}$ is the total derivative operator [3].

Thus, to find the variational symmetries of the systems of form (4), we assume initially that the components ξ, η and ζ of the vector field (8) are unknown functions of the variables s, x and z, and then write down the left-hand side of (9) by using expression (6) and the prolongation formulae (2.38) and (2.39) given in Ref. 3. Finally, equating the coefficients at the derivatives of the dependent variables x and z to zero we arrive at the following result (see also Refs. 4–8).

Proposition 2.1. *A system of form* (4) *admits a one-parameter Lie group $G_{\mathbf{v}}$ as a variational symmetry group if and only if the components ξ, η and ζ of its generator* (8) *are of the form*

$$\xi = 2\alpha_1 s + \alpha_0, \qquad \eta = \alpha_1 x + \alpha_2 z + \alpha_3, \qquad \zeta = \alpha_1 z - \alpha_2 x + \alpha_4 \tag{10}$$

where $\alpha_0, \ldots, \alpha_4 \in \mathbb{R}$, and the respective function $\mathcal{K}(x,z)$ is such that

$$\mathbf{v}(\mathcal{K}) + 2\alpha_1 \mathcal{K} = 0 \tag{11}$$

i.e., the function $\mathcal{K}(x,z)$ is either an invariant of the admitted group $G_{\mathbf{v}}$ (when $\alpha_1 = 0$) or an eigenfunction (when $\alpha_1 \neq 0$) of its generator $\mathbf{v}$.

3. Conservation Laws

Actually, the interest in the determination of the variational symmetries of a given system of differential equations is motivated by the fact that, in virtue of Bessel-Hagen's extension of Noether's theorem (see Refs. 3, 9 and 10), to each variational symmetry of the regarded system corresponds a conservation law (integral of motion) admitted by its smooth solutions.

Thus, if a given system of form (4) admits a vector field of form (8) as an infinitesimal divergence symmetry, then the aforementioned theorem implies that this system has an integral of motion of the form

$$\varepsilon = -\frac{1}{2}\xi\left(\dot{x}^2 + \dot{z}^2\right) + \eta\dot{x} + \zeta\dot{z} + h(x,z) \tag{12}$$

where the function $h(x,z)$ remains to be determined by the condition $\dot{\varepsilon} = 0$ on the solutions of the respective system.

In this setting, we can formulate the following results that unify and generalize the achievements of various authors [4–8].

Proposition 3.1. *Any system of form* (4) *admits the group* $G_{\mathbf{v}_0}$ *generated by the vector field* $\mathbf{v}_0 = \partial/\partial s$ *as a variational symmetry group and the corresponding integral of motion can be cast in the form* $v = \dot{x}^2 + \dot{z}^2$.

Proof. This statement is a straightforward consequence of Proposition 2.1 and formula (12) and has already been mentioned in the Introduction. □

Proposition 3.2. *Each system of form* (4) *that admits a nontrivial group* $G_{\mathbf{v}_1}$ *generated by a vector field* $\mathbf{v}_1$ *of the form* (8), (10) *with* $\alpha_0 = 0$, *i.e.*,

$$\mathbf{v}_1 = 2\alpha_1 s\frac{\partial}{\partial s} + (\alpha_1 x + \alpha_2 z + \alpha_3)\frac{\partial}{\partial x} + (\alpha_1 z - \alpha_2 x + \alpha_4)\frac{\partial}{\partial z} \tag{13}$$

as a variational symmetry group has two integrals of motion v *and* ε *that can be written as follows*

$$v = \dot{x}^2 + \dot{z}^2, \qquad \varepsilon = -\frac{1}{2}\xi\left(\dot{x}^2 + \dot{z}^2\right) + \eta\dot{x} + \zeta\dot{z} + h(x,z) \tag{14}$$

where the function $h(x,z)$ *is such that*

$$\frac{\partial h(x,z)}{\partial x} + \zeta\,\mathcal{K}(x,z) = 0, \qquad \frac{\partial h(x,z)}{\partial z} - \eta\,\mathcal{K}(x,z) = 0. \tag{15}$$

Proof. From Proposition 3.1 we already know that v is an integral of motion for any system of form (4). Moreover, under the above assumptions it is easy to ascertain also that $\dot{\varepsilon} = 0$ on the solutions of system (4) provided that relations (15) hold. This completes the proof but it should be noted that relation (11) is just the compatibility condition for equations (15). □

4. Group Classification

Propositions 2.1 and 3.2 allow us to characterize the systems of form (4) that have variational symmetries by representing the respective functions $\mathcal{K}$ in terms of the canonical co-ordinates [11] of the admitted Lie group $G_{\mathbf{v}_1}$ and to express the corresponding integrals of motion (14) by means of these functions. Below, we analyze in detail all possible cases.

First, suppose that a system of form (4) admits a vector field $\mathbf{v}_1$ specified by formula (13) as an infinitesimal divergence symmetry and that $\alpha_1 \neq 0$. In this case, we can introduce new dependent variables

$$u = \frac{1}{\alpha_1^2 + \alpha_2^2}\left[\alpha_1 \arctan\left(\frac{\zeta}{\eta}\right) + \alpha_2 \ln\sqrt{\eta^2 + \zeta^2}\right], \tag{16}$$

$$w = \frac{1}{\alpha_1^2 + \alpha_2^2}\left[\alpha_1 \ln\sqrt{\eta^2 + \zeta^2} - \alpha_2 \arctan\left(\frac{\zeta}{\eta}\right)\right] \tag{17}$$

which, as follows by relations (8), (10) and (13), obey

$$\mathbf{v}_1(u) = 0, \qquad \mathbf{v}_1(w) = 1$$

i.e., u and w are the canonical co-ordinates of the restriction of admitted variational symmetry group $G_{\mathbf{v}_1}$ on the space of the dependent variables. In these variables, condition (11) reads

$$\frac{\partial \mathcal{K}}{\partial w} + 2\alpha_1 \mathcal{K} = 0$$

and the latter can be easily solved to obtain

$$\mathcal{K} = K_1(u) \exp[-2\alpha_1 w] \tag{18}$$

where $K_1(u)$ is an arbitrary function of the indicated variable. At the same time, relations (15) take the form

$$\frac{\partial h}{\partial w} = 0, \qquad \frac{\partial h}{\partial u} - K_1(u) \exp[2\alpha_2 u] = 0$$

that immediately gives the following expression for the function h, namely

$$h = h_1(u), \qquad h_1(u) = \int K_1(u) \exp[2\alpha_2 u]\, \mathrm{d}u \tag{19}$$

up to an arbitrary real constant that can be omitted. Consequently, the two integrals of motion (14) read

$$v = \left(\dot{u}^2 + \dot{w}^2\right) \exp[2(\alpha_2 u + \alpha_1 w)], \tag{20}$$

$$\varepsilon = \dot{w} \exp[2(\alpha_2 u + \alpha_1 w)] - \alpha_1 s + h_1(u) \tag{21}$$

in which the first one is taken into account for the derivation of the second.

The following Proposition summarizes the above results.

Proposition 4.1. *A system of form* (4) *admits a Lie group* $G_{\mathbf{v}_1}$ *generated by a vector field* $\mathbf{v}_1$ *specified by formula* (13) *with* $\alpha_1 \neq 0$ *as a variational symmetry if and only if upon the transformation of the dependent variables* x *and* z *to* u *and* w, *given by formulae* (16) *and* (17), *the function* $\mathcal{K}$ *has the form* (18). *Each such system possesses two integrals of motion, given by expressions* (20) *and* (21), *where the function* $h_1(u)$ *has the form* (19).

Next, suppose that a system of form (4) admits a variational symmetry group $G_{\mathbf{v}_2}$ generated by a vector field of the form

$$\mathbf{v}_2 = (\alpha_2 z + \alpha_3)\frac{\partial}{\partial x} - (\alpha_2 x - \alpha_4)\frac{\partial}{\partial z} \tag{22}$$

with $\alpha_2 \neq 0$. In this case, without loss of generality one may set $\alpha_2 = 1$ and transform the dependent variables to the following ones

$$r = \sqrt{(x - \alpha_4)^2 + (z + \alpha_3)^2}, \quad \vartheta = \arctan\left(\frac{z + \alpha_3}{x - \alpha_4}\right) \tag{23}$$

which, together with the independent variable s, are nothing else but the canonical co-ordinates of the admitted symmetry group $G_{\mathbf{v}_2}$. Indeed, in view of relations (8), (10) and (22), we have

$$\mathbf{v}_2(s) = 0, \qquad \mathbf{v}_2(r) = 0, \qquad \mathbf{v}_2(\vartheta) = -1.$$

Then, according to Proposition 2.1, cf. condition (11), the function $\mathcal{K}$ is an invariant of the admitted symmetry group, i.e.,

$$\mathcal{K} = K_2(r) \tag{24}$$

where $K_2(r)$ is an arbitrary function of the basic invariant r. Successively, in the new dependent variables (23), equations (15) take the form

$$\frac{\partial h}{\partial \vartheta} = 0, \qquad \frac{\partial h}{\partial r} - rK_2(r) = 0$$

and can be readily solved to obtain

$$h = h_2(r) = \int rK_2(r)\,\mathrm{d}r \tag{25}$$

again up to an arbitrary real constant that can be omitted. Consequently, the integrals of motion (14) this time read

$$v = r^2\dot{\vartheta}^2 + \dot{r}^2, \tag{26}$$

$$\varepsilon = -r^2\dot{\vartheta} + h_2(r). \tag{27}$$

This results can be stated as follows.

Proposition 4.2. *A system of form* (4) *admits a Lie group* $G_{\mathbf{v}_2}$ *generated by a vector field* $\mathbf{v}_2$ *specified by the formula* (22) *with* $\alpha_2 \neq 0$ *as a variational symmetry if and only if upon the change of the dependent variables* x *and* z *to* r *and* ϑ, *given by formulae* (23), *the function* $\mathcal{K}$ *has exactly the form* (24). *Each such system possesses two integrals of motion which are given by expressions* (26) *and* (27).

Finally, let us characterize the systems of form (4) that admit a Lie group $G_{\mathbf{v}_3}$ generated by a vector field of the form

$$\mathbf{v}_3 = \alpha_3 \frac{\partial}{\partial x} + \alpha_4 \frac{\partial}{\partial z} \tag{28}$$

such that $\alpha_3^2 + \alpha_4^2 \neq 0$ as a variational symmetry group. In these cases, the canonical co-ordinates of the admitted variational symmetry group $G_{\mathbf{v}_3}$ are s and

$$p = \frac{\alpha_3 x + \alpha_4 z}{\alpha_3^2 + \alpha_4^2}, \qquad q = \frac{\alpha_4 x - \alpha_3 z}{\alpha_3^2 + \alpha_4^2} \tag{29}$$

since, according to expressions (8), (10) and (28), they satisfy the relations

$$\mathbf{v}_3(s) = 0, \qquad \mathbf{v}_3(p) = 1, \qquad \mathbf{v}_3(q) = 0.$$

Here again, according to Proposition 2.1, the function $\mathcal{K}$ is an invariant of the admitted symmetry group, i.e.,

$$\mathcal{K} = K_3(q) \tag{30}$$

where $K_3(q)$ is an arbitrary function of the basic invariant q. In the new dependent variables (29), equations (15) take the form

$$\frac{\partial h}{\partial p} = 0, \qquad \frac{\partial h}{\partial q} + \left(\alpha_3^2 + \alpha_4^2\right) K_3(q) = 0$$

which immediately produces

$$h = h_3(q) = -\left(\alpha_3^2 + \alpha_4^2\right) \int K_3(q)\, \mathrm{d}q \tag{31}$$

up to a real constant that is omitted. Thus, in terms of the variables p and q the integrals of motion (14) become

$$v = \left(\alpha_3^2 + \alpha_4^2\right)\left(\dot{p}^2 + \dot{q}^2\right), \tag{32}$$

$$\varepsilon = \left(\alpha_3^2 + \alpha_4^2\right)\left(\dot{p} - h_3(q)\right) \tag{33}$$

where relation (31) is taken into account.

The latter results can be formulated in the following way.

Proposition 4.3. *A system of form* (4) *admits a Lie group* $G_{\mathbf{v}_3}$ *generated by a vector field* $\mathbf{v}_3$ *such that* $\alpha_3^2 + \alpha_4^2 \neq 0$ *as a variational symmetry if and only if under the change of the dependent variables* x *and* z *to* p *and* q, *given by formulae* (29), *the function* $\mathcal{K}$ *has the form* (30). *Each such system possesses two integrals of motion given by expressions* (32) *and* (33).

It is worth noting that each one of the Frenet-Serret systems described in Propositions 4.1 – 4.3 admits a two-parameter variational symmetry group whose two-dimensional Lie algebra is spanned by the vector field $\mathbf{v}_0$ and one of the vector fields $\mathbf{v}_1$, $\mathbf{v}_2$ or $\mathbf{v}_3$. According to formulae (13), (22) and (28), the respective commutators are

$$[\mathbf{v}_0, \mathbf{v}_1] = 2\alpha_1 \mathbf{v}_0, \qquad [\mathbf{v}_0, \mathbf{v}_2] = 0, \qquad [\mathbf{v}_0, \mathbf{v}_3] = 0.$$

Thus, all possible cases for a two-dimensional Lie algebra, namely, to be abelian or isomorphic to the one with a basis satisfying the first of the above commutator relations (see Exercise 1.21 in Ref. 3), appear in our group classification problem.

5. Integrable Systems

Once the dynamical systems of form (4) admitting variational symmetries are identified and the corresponding integrals of motion are derived, we are at the position to examine the problem of their integrability by quadratures.

The systems characterized in Proposition 4.1 possess two functionally independent integrals of motion (20) and (21). Relation (21) can be cast in the form

$$\frac{\mathrm{d}}{\mathrm{d}s}\left(\frac{1}{2\alpha_1}\exp\left[2\alpha_1 w\right]\right) = \left(\varepsilon + \alpha_1 s - h_1\left(u\right)\right)\exp\left[-2\alpha_2 u\right]$$

which gives

$$w\left(s\right) = \frac{1}{2\alpha_1}\ln\left\{2\alpha_1 \int \left(\varepsilon + \alpha_1 s - h_1\left(u\left(s\right)\right)\right)\exp\left[-2\alpha_2 u\left(s\right)\right]\mathrm{d}s\right\}\cdot \tag{34}$$

In this way, one of the sought functions, i.e., $w\left(s\right)$, is expressed by means of the other one, $u\left(s\right)$, by a quadrature. It is clear however that the substitution of the expression (34) in the second integral of motion (20) or in the respective equations (4) leads to a rather complicated integro-differential equations for the function $u\left(s\right)$. These equations can hardly be solved by quadratures but, nevertheless, a separation of the variables for the systems of the considered type is actually achieved.

Next, let us specialize to the systems described in Proposition 4.2. Here, we have at disposition the two functionally independent integrals of motion (26) and (27). The second one readily gives the expression

$$\dot{\vartheta} = \frac{1}{r^2}\left(h_2(r) - \varepsilon\right) \tag{35}$$

which when substituted into the first yields

$$\dot{r} = \pm\frac{1}{r}\sqrt{vr^2 - \left(h_2(r) - \varepsilon\right)^2}. \tag{36}$$

Now, the integrability of the dynamical systems of the foregoing type by quadratures is obvious. Indeed, combining relations (35) and (36) we have

$$\vartheta = \pm\int \frac{\left(h_2(r) - \varepsilon\right)\mathrm{d}\, r}{r\sqrt{vr^2 - \left(h_2(r) - \varepsilon\right)^2}} + \vartheta_0 \tag{37}$$

where $\vartheta_0 \in \mathbb{R}$, which, together with equation (36), completes the problem.

Finally, consider the systems of form (4) specified in Proposition 4.3. In these cases, there are two functionally independent integrals of motion (32) and (33) to deal with. By solving (33) one finds

$$\dot{p} = \frac{\varepsilon}{\alpha_3^2 + \alpha_4^2} + h_3(q) \tag{38}$$

which substituted into (32) produces

$$\dot{q} = \pm\sqrt{\frac{v}{\alpha_3^2 + \alpha_4^2} - \left(\frac{\varepsilon}{\alpha_3^2 + \alpha_4^2} + h_3(q)\right)^2} \tag{39}$$

and hence the solution of our problem is given by the relations (39) and

$$p = \pm\int \frac{\left(\varepsilon + \left(\alpha_3^2 + \alpha_4^2\right) h_3(q)\right)\mathrm{d}q}{\sqrt{v\left(\alpha_3^2 + \alpha_4^2\right) - \left(\varepsilon + \left(\alpha_3^2 + \alpha_4^2\right) h_3(q)\right)^2}} + p_0, \qquad p_0 \in \mathbb{R} \tag{40}$$

which is obtained by combining the expressions (38) and (39).

6. Concluding Remarks

Two quite interesting classes of dynamical systems of the Frenet-Serret type (4) that are integrable by quadratures are identified in the present paper. These are the systems characterized in Propositions 4.2 and 4.3. In these cases the explicit parametrization of the corresponding trajectory curves can be obtained via direct integration of (37), respectively (40).

Two particular examples of plane curves associated with systems of the aforementioned types — the so-called Lévy's elasticae and the profile curves of classic Delaunay's surfaces — are considered in Ref. 12 where explicit parametric equations of the respective curves are derived to the very end.

Acknowledgments

The authors would like to thank Dr. Dimitar Trifonov and Dr. Stoil Donev for the valuable discussion concerning some of the problems considered in this paper. This research is partially supported by the National Science Fund of Bulgaria under the grant B–1531/2005 and the Inter Academy contracts # 23/2006 and # 35/2009 between the Bulgarian and the Polish Academies of Sciences.

References

1. J.D. Jackson, *Classical Electrodynamics*, Wiley, New York, 1999.
2. J. Liouville, Sur l'intégration des équations différentielles de la dynamique, *J. Math. Pure Appl.* (*Liouville*) **20**, 137 (1855).
3. P.J. Olver, *Applications of Lie Groups to Differential Equations*, 2nd Ed., Graduate Texts in Mathematics **107**, Springer, New York, 1993.
4. B. Dorizzi, B. Grammaticos, A. Ramani and P. Winternitz, Integrable Hamiltonian Systems with Velocity-Dependent Potentials, *J. Math. Phys.* **26**, 3070 (1985).
5. F. Haas and J. Goedert, Noether Symmetries for Two-Dimensional Charged Particle Motion, *J. Phys. A: Math. Gen.* **32**, 6837 (1999).
6. F. Haas and J. Goedert, Lie Symmetries for Two-Dimensional Charged-Particle Motion, *J. Phys. A: Math. Gen.* **33**, 4661 (2000).
7. G. Pucacco and K. Rosquist, On Integrable Hamiltonians with Velocity Dependent Potentials, *Celest. Mech. Dyn. Astron.* **88**, 185 (2004).
8. G. Pucacco and K. Rosquist, Integrable Hamiltonian Systems with Vector Potentials, *J. Math. Phys.* **46**, 012701 (2005).
9. E. Noether, Invariant Variationsprobleme, *Nachr. K. Ges. Wiss. Göttingen, Math.-Phys. Kl.*, Heft **2**, 235 (1918).
10. E. Bessel-Hagen, Über die Erhaltungssätze der Elektrodynamik, *Math. Ann.* **84**, 258 (1921).
11. G.W. Bluman and S. Kumei, *Symmetries and Differential Equations*, Springer, New York, 1989.
12. P. Djondjorov, V. Vassilev and I. Mladenov, Plane Curves Associated with Integrable Dynamical Systems of the Frenet-Serret Type, This volume pp 57–63.

AUTHOR INDEX